BASICS OF MECHANICS

AUTHOR (S)

Dr. Shubhankar Bhowmick
Professor, Department of Mechanical Engineering,
NIT Raipur, Raipur, Chhattisgarh

Dr. Neigapula Venkata Swamy Naidu
Professor, Department of Mechanical Engineering,
NIT Raipur, Raipur, Chhattisgarh

REVIEWER

Dr. Shrutidhara Sarma
Assistant Professor,
Mechanical Engineering, IIT Jodhpur, Rajasthan

KHANNA BOOK PUBLISHING CO. (P) Ltd.
4C/4344, Ansari Road, Darya Ganj, New Delhi, India

All India Council for Technical Education
Nelson Mandela Marg, Vasant Kunj, New Delhi, India

Dear Readers,

To prevent the piracy, this book is secured with HIGH SECURITY HOLOGRAM on the front title cover. In case you don't find the hologram on the front cover title, please write us to at contact@khannabooks.com or whatsapp us at +91-99109 09320 and avail special gift voucher for yourself.

Specimen of Hologram on front Cover title:

Moreover, there is a SPECIAL DISCOUNT COUPON for you with EVERY HOLOGRAM.

How to avail this SPECIAL DISCOUNT:

Step 1: Scratch the hologram

Step 2: Under the scratch area, your "coupon code" is available

Step 3: Logon to www.khannabooks.com

Step 4: Use your "coupon code" in the shopping cart and get your copy at a special discount

Step 5: Enjoy your reading!

ISBN: 978-93-5538-938-1

Book Code: UG415EN

Basics Of Mechanics

by Dr. Shubhankar Bhowmick and
Dr. Neigapula Venkata Swamy Naidu

First Edition: 2026

Originally Published by:
All India Council for Technical Education, Ministry of Education, GOI

Exclusively Published & Distributed by:
Khanna Book Publishing Co. (P) Ltd.
4C/4344, Ansari Road, Daryaganj,
New Delhi (INDIA)

Visit us at: **www.khannabooks.com**
Write us at: **contact@khannabooks.com**

To view complete list of books,
Please scan the QR Code:

KPH

Printed in Bharat.

प्रो. टी. जी. सीताराम
अध्यक्ष
Prof. T. G. Sitharam
Chairman

सत्यमेव जयते

आज़ादी का अमृत महोत्सव

अखिल भारतीय तकनीकी शिक्षा परिषद्
(भारत सरकार का एक सांविधिक निकाय)
(शिक्षा मंत्रालय, भारत सरकार)
नेल्सन मंडेला मार्ग, वसंत कुंज, नई दिल्ली–110070
दूरभाष : 011–26131498
ई–मेल : chairman@aicte-india.org

ALL INDIA COUNCIL FOR TECHNICAL EDUCATION
(A STATUTORY BODY OF THE GOVT. OF INDIA)
(Ministry of Education, Govt. of India)
Nelson Mandela Marg, Vasant Kunj, New Delhi-110070
Phone : 011-26131498
E-mail : chairman@aicte-india.org

FOREWORD

Engineers are the backbone of any modern society. They are the ones responsible for the marvels as well as the improved quality of life across the world. Engineers have driven humanity towards greater heights in a more evolved and unprecedented manner.

The All India Council for Technical Education (AICTE), have spared no efforts towards the strengthening of the technical education in the country. AICTE is always committed towards promoting quality Technical Education to make India a modern developed nation emphasizing on the overall welfare of mankind.

An array of initiatives has been taken by AICTE in last decade which have been accelerated now by the National Education Policy (NEP) 2020. The implementation of NEP under the visionary leadership of Hon'ble Prime Minister of India envisages the provision for education in regional languages to all, thereby ensuring that every graduate becomes competent enough and is in a position to contribute towards the national growth and development through innovation & entrepreneurship.

One of the spheres where AICTE had been relentlessly working since past couple of years is providing high quality original technical contents at Under Graduate & Diploma level prepared and translated by eminent educators in various Indian languages to its aspirants. For students pursuing 2nd year of their Engineering education, AICTE has identified 87 books, which shall be translated into 12 Indian languages - Hindi, Tamil, Gujarati, Odia, Bengali, Kannada, Urdu, Punjabi, Telugu, Marathi, Assamese & Malayalam. In addition to the English medium, books in different Indian Languages are going to support the students to understand the concepts in their respective mother tongue.

On behalf of AICTE, I express sincere gratitude to all distinguished authors, reviewers and translators from the renowned institutions of high repute for their admirable contribution in a record span of time.

AICTE is confident that these outcomes based original contents shall help aspirants to master the subject with comprehension and greater ease.

(Prof. T. G. Sitharam)

ACKNOWLEDGEMENT

The authors are grateful to the authorities of AICTE, particularly Prof. T. G. Sitharam, Chairman; Dr. Abhay Jere, Vice-Chairman; Prof. Rajive Kumar, Member-Secretary; Dr. Ramesh Unnikrishnan, Advisor-II and Dr. Sunil Luthra, Director, Training and Learning Bureau, for their planning to publish the books on ***Basics of Mechanics***. We sincerely acknowledge the valuable contributions of the reviewer of the book Dr. Shrutidhara Sarma, Assistant Professor, Indian Institute of Technology Jodhpur for making it student more appropriate and precise in terms of content so that students find the text easy to read and understand.

This book is an outcome of various suggestions of AICTE members, experts and authors who shared their opinion and thought to further develop the engineering education in our country. Acknowledgements are due to the contributors and different workers in this field whose published books, review articles, papers, photographs, footnotes, references and other valuable information enriched us at the time of writing the book.

Dr. Shubhankar Bhowmick

Dr. Neigapula Venkata Swamy Naidu

PREFACE

Many books have been written on Mechanics and the authors, during their studentships and well as in their professional life, have been greatly benefitted from these texts and books. The knowledge obtained from the peers, and these books together with the professional experience of the authors is penned here in this book titled "Basics of Mechanics". Following the great initiative of AICTE in coming out with textbooks for all engineering students, the authors have put up their effort to come out with this text that provides the maximum coverage of the listed topics in the syllabus along with additional information pertaining to each topic in a systematic and orderly manner.

The foremost priority in preparing this text is put upon the process of initiating the development of analytical capabilities of the student so that the student applies his problem-solving ability in simple and logical manner to the problems of Mechanics. Proper emphasis has been given to explanation of the fundamental concepts, leading examples with transition to more advanced theories, developing questions, multiple choice questions, questions of advanced level keeping in view the wide variety of students.

In unit 1, the idea of vectors, tensors, and vector and tensor algebra are covered in with in-depth discussions of Euler's theorem, Euler angles, coordinate systems, coordinate transformations, symmetric and anti-symmetric tensors, primary axes, eigenvalues, and eigenvectors, three-dimensional rotations. unit 2 presents various basic concepts of kinematic parameters, the range of the motion, velocities present in various motions with the various components of accelerations In unit 3, a variety of fundamental kinetics concepts, including Newton's Euler laws of motion, inertia and non-inertia frames, inertia tensor, angular momentum, polar moment of inertia, area moment of inertia, and mass moment of inertia are explained that shall be helpful to the readers in understanding the dynamics of rigid bodies. In unit 4, the readers shall acquire the in-depth knowledge of free body diagrams using various methods like method of Projections, Joints, Sections and moments. In the 5th unit, the readers shall be gaining the knowledge of general motion using various conditions like planar and 3-D motion by focusing on certain system and point of the body and analysing the forces acting on that system in both kinematic and kinetic parameters and to find out the velocities and angular motions, brief knowledge regarding the precession, "gyroscopic couple" and understanding through the various example and problems. Shear force and bending moments of beams is presented in unit 6 for different types of beams and beam loadings. The study and design of circular shafts subject to torques or twisting couples is presented in unit 7 wherein deformation and shear strains of the shafts subject to torque, shear stresses, twisting angle and power transmission of circular shafts have been discussed. In unit 8, types of friction, its laws, and its applications are explained.

Apart from illustrations and examples as required, the authors have meticulously enriched the text with numerous solved problems in every unit for proper understanding of the related topics in a constructive manner. The authors hope that the book serves the objective laid down by AICTE and inspires the students to learn and discuss the subject and contributes to the development of a solid foundation of the subject. We would be thankful to all beneficial comments and suggestions which will contribute to the improvement of the future editions of the book. It gives us immense pleasure to place this book in the hands of the teachers and students.

Dr. Shubhankar Bhowmick
Dr. Neigapula Venkata Swamy Naidu

OUTCOME BASED EDUCATION

For the implementation of an outcome based education the first requirement is to develop an outcome based curriculum and incorporate an outcome based assessment in the education system. By going through outcome based assessments evaluators will be able to evaluate whether the students have achieved the outlined standard, specific and measurable outcomes. With the proper incorporation of outcome based education there will be a definite commitment to achieve a minimum standard for all learners without giving up at any level. At the end of the programme running with the aid of outcome based education, a student will be able to arrive at the following outcomes:

PO1. **Engineering knowledge**: Apply the knowledge of mathematics, science, engineering fundamentals, and an engineering specialization to the solution of complex engineering problems.

PO2. **Problem analysis**: Identify, formulate, review research literature, and analyze complex engineering problems reaching substantiated conclusions using first principles of mathematics, natural sciences, and engineering sciences.

PO3. **Design / development of solutions**: Design solutions for complex engineering problems and design system components or processes that meet the specified needs with appropriate consideration for the public health and safety, and the cultural, societal, and environmental considerations.

PO4. **Conduct investigations of complex problems**: Use research-based knowledge and research methods including design of experiments, analysis and interpretation of data, and synthesis of the information to provide valid conclusions.

PO5. **Modern tool usage**: Create, select, and apply appropriate techniques, resources, and modern engineering and IT tools including prediction and modeling to complex engineering activities with an understanding of the limitations.

PO6. **The engineer and society**: Apply reasoning informed by the contextual knowledge to assess societal, health, safety, legal and cultural issues and the consequent responsibilities relevant to the professional engineering practice.

PO7. Environment and sustainability: Understand the impact of the professional engineering solutions in societal and environmental contexts, and demonstrate the knowledge of, and need for sustainable development.

PO8. Ethics: Apply ethical principles and commit to professional ethics and responsibilities and norms of the engineering practice.

PO9. Individual and team work: Function effectively as an individual, and as a member or leader in diverse teams, and in multidisciplinary settings.

PO10. Communication: Communicate effectively on complex engineering activities with the engineering community and with society at large, such as, being able to comprehend and write effective reports and design documentation, make effective presentations, and give and receive clear instructions.

PO11. Project management and finance: Demonstrate knowledge and understanding of the engineering and management principles and apply these to one's own work, as a member and leader in a team, to manage projects and in multidisciplinary environments.

PO12. Life-long learning: Recognize the need for, and have the preparation and ability to engage in independent and life-long learning in the broadest context of technological change.

COURSE OUTCOMES

After completion of the course the students will be able to:

CO-1: Describe planar, spatial and dynamical motion of rigid bodies with the help of Euler's equations

CO-2: Explain rotational motion characteristics from the idea of torque and angular momentum

CO-3: Apply vector calculus and Newton's laws to fundamental level problems of mechanics

CO-4: Apply Kepler's laws with the aid of other conservation laws to formulate central force problems

CO-5: Apply the concepts of Coriolis force in different environmental and geographical aspects.

CO-6: Analyze different types of oscillatory and vibratory motion problems in real life

Course Outcomes	Expected Mapping with Programme Outcomes (1- Weak Correlation; 2- Medium Correlation; 3- Strong Correlation)											
	PO-1	PO-2	PO-3	PO-4	PO-5	PO-6	PO-7	PO-8	PO-9	PO-10	PO-11	PO-12
CO-1	3	2	2	2	1	-	-	-	-	-	-	-
CO-2	3	1	2	1	-	-	-	-	-	-	-	-
CO-3	3	3	2	1	-	-	-	-	-	-	-	-
CO-4	3	3	3	2	1	-	-	-	-	-	-	-
CO-5	3	3	2	1	1	-	-	-	-	-	-	-
CO-6	3	3	3	2	1	-	-	-	-	-	-	-

GUIDELINES FOR TEACHERS

To implement Outcome Based Education (OBE) knowledge level and skill set of the students should be enhanced. Teachers should take a major responsibility for the proper implementation of OBE. Some of the responsibilities (not limited to) for the teachers in OBE system may be as follows:

- Within reasonable constraint, they should manoeuvre time to the best advantage of all students.
- They should assess the students only upon certain defined criterion without considering any other potential ineligibility to discriminate them.
- They should try to grow the learning abilities of the students to a certain level before they leave the institute.
- They should try to ensure that all the students are equipped with the quality knowledge as well as competence after they finish their education.
- They should always encourage the students to develop their ultimate performance capabilities.
- They should facilitate and encourage group work and team work to consolidate newer approach.
- They should follow Blooms taxonomy in every part of the assessment.

Bloom's Taxonomy

Level	Teacher should Check	Student should be able to	Possible Mode of Assessment
Create	Students ability to create	Design or Create	Mini project
Evaluate	Students ability to justify	Argue or Defend	Assignment
Analyse	Students ability to distinguish	Differentiate or Distinguish	Project/Lab Methodology
Apply	Students ability to use information	Operate or Demonstrate	Technical Presentation/ Demonstration
Understand	Students ability to explain the ideas	Explain or Classify	Presentation/Seminar
Remember	Students ability to recall (or remember)	Define or Recall	Quiz

GUIDELINES FOR STUDENTS

Students should take equal responsibility for implementing the OBE. Some of the responsibilities (not limited to) for the students in OBE system are as follows:

- Students should be well aware of each UO before the start of a unit in each and every course.
- Students should be well aware of each CO before the start of the course.
- Students should be well aware of each PO before the start of the programme.
- Students should think critically and reasonably with proper reflection and action.
- Learning of the students should be connected and integrated with practical and real life consequences.
- Students should be well aware of their competency at every level of OBE.

List of Figures

Unit 1 Units and Vectors

Unit 2 Kinematics of Rigid Body

Unit 3 Kinetics of Rigid Body

Unit 4 Free Body Diagram

Unit 8 Friction

LIST OF TABLES

CONTENTS

1 Units and Vectors

UNIT SPECIFICS

This unit discussed the following concepts:

- *Vectors and tensors and coordinate systems;*
- *Vector and tensor algebra;*
- *Symmetric and anti-symmetric tensors;*
- *Eigenvalues and Principal axes;*
- *Three-dimensional rotation- Euler's theorem;*
- *Axis-angle formulation and Euler angles;*
- *Coordinate transformation of vectors and tensors.*

RATIONALE

This unit on units and vectors helps students to understand about the vector and tensor algebra. The knowledge on the units and vectors is crucial for anyone studying in the fields of science and engineering. The coordinate systems, coordinate transformation of vectors and tensors, symmetric and anti-symmetric tensors, eigenvalues and principal axes knowledge is useful for the analysing directional quantities. Euler's theorem is influences various fields such as engineering, physics, and computer science by simplifying complex calculations, transforming differential equations into algebraic ones. The units and vectors are the backbone of many physical theories and technological applications, making their study and application essential for academic and professional success.

PRE-REQUISITES

Mathematics: Co-ordinate Systems (Class XII)
Physics: Mechanics (Class XII)

UNIT OUTCOMES

List of outcomes of this unit is as follows:

U1-O1: Describe basic vector operations
U1-O2: Describe the Cartesian, spherical and cylindrical coordinate system
U1-O3: Explain Vector and tensor algebra
U1-O4: Explain the symmetric, antisymmetric tensors and Eigenvalues and Principal axes
U1-O5: Apply Three-dimensional rotation- Euler's theorem and vectors to solve complex problems

Unit - 1 Outcomes	EXPECTED MAPPING WITH COURSE OUTCOMES (1- Weak Correlation; 2- Medium Correlation; 3- Strong Correlation)				
	CO-1	CO-2	CO-3	CO-4	CO-5
U1-O1	3	3	2	2	-
U1-O2	3	3	2	2	2
U1-O3	3	3	2	1	-
U1-O4	3	3	2	1	-
U1-O5	-	3	2	2	2

1.1 INTRODUCTION

Discovering the mysteries of nature through observations, measurements, as well as intuition and creativity, is the thrilling goal of physics. Its beauty resides in the fact that a few handfuls of fundamental principles can encompass a wide range of phenomena from atomic to cosmic scales. It is regarded as the most fundamental of the physical sciences and extensively relies on mathematics to express and establish hypotheses and principles. Physics' concepts and principles are applied in chemistry, geology, and astronomy. In addition to theory, physics also offers practical methods that are applied in every aspect of life. The richest source of human knowledge during the last century has been modern physical procedures.

Numerous complex events that at first glance can seem unrelated can all be explained by a single, basic physics law. When researching a complicated physical system, it is common practise to adopt a streamlined model of the system, focusing only on the key aspects of the system while ignoring the smaller effects. For instance, air resistance can be disregarded when analysing an object that is falling close to the earth's surface. Additionally, it is commonly believed that the earth is round and uniform. The earth is not homogeneous and is actually an ellipsoid. It can be considered that there is little discrepancy between these models' calculations.

The two branches of physics are "classical physics" and "modern physics". This book is primarily concerned with mechanics, a subfield of classical physics. Thermodynamics, electromagnetic, light and optics, and sound are further disciplines of classical physics. The foundation of classical physics is mechanics, which is the study of how objects move. On the other hand, theories created in the 20th century are included in modern fields of physics. The theories of relativity and quantum mechanics are the two basic ones. Many physical phenomena that classical physics cannot explain are described by modern physics.

1.2 UNITS

A physical quantity describes a physical phenomenon quantitatively. The physical quantity must be measured and then represented by a number in order to provide an accurate description. This measurement is performed by comparing the quantity to a standard, also known as a unit. For instance, mass is a physical term that describes the amount of matter that makes up an object. All the units should adhere to particular requirements, such as accuracy, accessibility, and stability over a wide range of times or environmental conditions.

The Metric System MKS was replaced in 1960 by the International System of Units (SI). The term "System International" is translated from French to form the acronym. The seven base fundamental units that make up the SI system are depicted in Table 1.1 and each one represents a quantity that is thought to be inherently independent. The radian, which measures plane angles, and the steradian, which measures solid angles, are two additional units that are included in the system. These base quantities serve as the foundation for all other physical quantities. For instance, the fundamental quantities length, mass, and time can be used to generate mechanical quantities as shown in Table 1.2. In addition, the greater and smaller values for each physical quantity specified in Table 1.3 are represented by powers of 10. The most recent definition for the SI units of length is the distance covered by light in a vacuum in a time interval of 1/299792458 seconds, knows as "the metre". The SI units of weight is a certain platinum-iridium alloy cylinder stored by the International Bureau of Weights and Measures is known as a kilogramme. The SI units of second is defined as radiation from cesium-133 atoms for 9192631770 cycles.

Table 1. 1 SI Base Units

Base Quantity	Unit Name	Unit Symbol
Length	meter	m
Mass	kilogram	kg
Time	second	s
Temperature	kelvin	K
Electric Current	ampere	A
Luminous Intensity	candela	cd
Amount of Substance	mole	mol

Table 1. 2 SI Derived Units

Derived Quantity	Unit Name	Unit Symbol	Equivalent SI Units
Force	Newton	N	$m.kg.s^{-2}$
Pressure	Pascal	Pa	$N.m^{-2}$
Energy	Joule	J	N.m
Power	Watt	W	$J.s^{-1}$
Frequency	Hertz	Hz	s^{-1}
Electric charge	Coulomb	C	s.A
Electric potential	Volt	V	$W.A^{-1}$

1.3 CONVERSION FACTORS

In addition to the SI units, there are two other important systems of units. The (CGS) system of units, which bases its measurements on the centimetre, gramme, and second, and the (FPS) system of units, which bases its measurements on the foot, pound, and second. The conversion factors between different systems of length, mass, and time units and the SI units are

- 1 m = 39.37 in = 3.281 ft = 6.214×10^{-4} mi
- 1 kg = 10^3 g = 0.0685 slug = 6.02×10^{26} u
- 1 s = 1.667×10^{-2} min = 2.778×10^{-4} h = 3.169×10^{-8} yr

Table 1. 3 SI Prefixes

Factor	Prefix	Symbol
10^{-24}	yocto	y
10^{-21}	zepto	z
10^{-18}	atto	a
10^{-15}	femto	f
10^{-12}	pico	p
10^{-9}	nano	n
10^{-6}	micro	μ
10^{-3}	milli	m
10^{-2}	centi	c
10^{-1}	deci	d
10^{1}	deka	da
10^{2}	hecto	h
10^{3}	kilo	k
10^{6}	mega	M
10^{9}	giga	G
10^{12}	tera	T
10^{15}	peta	P
10^{18}	exa	E
10^{21}	zetta	Z

Example 1.1 What is a tree's length in inches and feet if it is 12 metres long?

Solution: $12m = (12m)\left(\frac{39.37in}{1m}\right) = 472.44in$

$$12m = (12m)\left(\frac{3.281ft}{1m}\right) = 39.37ft$$

Example 1.2 What is the room's volume in cubic foot if it is 2700 cm^3?

Solution: $2700cm^3 = (2700cm^3)\left(\frac{1m}{100cm}\right)^3 = 0.0027m^3$

$$0.0027m^3 = (0.0027m^3)\left(\frac{3.281ft}{1m}\right)^3 = 0.0953ft^3$$

1.4 DIMENSIONAL ANALYSIS

The letters L, M, and T, respectively, are used to denote the dimensions of length, mass, and time. A technique for determining the correct expressions and validating equations is called Dimensional Analysis. Because they follow the laws of algebra, only dimensions that are the same can be added or subtracted. The terms on both sides of an equation must have the same dimension in order to validate it. Brackets [ ] are used to indicate a physical quantity's dimension. A physical quantity's dimension is merely an expression of its underlying base quantities. Dimensionally consistent equations are required for all equations that express physical laws or principles. This fact can be used to create new physical laws, assess the validity of proposed correlations between physical quantities, and make it easier to recall physical rules.

Example 1.3 Find the dimensions of physical quantities *s*, *v*, and *a*. Also analyse the dimensional consistency of the following equations: (a) $v = u + at$; (b) $s = vt^2 + at$; (c) $v = cos(at^2/s)$.

Solution: [s] = length units = L; [*v*] = velocity (m/s) = L/T = LT^{-1}; [*a*]=acceleration (m/s^2)= LT^{-2};

(a) $[v] = LT^{-1}$
$[u] = [v] = LT^{-1}$

$[at] = [a].[t] = LT^{-2}.T = LT^{-1}$

This equation is dimensionally consistent since each of the three terms has the same dimension.

(b) $[s] = L$

$[vt^2] = LT^{-1}.T^2 = LT$

$[at] = LT^{-2}.T = LT^{-1}$

This equation is dimensionally inconsistent because none of the three terms have the same dimension as any other.

(c) We must first confirm that the cos function's argument is dimensionless because this equation contains a trigonometric function:

$[at^2/s] = [a].[t]^2/[s] = LT^{-2}.T^2/L = 1$

The argument has no dimensions. All is well thus far. The two terms in the equation—the left expression and the right expression—need to know each have their dimensions verified:

$[v] = LT^{-1}$

$[cos(at^2/s)] = 1$

Because the dimensions of the two terms differ, the equation cannot be dimensionally consistent.

1.5 VECTORS

When investigating physical quantities observed in nature, it is discovered that certain of these quantities, such an object's mass or the interval between two events, may be fully described by providing a number and its unit. These are referred to as scalar quantities. It is also discovered that other quantities, such as the force acting on an object, can be fully represented by providing a number, its unit, and a specific direction. These quantities are referred to as vector quantities.

Scalar quantities follow the laws of common arithmetic and have magnitude but no direction. The following are a few examples: mass, temperature, energy, volume, and time intervals denoted by a letter, such as *m*, *T*, *E*, *v*, and *t*. Vector quantities follow the laws of vector algebra and have both magnitude and direction. Displacement, force, velocity, and acceleration are some examples. A bold face letter, such **A**, designates a vector analytically. This notation is typically used in printed materials, as it is in this book. The indication $\vec{A}$ is used when writing by hand.

In handwriting, the magnitude of A denoted as $|\vec{A}|$ or in print, denoted as $|A|$ or A. According to **Fig. 1. 1**, a scaled-down arrow ST serves as the geometric representation of a vector. No matter whether coordinate system is used, the length and direction of the arrow describe the vector's magnitude and direction, respectively. S is referred to as the

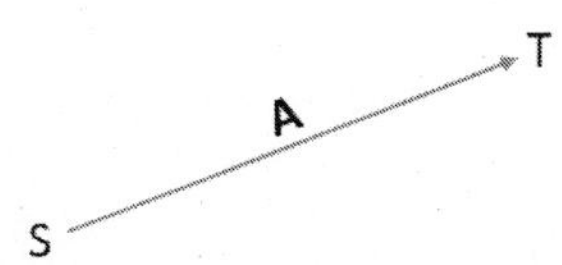

Fig. 1. 1 Geometrically, a vector is represented by an arrow ST

starting point (tail of A), while T is referred to as the terminal point (head of A).

1.6 VECTORS ALGEBRA

The use of mathematical operations on vectors will be covered in this section.

1.6.1 Equality of Two Vectors

Whether or not the initial points of the two vectors P and Q are same as shown in **Fig. 1. 2**, they are only said to be equal (P = Q) if their magnitude and direction are the same.

Fig. 1. 2 Equality of two vectors

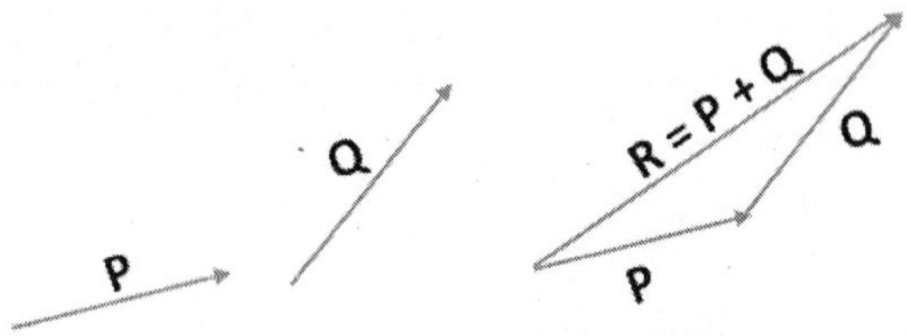

Fig. 1. 3 Geometric method for addition of two vectors

1.6.2 Addition

Vector addition can be done algebraically or geometrically. Here, we'll talk about the geometric approach, which can be used to solve problems without the use of a coordinate system. We'll talk about the algebraic approach later. Drawing a vector from the tail of P to the head of Q as shown in **Fig. 1. 3** will combine the two vectors P and Q using the geometric method. This technique is known as triangle approach.

Fig. 1. 4 illustrates an extension to add up more than two vectors. Fig. 1.5 illustrates a different method for adding vectors using the geometric approach. This technique, where R is the diagonal of a parallelogram with sides P and Q, is known as the parallelogram method. Analytically, Fig. 1.6 demonstrates that to find R,

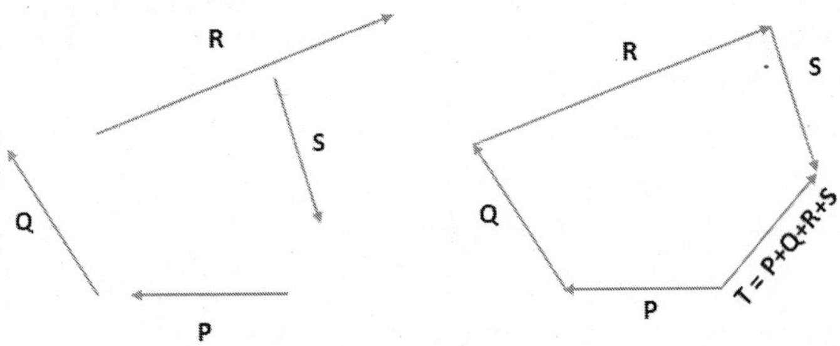

Fig. 1. 4 Numerous vectors addition using geometric method

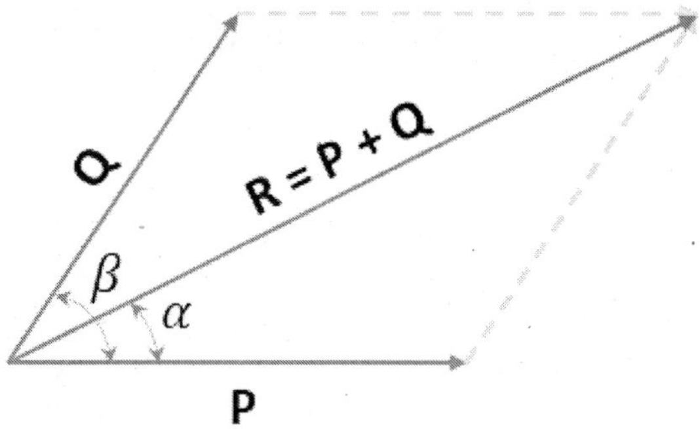

Fig. 1. 5 Adding two vectors using parallelogram method

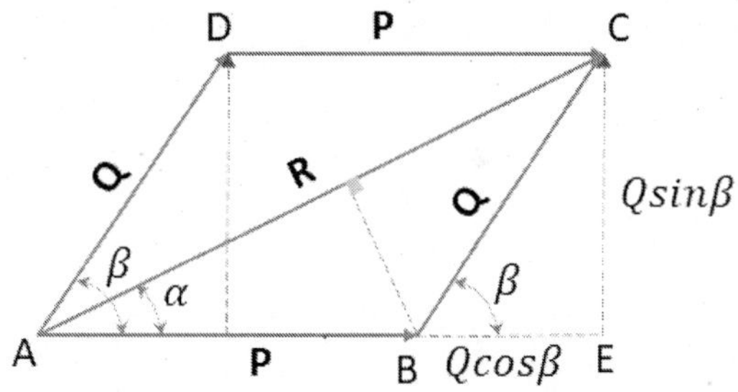

Fig. 1. 6 Determining the magnitude and direction of R

$$AC^2 = AE^2 + EC^2 \tag{1.1}$$

$$AE = AB + BE = P + Qcos\beta \tag{1.2}$$

Eq. (1.1) becomes,

$$R^2 = (P + Qcos\beta)^2 + (Qsin\beta)^2 = P^2 + Q^2 + 2PQcos\beta \tag{1.3}$$

Or

$$R = \sqrt{(P^2 + Q^2 + 2PQcos\beta)} \tag{1.4}$$

The direction of R is

$$tan\alpha = \frac{CE}{AE} = \frac{Qsin\beta}{P+Qcos\beta} \tag{1.5}$$

In contrast to the addition of scalar components, the "magnitude" of the resultant vector R is not always equal to P + Q. It is important to keep in mind that the "magnitude" of the resultant vector R is equal to P + Q only when P and Q are parallel.

Example 1.4 From her house, a jogger travels 0.8 km in a south direction before going 1.5 km west. Calculate the size and the direction of her resulting displacement.

Solution: We can observe from Fig. 1.7 that the resultant displacement's size is determined by

$$R = \sqrt{(0.8km)^2 + (1.5km)^2} = 1.7m$$

The direction of R is

$$\theta = \tan^{-1}\left(\frac{0.8m}{1.5m}\right) = 28.072^0$$

south of west

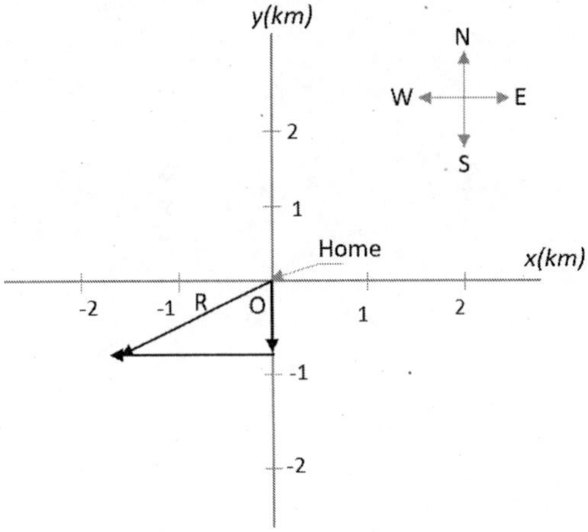

Fig. 1. 7 The total displacement of the jogger is the vector **R**

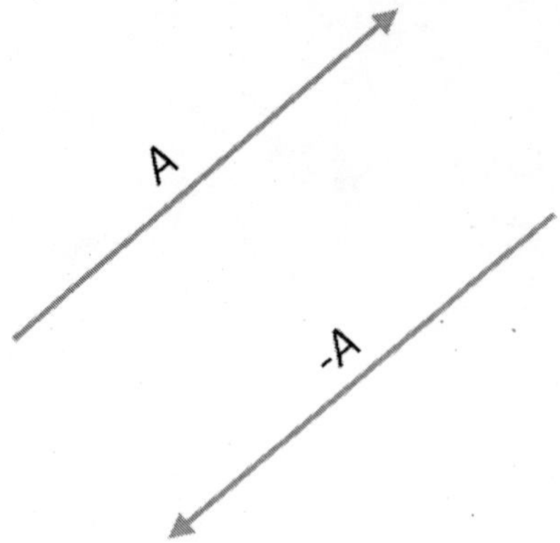

Fig. 1. 8 The vector A and -A are same in magnitude but opposite in direction

1.6.3 Negative of vector

The magnitude of vector A and negative vector of -A is same is same in magnitude but opposite in direction, as seen in Fig. 1.8.

1.6.4 Zero Vector

A zero vector is vector of zero in magnitude and having undefined direction.

1.6.5 Subtraction of Vectors

The vector P - Q is referred to as the vector that results from adding Q results to P. As illustrated in Fig. 1.9, P-Q is equivalently defined as the vector P added to the vector -P (P + (Q)).

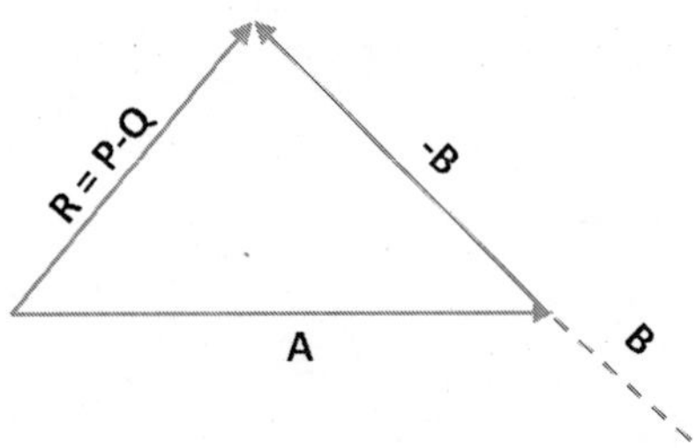

Fig. 1. 9 Subtraction of two vectors

1.6.6 Multiplication of a vector by a scaler

A vector *nP* or *Pn* is created when a vector P and a scalar n are combined. As seen in Fig. 1.10, it has the magnitude nP and moves in the same direction as P if n is positive and the opposite of P if n is negative.

1.6.7 Some Properties

- P+Q = Q+P (Commutative addition law) This is seen in Figure 1.11.
- (P+Q)+R = P+(Q+R), as may be observed in Fig. 1.12 (Associative law of addition).
- P+0 = P
- P+(-P) = 0
- m(nP) = (mn)P = n(mP) (m and n are scalers) (Associative law of association)
- m(P+Q) = mP+mQ (Distributive law)
- (m+n)P = mA+nA (Distributive law)
- 1·P = P, 0·A = 0 (Here, the direction of zero vector is the same.)

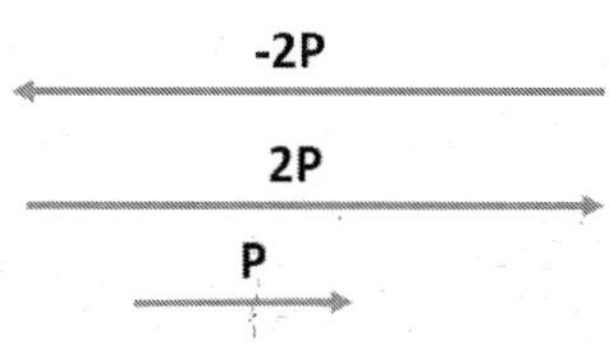

Fig. 1. 10 Scaler product of a vector

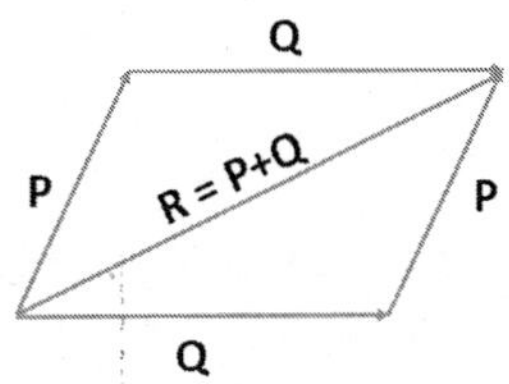

Fig. 1. 11 Commutative law addition

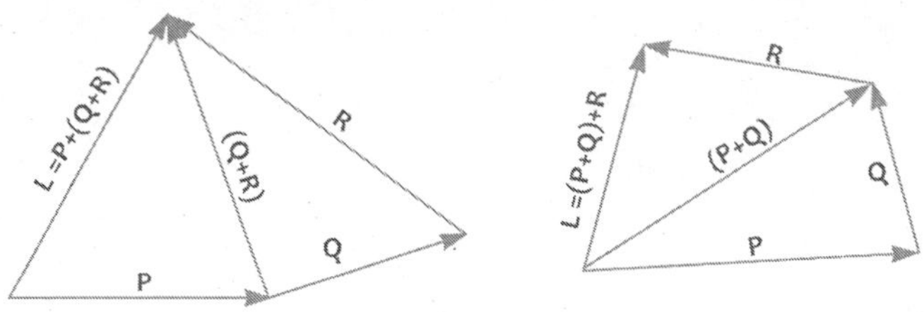

Fig. 1. 12 Associative law of addition

1.6.8 The Unit Vector

A vector with a magnitude of unity and the same direction as P, is the unit vector.
$\hat{P} = P/|P|$ is a unit vector when P = 1.

1.6.9 The scalar (Dot) Product

The scalar product is a scalar quantity with the formula P·Q = PQ cos θ, where θ is the smaller angle between P and Q ($0 \le \theta \le \pi$) (see Fig. 1.13).

1.6.9.1 Some Properties of scaler product

- P·Q = Q·P (Commutative law of scaler product)
- P·(Q+R) = P·Q + P· *R* (Distributive law)
- m(P·Q) = (mP) ·*Q = P(mQ) = (P·Q)m where m is a scalar*

1.6.10 The Vector (Cross) Product

A vector quantity with the definition R = P×Q (read P cross Q) and a magnitude of |P×Q| = PQ sin θ, ($0 \leq \theta \leq \pi$) is the vector product. The right-hand rule turned from P to Q, as shown in Fig. 1.14, can be used to determine the direction of R. R is parallel to the plane that is created by P and Q.

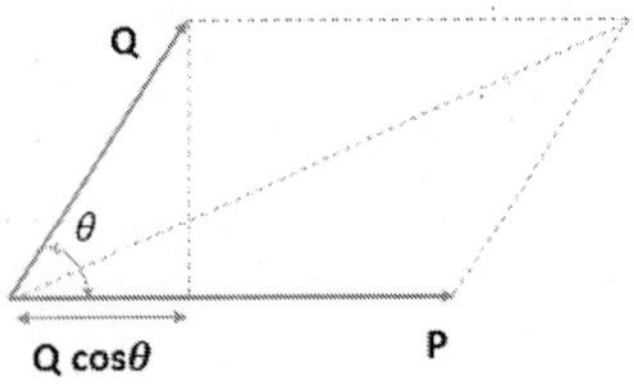

Fig. 1. 13 The scalar product of two vectors

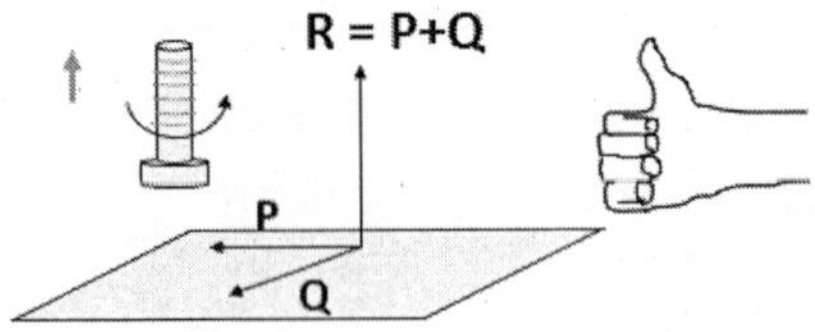

Fig. 1. 14 The vector product of two vectors

1.6.10.1 Some properties

- *P·P = P²,0·A = 0*
- *P×Q = - P×Q*
- $P\times (Q + R) = P \times Q + P \times R$ *(Distributive law)*
- *(P+Q) ×R = P ×Q + Q × R*
- *m(P × R) = (mP) ×Q = P ×(mQ) = (P ×Q)m* , where m is a scaler
- $|P \times Q|$= The area of a parallelogram that has sides P and Q as shown in Fig. 1.15

1.7 Coordinates System

It is necessary to employ a coordinate system in order to describe the position of a point in space. A set of marked axes plus a reference point known as the origin O, make up a coordinate system. Positive axes point in the direction of rising numbers, whereas negative axes go in the opposite direction. The polar coordinates of a point are shown in Figures 1.17 and 1.16, respectively. The rectangular coordinates x and y are connected to the polar coordinates r and by the following relations:

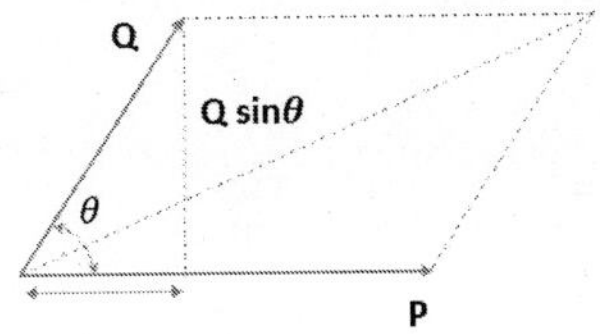

Fig. 1. 15 Area of a parallelogram that has sides P and Q

The magnitude of the vector product $|\boldsymbol{IP\ x\ QI}|$ = is the area of a parallelogram with sides P and Q.

$$x = r\,cos\theta \quad (1.6a)$$
$$y = r\,sin\,\theta \quad (1.6b)$$
$$tan\,\theta = \frac{y}{x} \quad (1.6c)$$
$$r = \sqrt{x^2 + y^2} \quad (1.6d)$$

Fig. 1.18 illustrates the system of cartesian coordinates in three dimensions. The cylindrical and spherical coordinates are other three-dimensional coordinate systems that are employed (Figs. 1.19 and 1.20).

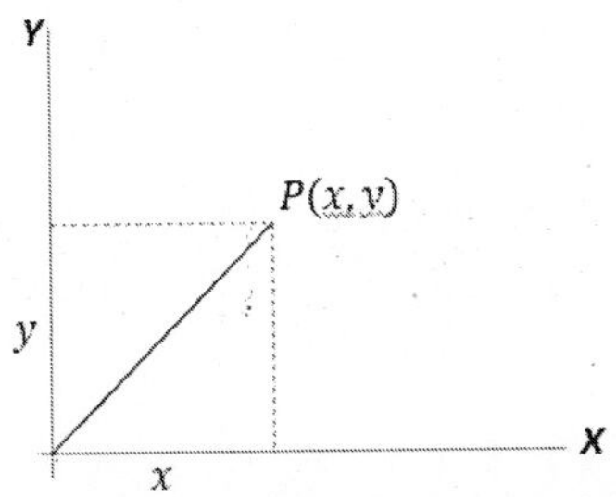

Fig. 1. 16 The rectangular (cartesian) coordinate system.

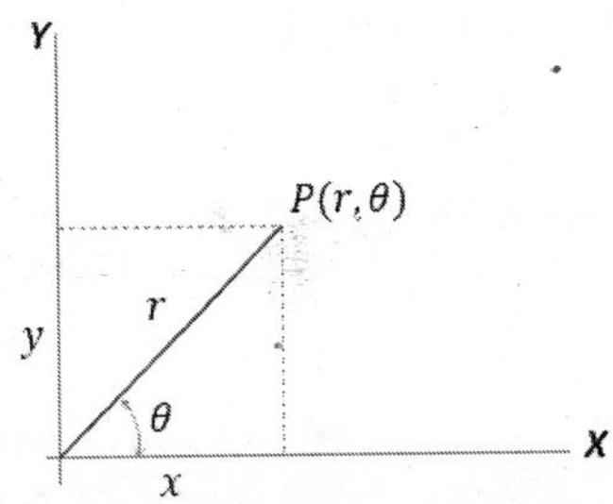

Fig. 1. 17 The polar coordinate system.

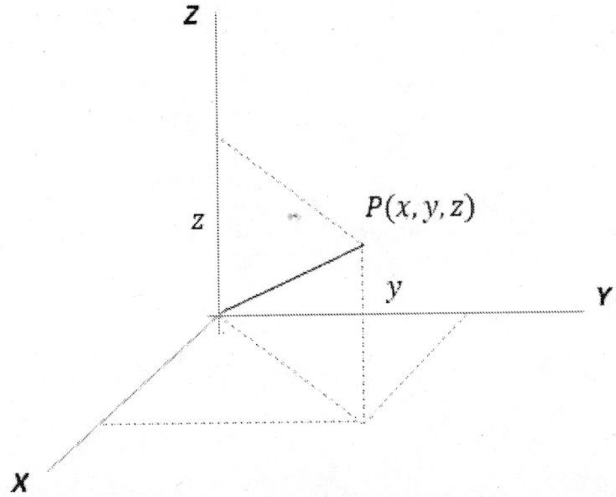

Fig. 1. 18 The cartesian coordinate system in three dimensions.

1.8 Vectors in terms of components

The vector P may be written as the combination of two additional vectors in two dimensions, where $P_x = P\cos\theta$ and $P_y = P\sin\theta$ are displayed in Fig. 1.21.

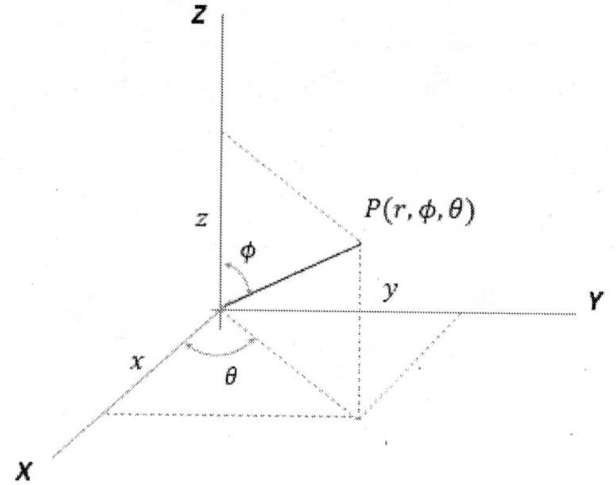

Fig. 1. 19 The spherical coordinate system

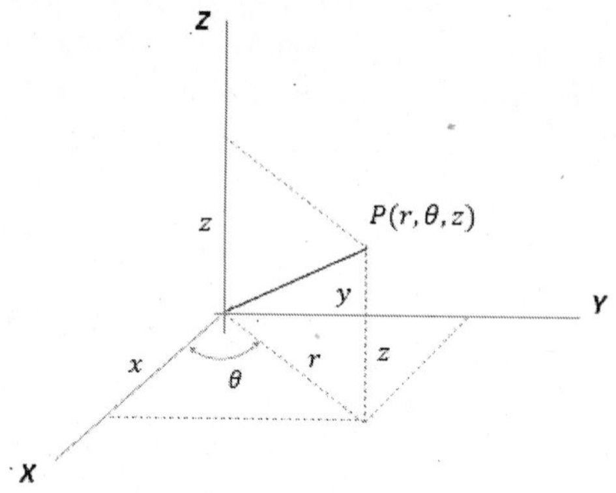

Fig. 1. 20 The cylindrical coordinate system

The rectangular components of P, known as components of P in the x and y directions, are P_x and P_y. the magnitude and direction of P through the expression:

$$P = \sqrt{P_x^2 + P_y^2} \quad (1.7a)$$

The magnitude of P in three dimensions (see Fig. 1.22) is given by

$$P = \sqrt{P_x^2 + P_y^2 + P_z^2} \quad (1.7b)$$

Direction is given by

$$\cos\alpha = \frac{P_x}{P}, \cos\beta = \frac{P_y}{P}, \cos\gamma = \frac{P_z}{P} \quad (1.7c)$$

In two dimensions (Fig. 1.21), the vector P can be expressed as the sum of two other vectors $P = P_x + P_y$, where $P_x = P\cos\theta$ and $P_y = P\sin\theta$

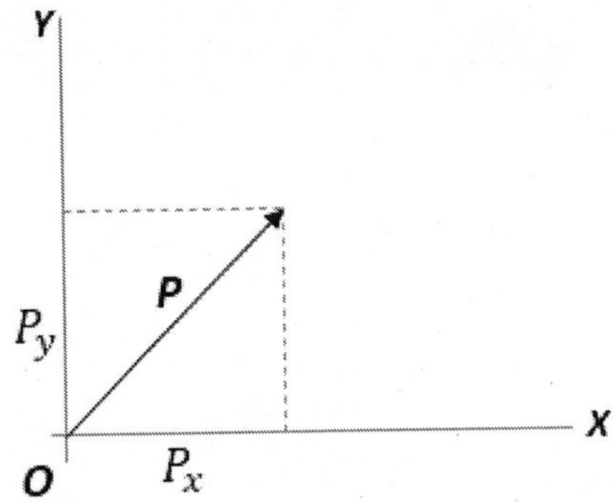

Fig. 1. 21 Vector P in two-dimensions

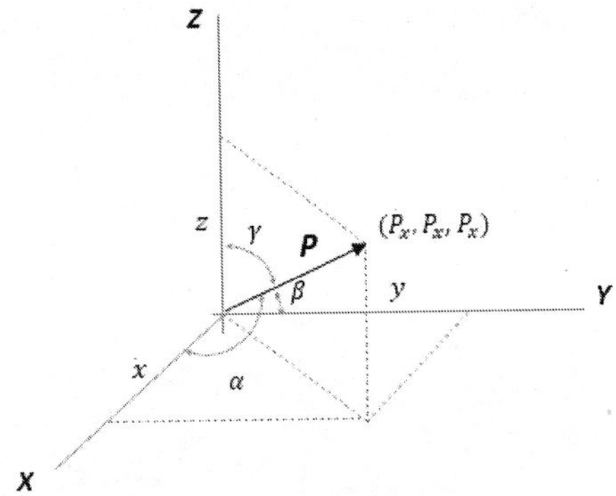

Fig. 1. 22 Vector P in three-dimensions

$$P = \sqrt{P_x^2 + P_y^2 + P_z^2} \tag{1.7d}$$

1.8.1 Rectangular Unit vectors

In the rectangular coordinate system shown in Fig. 1.23, the rectangular unit vectors e_1, e_2, and e_3 are unit vectors that are specified to be in the direction of the positive x-, y-, and z-axes, respectively. It should be noted that this method of axes marking creates a right-handed system. This term comes from the fact that a right-handed screw will progress in the positive z-direction when it is turned through 90 degrees from the x-axis into the y-axis. (Note that the right-handed coordinate system is employed throughout this book.) The unit vector representation of vector P is –

$$P = P_x e_1 + P_y e_2 + P_z e_3 \tag{1.8}$$

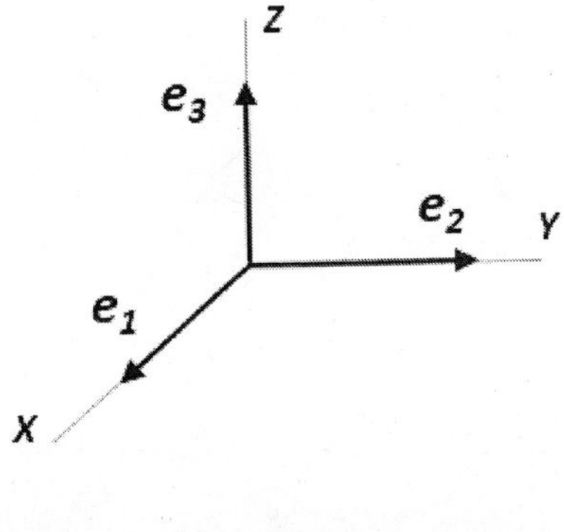

Fig. 1. 23 Representation of the rectangular unit vectors.

The rectangular unit vectors e_1, e_2 and e_3 are unit vectors defined to be in the direction of the positive x, y, and z axes respectively as shown in Fig. 1.23.

1.8.2 Component Method

Suppose we have $P = P_x e_1 + P_y e_2$ and $Q = Q_x e_1 + Q_y e_2$

1.8.2.1 Addition

The resultant vector C is given by

$$R = P + Q = (P_x + Q_x)e_1 + (P_y + Q_y)e_2 = R_x e_1 + R_y e_2 \tag{1.9a}$$

$$R_x = P_x + Q_x$$

$$R_y = P_y + Q_y$$

Magnitude of R is

$$R = \sqrt{R_x^2 + R_y^2}$$

Direction of R is

$$\tan\theta = \frac{R_y}{R_x} = \frac{P_y + Q_y}{P_x + Q_x}$$

For three dimensions -

$$R = (P_x + Q_x)e_1 + (P_y + Q_y)e_2 + (P_z + Q_z)e_3 = R_x e_1 + R_y e_2 + R_z e_3 \tag{1.9b}$$

Magnitude of R is

$$R = \sqrt{R_x^2 + R_y^2 + R_z^2}$$

And directions are

$$\cos\alpha = R_x / R,\ \cos\beta = R_y / R,\ \cos\gamma = R_z / R$$

It is simple to add any quantity of vectors using this component methodology.

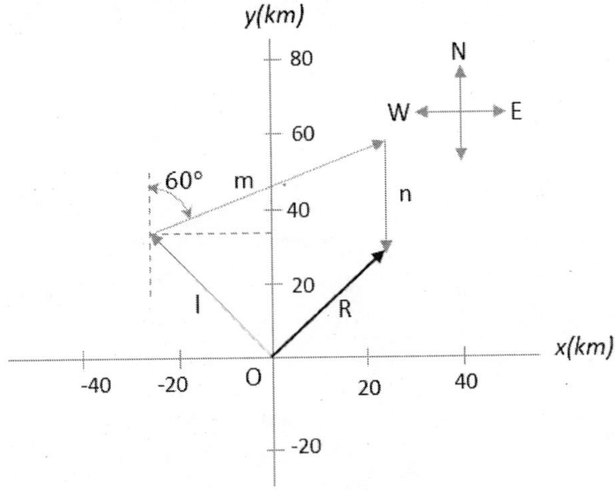

Fig. 1. 24 Representation of resultant vector R

Fig. 1.24 shown that the displacements are drawn to scale with the head of l placed at the tail of m and the head of m placed at the tail of n. the resultant vector R is the vector that extends from the tail of l to the head of n.

Example 1.5 A car goes 40 kilometres toward the northwest, 60 kilometres at 30 degrees north of east, and then 15 kilometres straight south. Calculate the displacement of car from its initial position, both visually and analytically, to determine its size and direction.

Solution 1.5: Drawing the displacements to scale graphically in Fig. 1.24, the head of A is positioned at the tail of B, and the head of B is positioned at the tail of C. The vector that runs from the tail of A to the head of C is the resulting vector, R. The magnitude of R is calculated to have a value of 34.8 km and a direction of 49.8° from the positive x axis using graph paper and a protractor. Analytically, we can see from Fig. 1.24 that

$$A_x = A\cos 135^0 = (40km)(-0.707) = -28.28km$$

$$A_y = A\sin 135^0 = (40km)(0.707) = 28.28km$$

$$B_x = B\cos 30^0 = (60km)(0.866) = 51.96km$$

$$B_y = B\sin 30^0 = (60km)(0.5) = 30km$$

$$C_x = C\cos 270^0 = (15km)(0) = 0$$

$$C_y = C\sin 270^0 = (15km)(-1) = -15km$$

$$R = A + B + C = (A_x + B_x + C_x)e_1 + (A_y + B_y + C_y)e_2 + (A_z + B_z + C_z)e_3 = 23.68e_1 + 43.28e$$

Thus, the magnitude of R is given by

$$R = \sqrt{R_x^2 + R_y^2} = \sqrt{(23.68km)^2 + (43.28km)^2} = 49.33\,km$$

and its direction is

$$\theta = \tan^{-1}\left(\frac{43.28km}{23.68km}\right) = 61.31^0$$

north of east.

1.8.2.2 Subtraction

$$R = P - Q = (P_x - Q_x)e_1 + (P_y - Q_y)e_2 + (P_z - Q_z)e_3 \qquad (1.9c)$$

The direction and magnitude of R are the same as in the case of addition. Except that the plus sign is swapped out for the negative sign,

1.8.2.3 Scalar Product

$$P{\bullet}Q = (P_x e_1 + P_y e_2 + P_z e_3){\bullet}(Q_x e_1 + Q_y e_2 + Q_z e_3) \qquad (1.9d)$$

We obtain nine terms by applying the distributive rule and the definition of scalar product:

since $\mathbf{e_1} \cdot \mathbf{e_1} = \mathbf{e_2} \cdot \mathbf{e_2} = \mathbf{e_3} \cdot \mathbf{e_3}$, and $\mathbf{e_1} \cdot \mathbf{e_2} = \mathbf{e_2} \cdot \mathbf{e_3} = \mathbf{e_3} \cdot \mathbf{e_1} = \mathbf{0}$, we get

$$P{\bullet}Q = P_x Q_x + P_y Q_y + P_z Q_z$$

The dot product of any vector (for example A) by itself is

$$P{\bullet}P = P^2 = P_x^2 + P_y^2 + P_z^2$$

1.8.2.4 The Angle Between Two Vectors

$$P{\bullet}Q = PQ\cos\theta = P_x Q_x + P_y Q_y + P_z Q_z \qquad (1.9e)$$

$$\cos\theta = \frac{P_x Q_x + P_y Q_y + P_z Q_z}{PQ}$$

Example 1.6: Two vectors **P** and **Q** are given by **P** = **2e₁** + **7e₂** – **4e₃** and **Q** = **3e₁** – **4e₂** + **7e₃**. Find the angle between them.

Solution 1.6

$$P \bullet Q = PQ\cos\theta = P_x Q_x + P_y Q_y + P_z Q_z$$

$$P = \sqrt{P_x^2 + P_y^2 + P_z^2} = \sqrt{4 + 49 + 16} = 8.3$$

$$Q = \sqrt{Q_x^2 + Q_y^2 + Q_z^2} = \sqrt{9 + 16 + 49} = 8.6$$

$$\cos\phi = \frac{P_x Q_x + P_y Q_y + P_z Q_z}{PQ} = \frac{6 - 28 - 21}{(8.3)(8.6)} = -0.7$$

$$\phi = 134.46^0$$

1.8.2.5 Perpendicular and Parallel Vectors

If **P×Q** = 0, nonzero vectors P and Q are parallel, and **P** · **Q** = 0 else they are perpendicular. P = mQ states that any two parallel vectors P and Q have the same direction if m > 0 and the opposite, if m < 0. We can also write.

$$\frac{P}{Q} = m$$

$$\frac{P_x}{Q_x} = \frac{P_y}{Q_y} = \frac{P_z}{Q_z} \quad (1.9f)$$

1.8.2.6 Vector Product

We can see from the definition of the vector product that

$$e_1 \times e_1 = e_2 \times e_2 = e_3 \times e_3 = 0$$

$$e_1 \times e_2 = e_3, e_2 \times e_3 = e_1, e_3 \times e_1 = e_2$$

$$e_2 \times e_1 = -e_3, e_3 \times e_2 = -e_1, e_1 \times e_3 = -e_2$$

The positive products are given by reading anticlockwise and the negative products are given by reading clockwise if the unit vectors are written in a circle as in Fig. 1.25. The coordinate system used for these results is right-handed. We have

$$P \times Q = \left(P_x e_1 + P_y e_2 + P_z e_3\right) \times \left(Q_x e_1 + Q_y e_2 + Q_z e_3\right) \tag{1.9g}$$

We obtain using the distributive rule and the unit vector relationships mentioned above

$$P \times Q = \left(P_y Q_z - P_z Q_y k\right) e_1 + \left(P_z Q_x - P_x Q_z\right) e_2 + \left(P_x Q_y - P_y Q_x\right) e_3$$

given that an order 2 determinant is defined as

$$\begin{vmatrix} a_1 & a_2 \\ b_1 & b_2 \end{vmatrix} = a_1 b_2 - a_2 b_1$$

Then, the expression above may be expressed as

$$A \times B = \begin{vmatrix} A_y & A_z \\ B_y & B_z \end{vmatrix} e_1 - \begin{vmatrix} A_x & A_z \\ B_x & B_z \end{vmatrix} e_2 + \begin{vmatrix} A_x & A_y \\ B_x & B_y \end{vmatrix} e_3 \tag{1.9h}$$

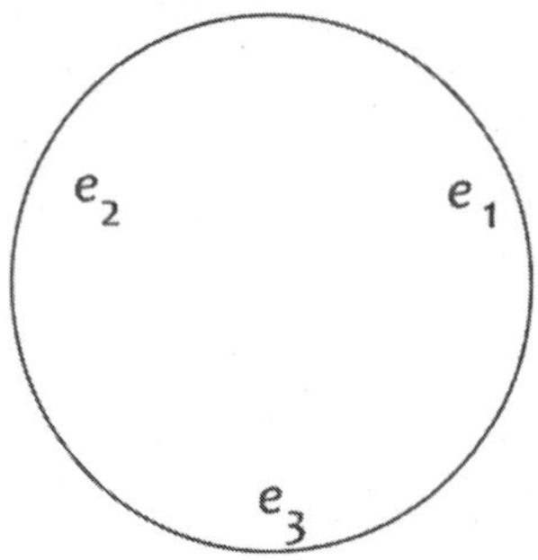

Fig. 1. 25 Sign convention of unit vector products

Fig. 1.25 represents the unit vectors around a circle, then reading counter clockwise gives the positive products and reading clockwise gives the negative products

Example 1.7 Two vectors **P** and **Q** are given by **P** = $-2e_1 + 5e_2$ and **Q** = $3e_1 + 2e_2$. Find: (a) the sum of **P** and **Q**, ;(b) − **Q** and **3P** ;(c) **P** · **Q** and **P** × **Q**.

Solution 1.7

(a)

$$R = P + Q = \left(P_x + Q_x\right) e_1 + \left(P_y + Q_y\right) e_2 = (-2+3) e_1 + (5+2) e_2 = e_1 + 7e_2$$

$$R_x = 1$$

$$R_y = 7$$

(b)

$$-Q = -3e_1 - 2e_2$$

$$3P = -6e_1 + 15e_2$$

(c)

$$P \cdot Q = (-2e_1 + 5e_2)(3e_1 + 2e_2) = -2e_1 \cdot 3e_1 - 2e_1 \cdot 2e_2 + 5e_2 \cdot 3e_1 + 5e_2 \cdot 2e_2 = -6 + 10 = 4$$

$$P \times Q = (-2e_1 + 5e_2) \times (3e_1 + 2e_2) = -2e_1 \times 2e_2 + 5e_2 \times 3e_1 = -4e_3 - 15e_3 = -19e_3$$

Example 1.8: To each of the vectors, discover a vector of magnitude 1 that is perpendicular. **P** = **3e$_1$ +2 e$_2$−7e$_3$** and **Q** = **8e$_1$+6e$_2$−3e$_3$**.

Solution 1.8: According to the definition of unit vector, we have

$$R = \frac{P \times Q}{|P \times Q|}$$

According to the definition of a unit vector, we may write as follows: where R is a unit vector perpendicular to the plane formed by P and Q. we have

$$P \times Q = \begin{vmatrix} e_1 & e_2 & e_3 \\ 3 & 2 & -7 \\ 8 & 6 & -3 \end{vmatrix} = 39e_1 + 46e_2 + 2e_3$$

$$|P \times Q| = \sqrt{(39)^2 + (46)^2 + (2)^2} = 60.44$$

$$R = \frac{39e_1 + 46e_2 + 2e_3}{60.44} = 0.64e_1 + 0.76e_2 + 0.033e_3$$

Example 1.9 Given that **X** = **4e$_1$ − 2e$_2$ − 7e$_3$**, **Y** = **5e$_1$ − 2e$_2$** and **Z** = **3e$_2$ − 5e$_3$**, find (a) **X × Y** (b) **(X × Y) × Z** (c) **X ·(Y × Z)**.

Solution 1.9

(a)

$$X \times Y = \begin{vmatrix} e_1 & e_2 & e_3 \\ 4 & -2 & -7 \\ 5 & -2 & 0 \end{vmatrix} = -14e_1 - 35e_2 + 2e_3$$

(b)

$$(X \times Y) \times Z = \begin{vmatrix} e_1 & e_2 & e_3 \\ -14 & -35 & 2 \\ 0 & 3 & -5 \end{vmatrix} = 169e_1 - 70e_2 - 42e_3$$

(c)

$$Y \times Z = \begin{vmatrix} e_1 & e_2 & e_3 \\ 5 & -2 & 0 \\ 0 & 3 & 5 \end{vmatrix} = -10e_1 + 25e_2 + 15e_3$$

$$X \cdot (Y \times Z) = (4e_1 - 2e_2 - 7e_3) \cdot (-10e_1 + 25e_2 + 15e_3) = -40 - 50 - 105 = 15$$

Example 1.10 Find the area of a triangle using the vector approach given the positions of its three vertices. are A(5, 3, 2) , B(4, 1, 3) , C(−2, 4, 1) .

Solution 1.10

$$AB = (5-4)e_1 + (3-1)e_2 + (2-3)e_3 = e_1 + 2e_2 - e_3$$

$$AC = (-2-5)e_1 + (4-3)e_2 + (1-2)e_3 = -7e_1 + e_2 - e_3$$

$$Area = \frac{1}{2}|AB \times AC| = \frac{1}{2}|(e_1 + 2e_2 - e_3) \times (-7e_1 + e_2 - e_3)| = \frac{1}{2}|(-e_1 - 7e_2 + 15e_3)|$$

$$= 2\sqrt{(-1)^2 + (-7)^2 + (15)^2} = 16.58$$

Fig. 1. 26 The triple scalar product is equal to the volume of a parallelepiped with sides P, Q, and R

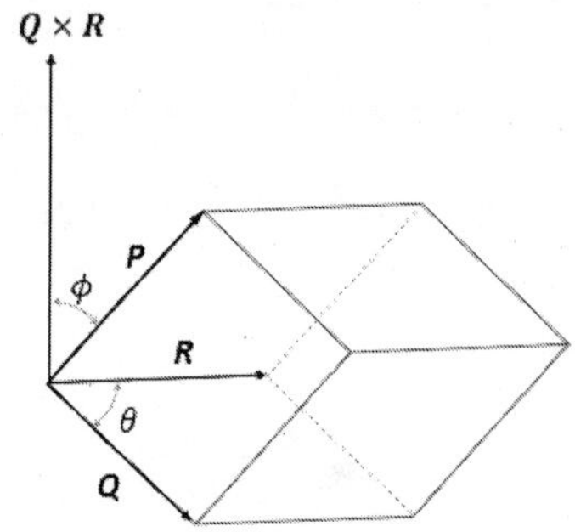

1.8.2.7 Triple Product

Scalar Triple Product

A scalar quantity known as the triple scalar product is defined as P·(Q × R). A determinant involving the elements of the vectors can be used to express this quantity.

$$P \cdot (Q \times R) = \begin{vmatrix} P_x & P_y & P_z \\ Q_x & Q_y & Q_z \\ R_x & R_y & R_z \end{vmatrix} \quad (1.9i)$$

where $P=P_xe_1+P_ye_2+P_ze_3$, $Q=Q_xe_1+Q_ye_2+Q_ze_3$, $R=R_xe_1+R_ye_2+R_ze_3$ Additionally, as shown in Fig. 1.26, the triple scalar product is equal to the volume of a parallelepiped with sides P, Q, and R. The triple scalar product can be represented as **P ·(Q × R)** or as **P ·(Q × R)** since any edges can be utilised. For a coordinate system that is right-handed, these products are positive and negative, respectively. As a result, there exist 6 equivalent triple scalar products, or 12 if the criteria of the type **(Q × R)· P** are taken into account. Three of these six products are positive, while the other four are negative. You may demonstrate by increasing the determinant that

$$P\cdot(Q\times R)=Q\cdot(R\times P)=R\cdot(P\times Q)=-P\cdot(R\times Q)=-Q\cdot(\mathrm{P}\times\mathrm{R})=-R\cdot(\mathrm{Q}\times\mathrm{P})$$

Vector Triple Product

A vector quantity with the definition of the triple vector product is P × (Q × R). By extending this equation, you can demonstrate that

$$\mathrm{P}\times(\mathrm{Q}\times\mathrm{R}) = (\mathrm{P}\cdot\mathrm{R})\mathrm{Q} - (\mathrm{P}\cdot\mathrm{Q})R \tag{1.9j}$$

Example 1.11 Given that $P=P_xe_1$, $Q=Q_xe_1+Q_ze_3$ and $R=R_ye_2$, show that the identity $P\times(Q\times R) = (P\cdot R)Q-(P\cdot Q)R$ is correct

Solution 1.11

$$(Q\times R)=\begin{vmatrix} e_1 & e_2 & e_3 \\ Q_x & 0 & Q_z \\ 0 & R_y & 0 \end{vmatrix}=-Q_zR_ye_1+Q_xR_ye_3$$

$$P\times(Q\times R)=\begin{vmatrix} e_1 & e_2 & e_3 \\ P_x & 0 & 0 \\ -Q_zR_y & 0 & Q_xR_y \end{vmatrix}=-P_xQ_xR_ye_2$$

$$(P\cdot R)Q = 0$$

$$-(\mathrm{P}\cdot\mathrm{Q})R=-(P_xQ_x)R=-P_xQ_xR_ye_2$$

The identity is therefore valid.

1.8.3 Position Vector

A vector that locates one point in relation to another in space is referred to as a position vector. Let's assume that O and P in Fig. 1.27 have the coordinates (0,0,0) and (x_P, y_P, and z_P), respectively. Point P's position vector with respect to point O can be expressed as, $\overrightarrow{r_P}$. It can be written as $\overrightarrow{r_{OP}}$ or $\overrightarrow{r_{P/O}}$ and given by

$$\overrightarrow{r_P} = x_P e_1 + y_P e_2 + z_P e_3$$

Once more, if Q is a different point with the coordinates (x_Q, y_Q, and z_Q), then $\overrightarrow{r_{QP}}$ the position vector of P with respect to Q,

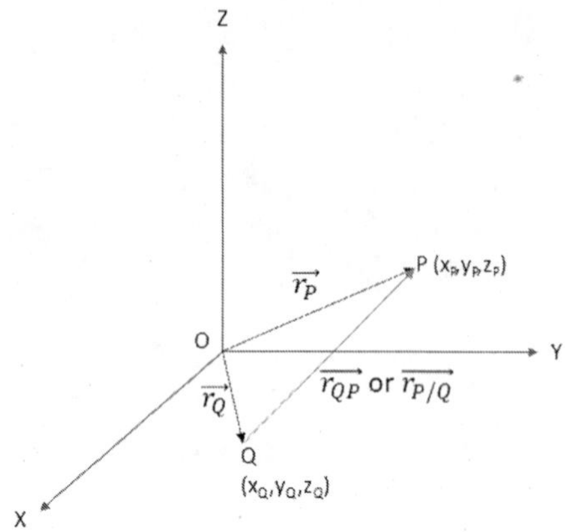

Fig. 1. 27 Position Vector.

$$\overrightarrow{r_P} - \overrightarrow{r_Q} = (x_P e_1 + y_P e_2 + z_P e_3) - (x_Q e_1 + y_Q e_2 + z_Q e_3)$$

Or

$$\overrightarrow{r_{QP}} = (x_P - x_Q)e_1 + (y_P - y_Q)e_2 + (z_P - z_Q)e_3$$

Hence,

$$\left|\overrightarrow{r_{QP}}\right| = \sqrt{(x_P - x_Q)^2 + (y_P - y_Q)^2 + (z_P - z_Q)^2} \quad (1.10)$$

However, we designate it as $\overrightarrow{r_{P/Q}}$ if position P is referred to with respect to Q. Recall that $\overrightarrow{r_{Q/P}}$ =- $\overrightarrow{r_{P/Q}}$.

1.9 Derivatives of Vectors

1.9.1 Some Rule

If A(t) is a vector function of the scalar variable t, then t can be

$$P(t) = P_x(t)e_1 + P_y(t)e_2 + P_z(t)e_3 \quad (1.11a)$$

Then

$$\frac{dP(t)}{dx} = \frac{dP_x(t)}{dx}e_1 + \frac{dP_y(t)}{dx}e_2 + \frac{dP_z(t)}{dx}e_3$$

1.9.2 Some Rule

If ϕ(t) is a scalar function and P(t), Q(t) are vector functions, then

$$\frac{d}{dt}(\phi P) = \phi\frac{dP}{dt} + \frac{d\phi}{dt}P$$

$$\frac{d}{dt}(P\cdot Q)= \text{Q}\frac{dP}{dt}+\frac{dP}{dt}\cdot Q$$

$$\frac{d}{dt}(P\times Q)= P\times\frac{dQ}{dt}+\frac{dP}{dt}\times Q$$

Example 1.12 Two vectors p_1 and p_2 are given by $p_1 = 3t^2 e_1 + 2\cos t\, e_2 + 5e_3$, $p_2 = \sin t\, e_1 - \cos t\, e_2$, find at $t = 0$ (a) $\frac{d^2 p_1}{dt^2}$ and (b) $\frac{d(p_1 \cdot p_2)}{dt}$.

Solution 1.12

(a) $$\frac{dp_1}{dt} = 6te_1 - 2\sin t\, e_2$$

$$\frac{d^2 p_1}{dt^2} = 6e_1 - 2\cos t\, e_2$$

At t = 0

$$\frac{d^2 p_1}{dt^2} = 6e_1 - 2e_2$$

(b)

$$\frac{d(p_1 \cdot p_2)}{dt} = \frac{d\{(3t^2 e_1 + 2cost\, e_2 + 5e_3)(sinte_1 - cos\,t\, e_3)\}}{dt} =$$

$$\frac{d(3t^2\ sin\ t\ \text{-}\ 5\ cost)}{dt} = 6tsin\ t + 3t^2 cost + 5\ sin\ t = (6t+5)sin\ t + 3t^2\ cost$$

At t = 0

$$\frac{d(p_1 \cdot p_2)}{dt} = 0$$

1.9.3 Gradient, Divergence and Curl

If ϕ (t) is a scalar function, P(t), and Q(t), respectively, are vector functions P(x, y, z) is known as a vector field if P = P(x, y, z) is a vector function of the three variables x, y, and z. Scalar field is the term used to describe the scalar function φ(x, y, z).

1.9.3.1 Del

The definition of the vector differential operator del is

$$\nabla = e_1\frac{\partial}{\partial x} + e_2\frac{\partial}{\partial y} + e_3\frac{\partial}{\partial z} \quad (1.11b)$$

1.9.3.2 Gradient

$$\nabla\phi = \left(e_1\frac{\partial}{\partial x}+e_2\frac{\partial}{\partial y}+e_3\frac{\partial}{\partial z}\right)\phi = e_1\frac{\partial\phi}{\partial x}+e_2\frac{\partial\phi}{\partial y}+e_3\frac{\partial\phi}{\partial z} \quad (1.11c)$$

The gradient of the vector $\nabla\phi$ is denoted by the symbol grad ϕ

1.9.3.3 Divergence

$$\nabla\cdot P = \left(e_1\frac{\partial}{\partial x}+e_2\frac{\partial}{\partial y}+e_3\frac{\partial}{\partial z}\right)\cdot\left(P_x e_1+P_y e_2+P_z e_3\right)$$

$$=\frac{\partial P_x}{\partial x}+\frac{\partial P_y}{\partial y}+\frac{\partial P_z}{\partial z} \quad (1.11d)$$

A is referred to as the divergence of A. (written div A).

1.9.3.4 Curl

$$\nabla\times P = \left(e_1\frac{\partial}{\partial x}+e_2\frac{\partial}{\partial y}+e_3\frac{\partial}{\partial z}\right)\times\left(P_x e_1+P_y e_2+P_z e_3\right)$$

$$\begin{vmatrix} e_1 & e_2 & e_3 \\ \frac{\partial}{\partial x} & \frac{\partial}{\partial y} & \frac{\partial}{\partial z} \\ P_x & P_y & P_z \end{vmatrix} = \left(\frac{\partial P_z}{\partial y}-\frac{\partial P_y}{\partial z}\right)e_1+\left(\frac{\partial P_x}{\partial z}-\frac{\partial P_z}{\partial x}\right)e_2+\left(\frac{\partial P_y}{\partial x}-\frac{\partial P_x}{\partial y}\right)e_3 \quad (1.11e)$$

A is referred to as the curl of A. (written curl A).

1.9.3.5 Some Identities

- $div\,curl\,A = \nabla\cdot(\nabla\times A) = 0$
- $curl\,grad\phi = \nabla\times(\nabla\phi) = 0$

Example 1.13: There are two fields: a scalar field q and a vector field p. $p = 2xy^2e_1+3x^2e_2$ and $q = (3y^2-z)e_1+5y^2e_2$, Find at the point (2,-1) (a) $\nabla\cdot p$ (b) $\nabla\times p$ (c) ∇q.

Solution 1.13

(a) $$\nabla\cdot p = \frac{\partial p_x}{\partial x}+\frac{\partial p_y}{\partial y} = 2y^2$$

At (2,-1),

$$\nabla\cdot p = 5$$

(b)

$$\nabla \times p = \begin{vmatrix} e_1 & e_2 & e_3 \\ \dfrac{\partial}{\partial x} & \dfrac{\partial}{\partial y} & \dfrac{\partial}{\partial z} \\ 2xy^2 & 3x^2 & 0 \end{vmatrix} = (6x - 4xy)\, e_3$$

At (2,-1),

$$\nabla \times p = 20 e_3 .$$

(c)

$$\nabla q = \frac{\partial q_x}{\partial x} e_1 + \frac{\partial q_y}{\partial y} e_2 + \frac{\partial q_z}{\partial z} e_3 = 10 y\, e_2$$

At (2,-1),

$$\nabla q = -10 e_2 .$$

1.10 Derivatives of vector

If $P(t) = P_x(t)e_1 + P_y(t)e_2 + P_z(t)e_3$, the indefinite integral has the following definition, where t is a scalar variable:

$$\int P(t)dt = e_1 \int P_x(t)dt + e_2 \int P_y(t)dt + e_3 \int P_z(t)dt \qquad (1.12a)$$

If $P(t) = dQ(t)/dt$, then

$$\int P(t)dt = \int \frac{d}{dx}\{Q(t)\}dt = Q(t) + R$$

where constant vector C can be used. Between the boundaries of t = a and t = b, the definite integral is defined as

$$\int_a^b P(t)dt = \int_a^b \frac{d}{dt}\{Q(t)\}dt = Q(t) + R\Big|_a^b = Q(b) - Q(a)$$

1.10.1 Line integrals

An integral that follows a line or curve is referred to as a line integral. Either this curve is open or closed. Three alternative versions of the line integral are possible, as indicated by $\int_c \phi dr$,

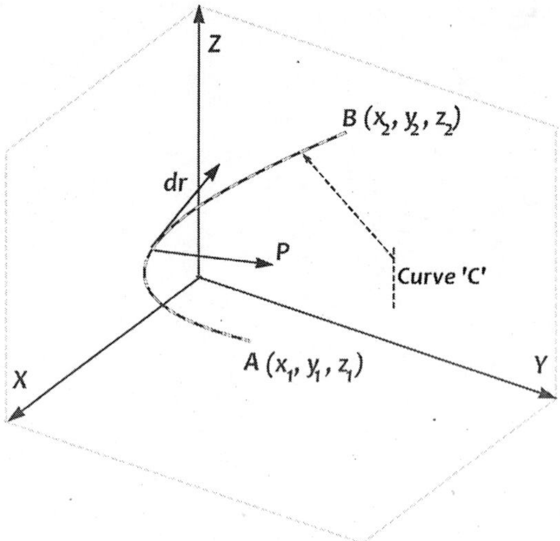

Fig. 1. 28 The line integral.

$\int_c P.dr$, and $\int_c P \times dr$. The second, which is the most usual, will be utilised throughout this book. The position vector of any point '***P***' along curve ***C*** (see Fig. 1.28), which stretches from ***A*** (x_1, y_1, z_1) at t_1 to ***B*** (x_2, y_2, z_2) at t_2 is given by

$$R(t) = x(t)e_1 + y(t)e_2 + z(t)e_3$$

Suppose that $P = P(x, y, z) = P_x e_1 + P_y e_2 + P_z e_3$ is a vector field, the line integral of ***P*** is given by:

$$\int_A^B P.dr = \int_C P.dr = \int_C (P_x dx + P_y dy + P_z dz) \tag{1.12.b}$$

As one can see, the tangential component of P along C is designated as $P.r$. The line integral is represented as if C is a simple closed curve (does not overlap with itself) and is written as

$$\oint_C P.dr = \oint_C (P_x dx + P_y dy + P_z dz)$$

1.10.2 Independence of Path

In general, the line integral relies on the path, however this is not always the case. Instead, it is not dependent on the curve itself but simply on the coordinates of the end points (path) of curve. The line integral in Eq. (1.10.1) joins the points ***A*** and ***B*** if and only if $P = \nabla \phi$, or equivalently $\nabla \times P = 0$ is correct, irrespective of the path. Thus, the value of Eq. (1.10.1) is provided by

$$\int_A^B P \cdot dr = \int_A^B d\phi = \phi(A) - \phi(B) = \phi(x_2, y_2, z_2) - \phi(x_1, y_1, z_1)$$

Note that $\phi(x, y, z)$ has continuous partial derivatives. Furthermore, if the line integral of A is independent of the path, then the line integral of A about any closed path is equal to zero:

As you can see $\phi(x, y, z)$ has continuous partial derivatives. Additionally, if P is line integral and it is independent of the path, then line integral of P about any closed path is equivalent to zero:

$$\oint_C P \cdot dr = 0$$

Example 1.14 A force field is given by $F = \left(4x^2y^3 + 2z^2\right)e_1 + \left(3x^2y^2\right)e_2 + \left(2xz^2 - 1\right)e_3$.

(a) Show that $\nabla \times F$,

(b) Find a scalar function ϕ such that $F = \nabla \phi$

Solution 1.14

(a)
$$\nabla \times F = \begin{vmatrix} e_1 & e_2 & e_3 \\ \dfrac{\partial}{\partial x} & \dfrac{\partial}{\partial y} & \dfrac{\partial}{\partial z} \\ \left(4x^2y^3 + 2z^2\right) & \left(3x^2y^2\right) & \left(2xz^2 - 1\right) \end{vmatrix} = \left(4z - 2z^2\right)e_2 + \left(12x^2y^2 - 6xy^2\right)e_3 = 0$$

(b)

$$F \cdot dr = \nabla \phi \cdot dr = \frac{\partial \phi}{\partial x}dx + \frac{\partial \phi}{\partial y}dy + \frac{\partial \phi}{\partial z}dz = d\phi$$

$$d\phi = \left(4x^2y^3 + 2z^2\right)dx + \left(3x^2y^2\right)dy + \left(2xz^2 - 1\right)dz$$

Hence

$$\phi = \left(\frac{4}{3}x^3y^2 + 2z^2x\right) + \left(y^3x^2\right) + \left(\frac{2}{3}xz^3 - z\right).$$

Example 1.15 A vector **F** is given by $F = 6xy^2e_1 - \left(2y^2 + y\right)e_2$. Compute $\int_c F \cdot dr$ along each of the following paths:

(a) The straight lines from (0, 0) to (0, 1) and then to (1, 1).

(b) Along the straight line $y = x$.

(c) Along the curve $x = t$, $y = t^2$.

Solution 1.15

(a) Along the straight line from (0,0) to (0,1) we have $x = 0$, and $dx = 0$, therefore

$$\int_c F \cdot dr = \int_c 6xy^2dx - \left(2y^2 + y\right)dy = \int_{y=0}^{1} -(2y^2 + y)dy = -\frac{2}{3}y^3\Big|_0^1 - \frac{1}{2}y^2\Big|_0^1 = -\frac{7}{6}$$

Along the straight line from (0, 1) to (1, 1) we have $y = 1$, $dy = 1$, hence

$$\int_c F \cdot dr = \int_{x=0}^{1} 6xy^2dx = 3x^2\Big|_0^1 = 3$$

Thus, we have for the total path

$$\int_c F \cdot dr = -\frac{7}{6} + 3 = \frac{11}{6}$$

(b) Along the straight line $y = x$, we have $dy = dx$,

$$\int_c F \cdot dr = \int_c 6xy^2dx - \left(2y^2 + y\right)dy = \int_{x=0}^{1} (6x^3 - 2x^2 + x)\,dx$$

$$= \left(\frac{2}{3}x^4 - \frac{2}{3}x^3 + \frac{1}{2}x^2\right)\Big|_0^1 = 1/2$$

(c) Finally along the curve $x = t$, $y = t^2$, we have $dx = dt$, $dy = 2tdt$, furthermore the points (0, 0) and (1, 1) corresponds to $t = 0$and $t = 1$, respectively. He

$$\int_c F \cdot dr = \int_c 6xy^2dx - \left(2y^2 + y\right)dy = \int_{t=0}^{1} 6t^3dt - 2t(2t^2 + t)dt$$

$$= 2/3t^4 - t^4 - 2/3t^3\Big|_0^1 = -1/3$$

Example 1.16 If a vector **A** is given by $A = x^2yi - x^3j$, find the line integral $\int_c A \cdot dr$ along the circular arc shown in Fig. 1.28.

Solution 1.16 By using the polar coordinates, we have $x = \cos\theta$ and $y = \sin\theta$ (since r = 1), $dx = -\sin\theta d\theta$ and $dy = \cos\theta d\theta$, also $x^2 + y^2 = r^2 = 1$, therefore we have

$$\int_c A \cdot dr = \int_{\theta=\pi}^{-\pi/4} -\cos^2\theta \sin^2\theta d\theta - \cos^4\theta d\theta = \int_{\theta=\pi}^{-\pi/4} -\cos^2\theta\left(\sin^2\theta + \cos^2\theta\right)d\theta$$

$$= \int_{\theta=\pi}^{-\pi/4} -\cos^2\theta d\theta = -\frac{\sin 2\theta}{4} + \frac{x}{2}\Bigg|_{\pi}^{-\pi/4} = -0.25 - \frac{5\pi}{8}$$

(a)

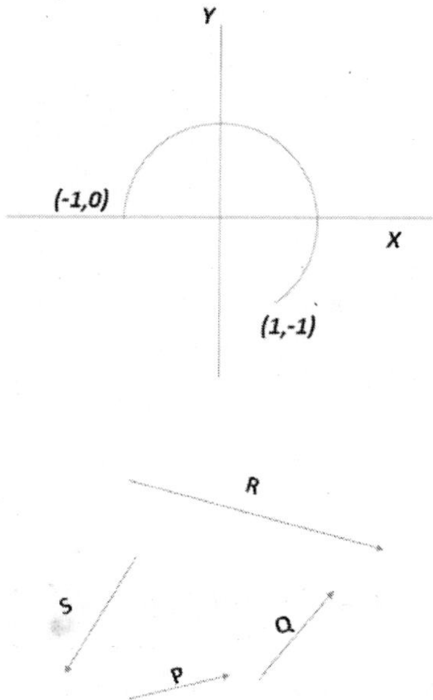

(b)

Fig. 1. 29 (a) The line integral along the curve using polar coordinates, **(b)** Vectors **P**, **Q**, **R** and **S**.

1.11 TENSOR

Tensors are merely mathematical entities that, like scalars and vectors, that can be employed to characterise physical qualities. In reality, tensors are simply a generalisation of "scalars and vectors"; a vector is a first rank tensor and a scalar is a zero-rank tensor. Scalars and vectors, are the two fundamental types of tensors, as well as "dual vectors", multilinear maps across vector spaces, and even some mathematical operations like the dot product, are just a few of the many different types of tensors. Tensors are defined independently of any basis, despite the fact that they are frequently identified by their "components" in a basis connected to a certain coordinate system. It is standard practise in applications to investigate scenarios where a different tensor can exist at each position of an item; for instance, the tension within an object may change from one spot to another. The stress experienced by the fluid (or solid) is expressed by the stress tensor σ_{ij}, which is a physical quantity; this stress is constant regardless of the coordinate system.

1.11.1 Newton's Second Law Equation using Tensor Notation

Newton's second law equation is given below

$$m\frac{d^2\vec{x}}{dt^2} = \vec{F} \tag{1.13a}$$

The Eq. (1.13a) has three components with reference to any coordinate system and it can be written in the component form as shown in equation (1.13b)

$$m\frac{d^2x_1}{dt^2} = F_1;\ m\frac{d^2x_2}{dt^2} = F_2;\ m\frac{d^2x_3}{dt^2} = F_3 \tag{1.13b}$$

The above Eq. (1.13b) can be written in tensor notation as given below

$$m\frac{d^2x_i}{dt^2} = F_i \tag{1.13c}$$

The Eq. (1.13c) should be applied in each coordinate direction if an index (also known as a free index) appears once in each term.

1.11.2 Divergence Equation using Tensor Notation

The divergence equation is written as

$$\nabla . U = 0 \tag{1.13d}$$

The Eq. (1.13d) reads as

$$\frac{\partial U_1}{\partial x_1} + \frac{\partial U_2}{\partial x_2} + \frac{\partial U_3}{\partial x_3} = 0 \Rightarrow \sum_{i=1}^{3} \frac{\partial U_i}{\partial x_i} = 0 \tag{1.13e}$$

The following rule is added in tensor notation: summation should be applied over an index if it appears twice (a "dummy index") within a term. The summation is often calculated using the three coordinate indices, 1, 2, and 3. When using a 2D coordinate system, the summation can only go from 1 to 2. Thus, Eq. (1.13d) is expressed as

$$\frac{\partial U_i}{\partial x_i} = 0 \ or \ U_{i,i} = 0 \ or \ \partial_i U_i = 0 \tag{1.13f}$$

It should be noted that the dummy index can be substituted for any index because it indicates a summation over each term.

i.e.

$$\frac{\partial U_j}{\partial x_j} = 0$$

1.11.3 Symmetric and Antisymmetric Tensors

If $p_{ij} = p_{ji}$, a tensor p_{ij} is said to be symmetric.

If $q_{ij} = -q_{ji}$, a tensor q_{ij} is said to be antisymmetric. Therefore, all diagonal components of an antisymmetric tensor must be zero (For example, $q_{22} = -q_{22} \Rightarrow q_{22} = 0$). A symmetric and an antisymmetric tensor's (inner) product is always zero. As evidence, consider the following:

$$p_{ij}q_{ij} = p_{ji}q_{ij} = -p_{ji}q_{ji} = -p_{ij}q_{ij} \tag{1.13g}$$

where the indices i and j were modified because they are dummy indices, followed by the facts that $q_{ij} = -q_{ji}$ (antisymmetric) and $p_{ij} = p_{ji}$ (symmetric). So, the product must be equal zero. This can also be demonstrated by writing the letters $p_{ij}q_{ij}$ in component form, i.e.

$$p_{ij}q_{ij} = p_{11}q_{11} + p_{12}q_{12} + p_{13}q_{13} + \ldots + p_{32}q_{32} + p_{33}q_{33} = 0 \tag{1.13h}$$

By inserting the following symmetric and antisymmetric components

$p_{12} = p_{21}, p_{13} = p_{31}, p_{23} = p_{32}$;

$q_{11} = q_{22} = q_{33} = 0$; $q_{12} = -q_{21}, q_{13} = -q_{31}, q_{23} = -q_{32}$

in the Eq. (1.7.3.2), i.e., $p_{ij}q_{ij} = 0$, is verified.

1.11.4 Second-order Tensor

The stress forces that a material is under at a specific position are represented by the second-order Cauchy stress tensor (**T**). The stress forces that a material would experience at the place that the stress tensor describes, along a plane perpendicular to ***v***, are characterised by the vector **T**.***v***, which is the product of the stress tensor and a unit vector v pointing in a specific direction. Three perpendicular stress vectors are depicted in the Fig. 1.30, with each direction being represented by a cube face. The stress tensor is a "second-order tensor" since it provides a mapping with a single vector as input and a single vector as output.

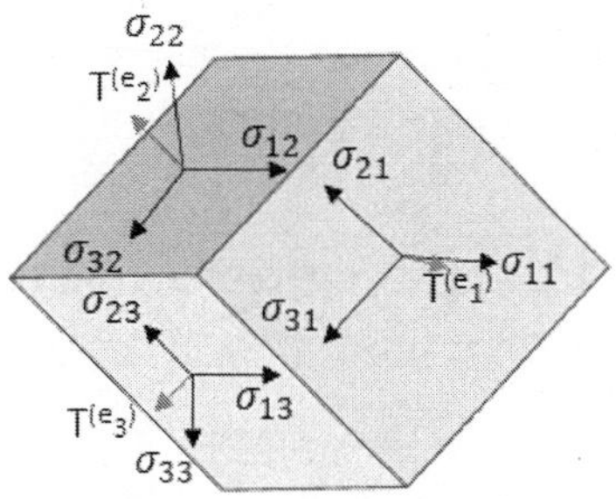

Fig. 1. 30 Second-order stress tensor.

1.12 TRANSFORMATION OF AXES

Every tensor, like a vector, is characterised about a basis, and if we select a new basis or angle from which to see the issue, the physical meaning remains the same but the tensor's component values will vary. Due to the geometry of the issue or the physical characteristics of the scenario, some orientations are simpler to work in than others. We must develop the ability to transfer our issue from one frame to another. We need pure rotational transformation matrices, hence they are denoted by the letter **R**.

Take the 2-dimensional simplification into consideration initially.

The basis with basis vectors x and y is being rotated into a new basis with basis vectors x' and y' as shown in Fig. 1.31. The resolution of the vectors x' = xcosα + ysinα and y' = xsinα+ ycosα results in a matrix that may be represented as follows:

$$\begin{Bmatrix} x' \\ y' \end{Bmatrix} = \begin{bmatrix} cos\alpha & sin\alpha \\ -sin\alpha & cos\alpha \end{bmatrix} \begin{Bmatrix} x \\ y \end{Bmatrix}$$

The cosines of the involved angles (known as direction cosines) make up this rotation matrix's. component, R. The component r_{ij}, or the part of x_j resolved along x_i', is the cosine of the angle between x_j (the old basis) and x_i' (the new basis).

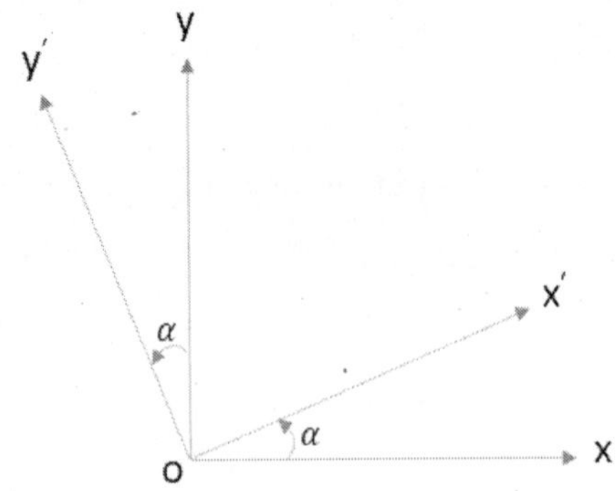

Fig. 1. 31 Rotation of Axes

As an illustration, we can generally state:

$x_i' = r_{i1}x_1 + r_{i2}x_2 + r_{i3}x_3$

We can write: $r_{ij} = x_i'.x_j$ since the scalar product for unit vectors is just the "cosine of the angle" between the two vectors. Completely written, we get:

$$R = \begin{bmatrix} x_1'.x_1 & x_1'.x_2 & x_1'.x_3 \\ x_2'.x_1 & x_2'.x_2 & x_2'.x_3 \\ x_3'.x_1 & x_3'.x_2 & x_3'.x_3 \end{bmatrix} \qquad (1.14a)$$

The transformation matrix shown here will change the basis from the old to the new. It is clear from the matrix that the aforementioned matrix has been transposed to transition from "new to old basis". In other words, "rotating and then rotating back" results in: $\mathbf{RR^T} = \mathbf{I}$, which is the initial result. As a result, the rotation matrix's transpose is its inverse.

1.12.1 Vector Transformation

Look at the vector: $\boldsymbol{b} = b_1\boldsymbol{y_1} + b_2\boldsymbol{y_2} + b_3\boldsymbol{y_3} = b_1'.\boldsymbol{y_1'} + b_2'.\boldsymbol{y_2'} + b_3'.\boldsymbol{y_3'}$

Knowing that a resolved onto the new 1 axis has the following component:

$$b_1' = \boldsymbol{b}.\boldsymbol{y_1'} = b_1r_{11} + b_2r_{12} + b_3r_{13}$$

The above outcome can be achieved by similarly resolving the other components. As a result, we use the following equation to rotate a vector: **b'** = **Rb**

1.12.2 Second Rank Tensor Transformation

Let's have a look at the general tensor equation, which is written as m = Tn (in the old basis) and m' = T'n', to get the transformation law for second rank tensors (in the new basis). We utilise the transformations m' = Rm and n' = Rn for the vectors.

Given the information given, we can easily substitute certain terms to determine that: m' = T'n' and that m' = Rm = RTn = RTR$^{-1}n'$ = RTR$^T n'$.

Hence: T' = RTR^T

This can be expressed as the final transformation law in suffix notation: $T_{ij} = r_{ip}r_{jq}T_{pq}{}'$, or vice versa, $T_{ij}{}' = r_{ip}r_{jq}T_{pq}$.

Remember that each of them represents 9 equations, one for each tensor component.

1.12.3 n^th^ Rank Tensor Transformation

The transformation law is as follows for an nth rank tensor. There are *n* different transformation matrices, and $T'_{ijk...} = r_{ip}r_{jq}r_{ks...}T_{pqs...}$. The benefit of the transformation laws is that they enable us to define tensors mathematically as "an object whose coefficients transform according to the aforementioned criteria."

As a result, an object keeps its physical significance regardless of the premise employed to describe it.

1.13 PRINCIPAL AXES AND EIGENVALUES

As demonstrated, a general second rank tensor has the following form:

$$\begin{bmatrix} T_{11} & T_{12} & T_{13} \\ T_{21} & T_{22} & T_{23} \\ T_{31} & T_{32} & T_{33} \end{bmatrix}$$

However, this takes a simpler form on a particular basis:

$$\begin{bmatrix} T_{11} & 0 & 0 \\ 0 & T_{22} & 0 \\ 0 & 0 & T_{33} \end{bmatrix}$$

i.e., all off-diagonal elements have a value of zero. These basis vectors are referred to as the principal axis (or directions), and the principal values are the non-zero tensor components. Keep in mind that the main axes for a given attribute typically do not match the crystal axes.

Principal axes make the mathematics easier to understand and draw attention to the situation's symmetry. Let's return to the example of electrical conductivity. When using an arbitrary basis, the equations have the following form:

$$j_1 = \sigma_{11}E_1 + \sigma_{12}E_2 + \sigma_{13}E_3$$

$$j_2 = \sigma_{21}E_1 + \sigma_{22}E_2 + \sigma_{23}E_3$$

$$j_3 = \sigma_{31}E_1 + \sigma_{32}E_2 + \sigma_{33}E_3$$

They primarily take the following forms:

$$j_1 = \sigma_{11}E_1;\ j_2 = \sigma_{22}E_2;\ j_3 = \sigma_{33}E_3$$

i.e. An action directed along a principal axis will also have an effect along that axis ("the conductivities along each principal axis are of course different from each other").

As we have seen, in the primary basis, the component equations change to Tx = λx, where x is a proportionality constant. Here, three alternative linear equations with three possible λ values are represented (the principal values).

$$T_{11}x_1 + T_{12}x_2 + T_{13}x_3 = \lambda x_1$$

$$T_{21}x_1 + T_{22}x_2 + T_{23}x_3 = \lambda x_2$$

$$T_{31}x_1 + T_{32}x_2 + T_{33}x_3 = \lambda x_3$$

This can be solved effectively when |T -λ I| = 0, that is, when:

$$\begin{vmatrix} T_{11}-\lambda & T_{12} & T_{13} \\ T_{21} & T_{22}-\lambda & T_{23} \\ T_{31} & T_{32} & T_{33}-\lambda \end{vmatrix} = 0 \tag{1.14b}$$

As a result, the secular equation, a cubic equation, is obtained in λ. This equation must be solved for λ in order to determine the principal values. We identify the vector x that resolves the problem for every one of the three solutions for λ. One of the primary axes is parallel to each of these solutions for x. If the vector points along the major axis, it can be of any length. Therefore, we often scale the vector to be of "unit length," giving us an "orthonormal basis."

It is important to note that the "unit vectors" along the primary axis are its eigenvectors and that the principal values are known as the eigenvalues of the matrix expressing T. The basic process of identifying them is helpful not only for simplifying tensors but is utilised frequently in physics and chemistry, such as when investigating modes of vibration and computing energies in "quantum mechanics."

Example 1.17: Determine the principal values (eigenvalues) and Principal axes (eigenvectors) for the given tensor.

$$T = \begin{bmatrix} 2 & -1 & 1 \\ -1 & 2 & -1 \\ 1 & -1 & 1 \end{bmatrix}$$

Solution 1.17: The eigenvector ***x***, which corresponds to the eigenvalue λ, is defined as follows:

T***x*** = λ***x***

Then (T-λI).***x*** = 0

If and only if, the equation has a nonzero solution |T- λI|=0

$$\begin{vmatrix} 2-\lambda & -1 & 1 \\ -1 & 2-\lambda & -1 \\ 1 & -1 & 1-\lambda \end{vmatrix} = 0 \Rightarrow -\lambda^3 + 5\lambda^2 - 5\lambda + 1 = 0 \Rightarrow -(\lambda-1)(\lambda^2 - 4\lambda + 1) = 0$$

$$-(\lambda-1)(\lambda+(\sqrt{3}\text{-}2))(\lambda-(\sqrt{3}+2)) = 0$$

$\lambda_1 = 1;\ \lambda_2 = -(\sqrt{3}\text{-}2);\ \lambda_3 = (\sqrt{3}+2)$ are the principal values (eigenvalues) from the equation

T***x*** = λ***x***

Find the individual vectors for each λ from the following equation:

$$\begin{bmatrix} 2-\lambda & -1 & 1 \\ -1 & 2-\lambda & -1 \\ 1 & -1 & 1-\lambda \end{bmatrix} \begin{Bmatrix} x_1 \\ x_2 \\ x_3 \end{Bmatrix} = \begin{Bmatrix} 0 \\ 0 \\ 0 \end{Bmatrix}; \text{ find the principal axis } \begin{Bmatrix} x_1 \\ x_2 \\ x_3 \end{Bmatrix}$$

We can determine a relationship between the components by expanding the equations and applying each principal value to the aforementioned equation. The vector can then be scaled to have a length of one.

$$\text{Try } \lambda_1 = 1;\ \begin{bmatrix} 2-1 & -1 & 1 \\ -1 & 2-1 & -1 \\ 1 & -1 & 1-1 \end{bmatrix} \begin{Bmatrix} x_1 \\ x_2 \\ x_3 \end{Bmatrix} = \begin{Bmatrix} 0 \\ 0 \\ 0 \end{Bmatrix} \Rightarrow \begin{bmatrix} x_1 - x_2 + x_3 \\ -x_1 + x_2 - x_3 \\ x_1 - x_2 + 0 \end{bmatrix} = \begin{Bmatrix} 0 \\ 0 \\ 0 \end{Bmatrix} \Rightarrow x_1 = x_2;\ x_3 = 0$$

By making the vector's length equal to one and the scalar product with itself equal to 1, we can find that $x_1 = x_2 = 1$ and obtain the primary axis.

Principal axis or eigenvector for the principal value or eigenvalue of $\lambda_1 = 1$ is

$$x_1 = \begin{Bmatrix} 1 \\ 1 \\ 0 \end{Bmatrix}$$

Similarly,

$$x_2 = \begin{Bmatrix} \frac{-\sqrt{3}+1}{2} \\ \frac{\sqrt{3}-1}{2} \\ 1 \end{Bmatrix} for\ \lambda_2 = -\sqrt{3} + 2\ and\ x_3 = \begin{Bmatrix} \frac{\sqrt{3}+1}{2} \\ \frac{-\sqrt{3}-1}{2} \\ 1 \end{Bmatrix}\ for\ \lambda_2 = \sqrt{3} + 2$$

1.14 Three-dimensional rotation- Euler's theorem

A rotation is a "transformation" in which the length, volume (with sign), and angle between two vector pairs are all preserved. The length preservation property is represented as follows if vector $\boldsymbol{p}$ is rotated into vector $\boldsymbol{q}$:

$$|\boldsymbol{p}| = |\boldsymbol{q}| \tag{1.15a}$$

Additionally, due to the feature of volume preservation, if vectors $\boldsymbol{p_1}$, $\boldsymbol{p_2}$, and $\boldsymbol{p_3}$ are transformed into vectors $\boldsymbol{q_1}$, $\boldsymbol{q_2}$, and $\boldsymbol{q_3}$, respectively, by the same exact rotation

$$\boldsymbol{p_1}.\boldsymbol{p_2} \times \boldsymbol{p_3} = \boldsymbol{q_1}.\boldsymbol{q_2} \times \boldsymbol{q_3} \tag{1.15b}$$

Lastly, we have the "property of angle" preservation between vectors

$$cos\alpha_{p_1,p_2} = \boldsymbol{p_1}.\boldsymbol{p_2} = \boldsymbol{q_1}.\boldsymbol{q_2} = cos\alpha_{q_1,q_2} \tag{1.15c}$$

where the angle between vectors i and j is denoted by $\alpha_{i,j}$. Rotations are known as rigid transformations due to these three characteristics. Rotations are used to express relations between coordinate systems in mechanics.

The majority of the instances and uses we have thought about involved rigid bodies rotating around a fixed axis. But there are several instances of rigid bodies rotating along an axis that is shifting. A spinning top, a gyroscope, a bicycle wheel moving around its axis, and a coin rolling across a table are all instances of this kind of motion. These motions can be extremely intricate and challenging to comprehend.

1.14.1 Euler's Theorem

According to Euler's Theorem, a single resultant rotation along an axis passing through a point is equivalent to two "component" rotations around various axes. When more than two "rotations" are used, they can be reduced to pairs, and each pair can then be reduced to a single rotation.

The foundation of the axis-angle representation is Euler's rotation theorem, which also states that every rotation of a rigid body or series of rotations of that body in three dimensions is equivalent to a "pure rotation" about a single fixed axis.

Any rotation can be explained using three angles, according to "Euler's rotation theorem." A general rotation R can be described as if the rotations are expressed in terms of the "rotation matrices" S, Q, and P (Fig. 1.32), R = P Q S.

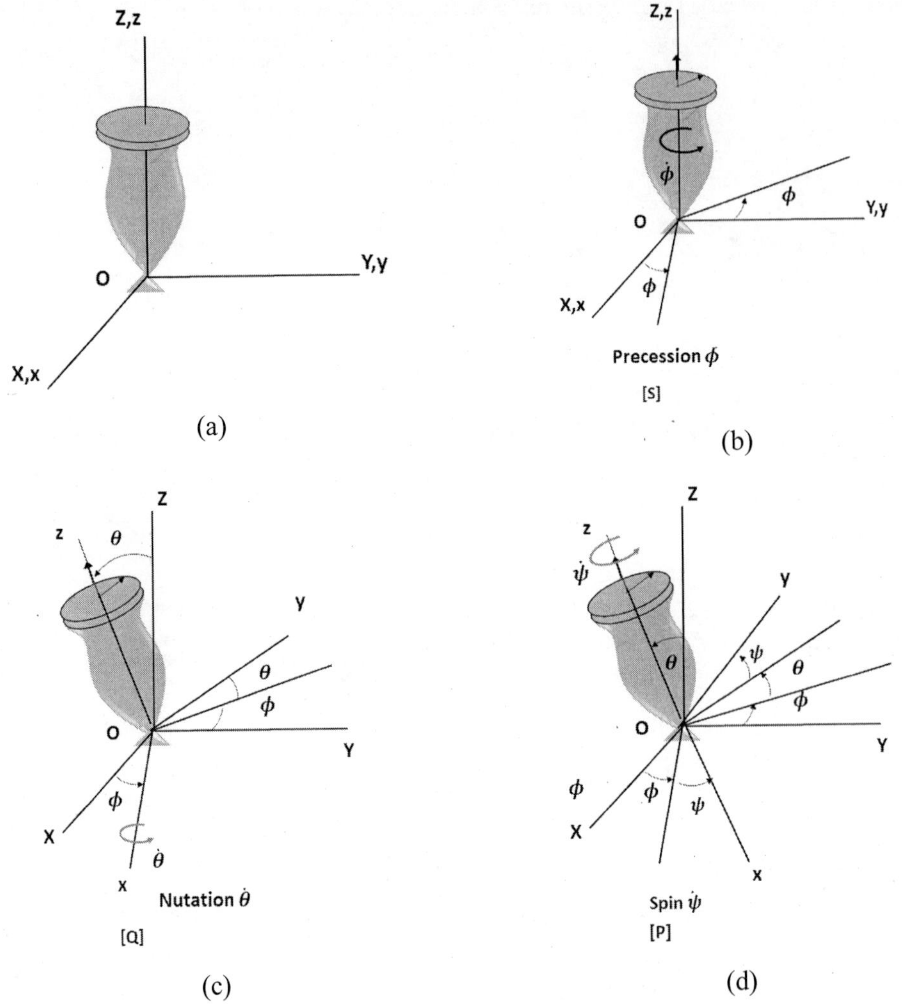

Fig. 1. 32 Euler Angles.

The three angles are known as Euler angles, and they provide the three "rotation matrices." Depending on the axis around which the "rotations" are performed, there are many Euler angle conventions. The matrix R as

$$R \equiv \begin{bmatrix} r_{11} & r_{12} & r_{13} \\ r_{21} & r_{22} & r_{23} \\ r_{31} & r_{32} & r_{33} \end{bmatrix} \tag{1.15d}$$

The most typical definition is the so-called "x-convention," which is depicted above. The rotation in this convention is determined by the Euler angles $(\emptyset, \theta, \psi)$, where

i. the first rotation uses S to rotate at an angle of $\emptyset$ about the z-axis
ii. the old x-axis (now x') is rotated by an angle $\theta \in [0,\pi]$ in the second rotation, and
iii. the old z-axis (now z') is rotated by an angle of ψ in the third rotation, which makes use of P.

The component rotations in the x-convention are then provided by

$$S \equiv \begin{bmatrix} \cos\emptyset & \sin\emptyset & 0 \\ -\sin\emptyset & \cos\emptyset & 0 \\ 0 & 0 & 1 \end{bmatrix}; Q \equiv \begin{bmatrix} 1 & 0 & 0 \\ 0 & \cos\theta & \sin\theta \\ 0 & -\sin\theta & \cos\theta \end{bmatrix}; P \equiv \begin{bmatrix} \cos\psi & \sin\psi & 0 \\ -\sin\psi & \cos\psi & 0 \\ 0 & 0 & 1 \end{bmatrix}$$

So

$$R \equiv \begin{bmatrix} (\cos\psi\cos\emptyset - \cos\theta\sin\emptyset\sin\psi) & (\cos\psi\sin\emptyset + \cos\theta\cos\emptyset\sin\psi) & \sin\psi\sin\theta \\ (-\sin\psi\cos\emptyset - \cos\theta\sin\emptyset\cos\psi) & (-\sin\psi\sin\emptyset + \cos\theta\cos\emptyset\cos\psi) & \cos\psi\sin\theta \\ \sin\theta\sin\emptyset & -\sin\theta\cos\emptyset & \cos\theta \end{bmatrix}$$

For a matrix, keep in mind that you can get the parts of the "angular velocity ω" in the body axes, $R \equiv [\boldsymbol{R_1} \quad \boldsymbol{R_2} \quad \boldsymbol{R_3}]$, i.e.,

$$\begin{bmatrix} r_{11} & r_{12} & r_{13} \\ r_{21} & r_{22} & r_{23} \\ r_{31} & r_{32} & r_{33} \end{bmatrix} \begin{Bmatrix} \omega_x \\ \omega_y \\ \omega_z \end{Bmatrix} = \begin{Bmatrix} r_{11}\omega_x + r_{12}\omega_y + r_{13}\omega_z \\ r_{21}\omega_x + r_{22}\omega_y + r_{23}\omega_z \\ r_{31}\omega_x + r_{32}\omega_y + r_{33}\omega_z \end{Bmatrix} = \begin{Bmatrix} r_{11} \\ r_{21} \\ r_{31} \end{Bmatrix} \omega_x + \begin{Bmatrix} r_{12} \\ r_{22} \\ r_{32} \end{Bmatrix} \omega_y + \begin{Bmatrix} r_{13} \\ r_{23} \\ r_{33} \end{Bmatrix} \omega_z$$

$$= \left[\boldsymbol{R_1}\omega_x + \boldsymbol{R_2}\omega_y + \boldsymbol{R_3}\omega_z\right]$$

Now, if we examine the ω_z component of R ω, we can see that it corresponds to rotation about the ϕ axis,

$$\omega_\emptyset = \boldsymbol{R_3}\omega_z = \begin{Bmatrix} \sin\psi\sin\theta \\ \cos\psi\sin\theta \\ \cos\theta \end{Bmatrix} \dot{\emptyset}$$

Look at the ω_x component of P ω to see how the line of nodes translates into a θ rotation about the x-axis,

$$\omega_\theta = \boldsymbol{P_1}\omega_x = \begin{Bmatrix} \cos\psi \\ -\sin\psi \\ 0 \end{Bmatrix} \dot{\theta}$$

Similarly, examine the ω_ψ component of P ω to determine rotation by ψ about the remaining axis,

$$\omega_\psi = \boldsymbol{P_3}\omega_z = \begin{Bmatrix} 0 \\ 0 \\ 1 \end{Bmatrix} \dot{\psi}$$

Combining the parts results in

$$\omega = \begin{Bmatrix} \sin\psi\sin\theta\,\dot{\emptyset} + \cos\psi\,\dot{\theta} \\ \cos\psi\sin\theta\,\dot{\emptyset} + \sin\psi\,\dot{\theta} \\ \cos\theta\,\dot{\emptyset} + \dot{\psi} \end{Bmatrix}$$

UNIT SUMMARY

The many ideas of vectors, tensors, and vector and tensor algebra were covered in this unit. In-depth discussions of Euler's theorem, Euler angles, coordinate systems, coordinate transformations, symmetric and anti-symmetric tensors, primary axes, eigenvalues, and eigenvectors, three-dimensional rotations, are also included in this unit. The skills learned in this unit will undoubtedly be able to handle the numerous vector mechanics problems.

EXERCISES

Multiple Choice Questions

1. The generalisation of "vector analysis" is called---------------
   a) Real analysis b) Vector analysis c) Tensor analysis d) Scalar analysis
2. The 0th rank tensor is known as-----------------------
   a) Scalar b) Vector c) Matrix d) None of the above
3. The 1st rank tensor is known as-----------------------
   a) Scalar b) Vector c) Matrix d) None of the above
4. Which of the following variables is a scalar value?
   a) Velocity b) Acceleration c) Force d) Mass
5. Which of the following variables is a vector?
   a) Velocity b) Acceleration c) Force d) Mass
6. In order for the summation convention to apply, an index must be repeated in the given expression--------------------
   a) Dummy index b) free index c) Dummy and free index d)None of the above
7. In a given expression, an index that only appears once is referred to as----------
   a) Dummy index b) free index c) Dummy and free index d)None of the above
8. If the tensor is symmetric, then----------
   a) $A_{ij} = A_{ii}$ b) $A_{ij} = A_{ji}$ c) $A_{ij} = -A_{ji}$ d) $A_{ij} = A_{jj}$
9. If the tensor is antisymmetric, then----------
   a) $A_{ij} = A_{ii}$ b) $A_{ij} = A_{ji}$ c) $A_{ij} = -A_{ji}$ d) $A_{ij} = A_{jj}$
10. The sum of the three vectors **P**, **Q**, and **R** is 0. Decide which is false.
    a) (P X Q) X R is not zero unless Q, R are parallel
    b) (**P** X **Q**).**R** is not zero unless **Q**, **R** are parallel
    c) If **P**, **Q**, **R** define a plane, (**P** X **Q**) X **R** is in that plane
    d) (**P** X **Q**).**R** = |**P**| |**Q**| |**R**| → $R^2 = P^2 + Q^2$

Answers of Multiple Choice Questions

1) c 2) a 3) b 4) d 5) a 6) a 7) b 8) b 9) c 10) b

Short and Long Answer Type Questions

1. What is tensor explain with example?
2. What differentiates tensors from vectors?
3. How many different coordinate systems may a vector be defined in? Explain each.
4. Identify the unit vector in the direction $\overline{PQ}$, where P is equal to (2,3,4) and Q is equal to (5,6,7).
5. Determine the angle between the vectors: $\vec{P} = e_1 + e_2 - e_3;\ \vec{Q} = e_1 - e_2 + e_3$
6. Determine $\vec{P} \times (\vec{Q} \times \vec{R}), if \vec{P} = 3e_1 + e_2 - 4e_3;\ \vec{Q} = e_1 - 2e_2 + 3e_3;\ \vec{R} = e_1 + e_2 + 3e_3$
7. Two points X and Y have position vectors of $3e_1+e_2$ and e_1-3e_2, respectively. The point Z that externally divides the line segment XY in a ratio of 2:1 should have its position vector written down.
8. Calculate the distance, scalar products, and angles between the three vectors **P**, **Q**, and **R**, which have the components {1, 1, 0}, {0, 1, 1} and {1, 1, 1} respectively.
9. How are the Eigenvalues of a square matrix determined?
10. How is the Euler theorem solved?

Numerical Problems

1. For the vectors p: {1, 0, 0} and q: {0, 1, 0}, write the symmetric traceless and the antisymmetric components of the dyadic tensor $A_{\mu\nu} = a_\mu b_\nu$ in matrix form. Calculate the symmetrical and antisymmetrical components' norm squared, then compare the results to $A_{\mu\nu}A_{\mu\nu}$ and $A_{\mu\nu}\ A_{\nu\mu}$.
2. The diagonals of three vectors that are along the adjacent faces of a cube and have the magnitudes *p*, 2*p*, and 3*p* come together at a point. Identify the resultant.
3. Identify and demonstrate the coplanarity of the points with the coordinates (2, -1 1), (1, -3,-5), and (3 -4 -4).
4. Determine the values of *l* and *m* to make **P** = $3e_1+le_2-e_3$ and q = $2e_1+e_2+me_3$ equal magnitude, mutually perpendicular vectors.
5. Assume that vectors **a** = e_2-e_3 and **b** = $e_1-e_2-e_3$. Determine the vector **b** if the vectors fulfilling **a** x **b** + **c** = 0 and **a. b**=3.
6. Let the vectors **a** = $e_1-2e_2+e_3$ and **b** = $e_1-e_2+e_3$. Find the c.b, if vector c is satisfying the **b** x **c** = **b** x **a** and **c.a** = 0.
7. The vector **P**= $e_1+2e_2+2e_3$ makes a right turn while travelling across the positive x-axis. Determine the vector in its new location.
8. Find the eigenvalues and corresponding eigenvectors of the matrix A.

$$A = \begin{bmatrix} 2 & 2 & 2 \\ 2 & 3 & 2 \\ 2 & 2 & 4 \end{bmatrix}$$

9. Determine **P** × (**Q** × **R**) and **P.** (**Q** × **R**) for the vectors **P** = $2e_1 + 3e_2 + 5e_3$, **Q** = $3e_1 - 4e_2 + 2e_3$ and **R** = $3e_1 + e_2 + 5e_3$ respectively.
10. The object is moving in such a way that, at the instant depicted in Fig. P1.1, it is rotating about the Z-axis at ω_1 = 0.8 rad/s and spinning at ω_2 = 10 rad/s. As of this moment, ascertain the object's angular velocity.

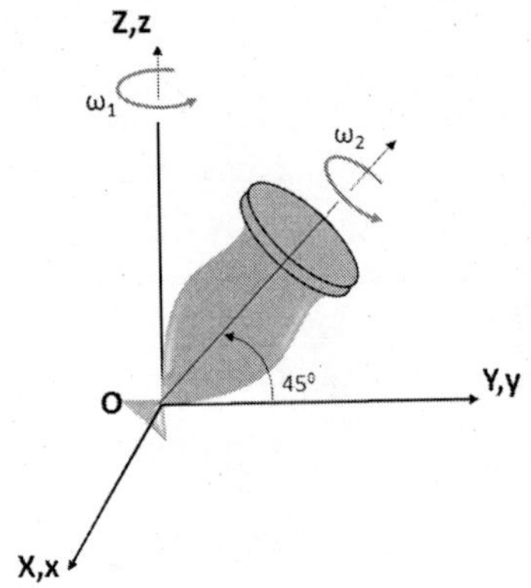

Fig. P1.1

PRACTICAL

Aim: To determine the weight of a given object (a wooden block) using the "parallelogram law of vectors".

Materials Required: Parallelogram apparatus, two slotted weights with hangers, a wooden block with a hook, a spring balance, a mirror strip, a roll of cotton thread, and a drawing pin.

Theory: The Gravesand device, also called the "Parallelogram apparatus", is shown in Fig. A1. It is made up of two pillars and a wooden board A fastened vertically. At the top of the board, two pulleys P and Q are installed at the same level. The device comes with three sets of slotted weights that can be used to test the "parallelogram law of vectors". In order to apply two forces P and Q, a thread containing hangers for the insertion of slotted weights is made to travel through the pulleys. A third force W is applied by suspending the provided object, the weight of which must be calculated, in the centre of the thread.

If the two adjacent sides of a "parallelogram" drawn from a point are used to represent two vectors working concurrently on a particle's magnitude and direction, then the diagonal of that parallelogram created from that point is used to represent the resultant's magnitude and direction completely.

As shown in Fig. A2, let's assume that two vectors **P** and **Q** are acting concurrently at an angle on a particle O. The opposite sides OA and OB of an OACB parallelogram drawn from a point O serve as their representations in terms of magnitude and direction. In terms of magnitude and direction, the diagonal OC will then reflect the resulting **R**. In terms of math, we can state,

$$\vec{R} = \vec{P} + \vec{Q}$$

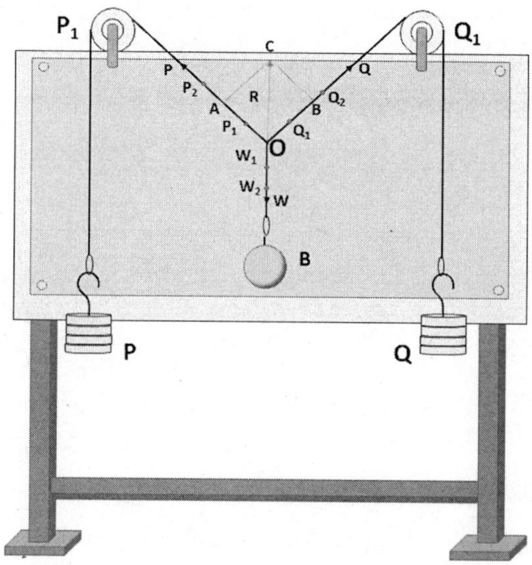

Fig. A1

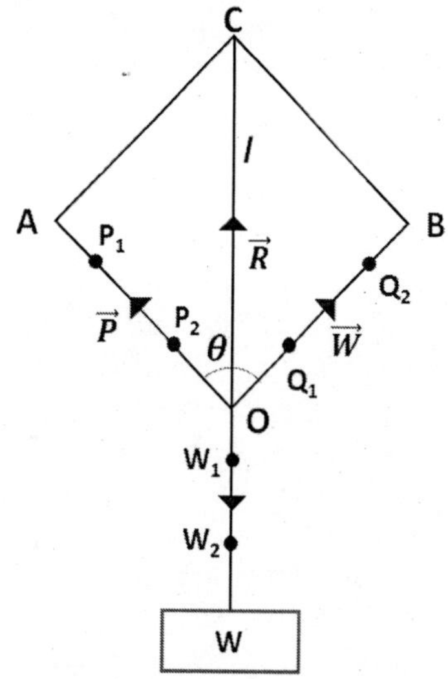

Fig. A2

The three vectors **P**, **Q**, and **W** are in equilibrium if a body of unknown weight W is hanged from the middle hanger while balancing weights **P** and **Q** are strung from the other two hangers. When everything is balanced, |**W**| = |**R**|. A body's weight is a force.

$$\vec{W} = \vec{P} + \vec{Q}$$

$$W = \sqrt{P^2 + Q^2 + 2PQ\cos\theta} \quad \text{(i)}$$

Percentage of error can be determined using below equation,

$$\%error = \frac{Actual\ Weight - Calculated\ Weight}{Actual\ Weight} \times 100 \quad \text{(ii)}$$

Procedure:

1. Using a plumb line, align the Gravesand apparatus' board in the vertical position.
2. With the use of drawing pins, secure a sheet of drawing paper to the board.
3. Take a long piece of thread and secure two hangers to either end.
4. Tie one end of a second, shorter thread to the specified body B, the weight of which is to be determined.
5. Form a tiny knot by tying the opposite end of the shorter thread in the middle of the longer thread, O. The intersection of the three threads is formed by this knot.
6. Place two equal slotted weights P and Q on the hangers, then pass the longer threads over the apparatus' two pulleys P_1 and P_2.
7. With the knot O placed someplace on the drawing paper, the body B is designed to hang vertically.
8. To keep the knot O at a position immediately below the paper's centre, adjust weights P and Q. O is now in its equilibrium position. Make sure that none of the weights touch the board and that they all hang freely.
9. On the white page, make a pencil mark at junction O.

10. Make sure that the two pulleys P_1 and P_2 are friction-free. In order to achieve this, slightly upset P and Q's locations before leaving them. If the knot doesn't return to its original equilibrium position, lubricate the pulleys to reduce friction.
11. Make a pencil mark on the paper sheet in the location of the knot's equilibrium (O).
12. Under one of the threads on the paper, spread the thin mirror strip out in a breadthwise direction. In the mirror, observe the thread's reflection. The strip and the thread should be at a straight angle in your line of sight. Right now, it is unable to see the thread's image (meaning, parallax is removed). On the paper, encircle the two locations P_1 and P_2 that are precisely behind the thread on either side of the strip.
13. Mark points Q_1, Q_2 and W_1, W_2 in a similar manner for the other two threads.
14. Remove the hangers and threads from the apparatus, and then write in table A1 the weights of the two hangers, H_1 and H_2, as well as the slotted weights. Let P and Q stand for these two weights.
15. With the use of a ruler, link P_1 and P_2, Q_1 and Q_2, and W_1 and W_2 to create three lines that should all intersect at O. Remove the paper from the board. The three lines shown here reflect the three forces' directions.
16. Select a reasonable scale to represent the vector, such as 10 gm (0.1N) = 1 cm.
17. To illustrate the forces and, correspondingly, cut the lines OA and OB. For instance, OA and OB are 6 cm and 6 cm, respectively, if P and Q are both 60 grammes.
18. Finish the parallelogram OACB by adding the adjacent sides OA and OB as illustrated in Fig. A2.
19. O and C invite you to join them. The resultant vector with the unknown weight W is represented by the length OC.
20. To determine the value of R, multiply the OC (l) measurement by the scale (10 g).
21. Using a protractor, determine θ and then use equation to determine W. (i).
22. Use new paper for the experiment and hangers that are different (in weight) for P and Q.
23. Calculate the weight R and record the result.
24. Equation is used to determine the percentage error after utilising the spring balance to measure the actual weight (S) of the unknown weight (W) (ii).

Observation:

Least count of spring balance = _______________ gm

Weight of B by spring balance = _______________gm

Scale factor: Let ______ cm = ___________gm

Table A1 Measurement of weight of given body

S. No	Forces		Length of the sides			$\angle AOB$ θ	Resultant Force, R = OC X Scale (gm wt.)	Measured Weight, W = R (gm wt.)	% error
	P (gm wt.)	Q (gm wt.)	OA (cm)	OB (cm)	OC (cm)				
1									
2									
3									

Result:

For equal weights

Unknown body weight by observation, W: ....

Unknown body weight of by calculation, W': ....

For un-equal weights

Unknown body weight by observation, W: ....

Unknown body weight of by calculation, W': ....

KNOW MORE

Gregorio Ricci-Curbastro

Tensor calculus, developed by the mathematician, was essential in the development of general relativity.

Mathematician Gregorio Ricci-Curbastro was born on January 12, 1853, in Lugo, present-day Italy. He is best known for developing the tensor calculus. Ricci received his education at home from private tutors. He then transferred to the Scuola "Normale Superiore" in Pisa, a significant hub for mathematical research, after studying at the universities of Rome and Bologna. He obtained a fellowship to study at the "Technische Hochschule" in Munich after receiving his PhD in 1875. Felix Klein, a well-known German mathematician, taught classes there that Ricci took.

Ricci started working as a professor at the University of Padua in 1880, and he remained there until he passed away in 1925. Early works by Ricci focused on differential equations and the principles governing "electric circuits", in particular. Ricci created his renowned theory of tensor calculus, an application of "vector calculus to tensor fields", by the year 1900. With the assistance of

"Tullio Levi-Civita", one of his students, he would continue to develop the idea. Later, Albert Einstein was able to construct his theory of general relativity because to this hypothesis. Even Einstein went to Padua in 1921 to meet Ricci. For his contributions to mathematical physics, Ricci was recognised with numerous medals and awards. He was also admitted to membership in a number of academies, including the "Veneto Institute of Science", where he presided from 1916 to 1919. Additionally engaged in politics, he held council positions in Padua and his hometown of Lugo.

REFERENCES AND SUGGESTED READINGS

- Meriam, J.L. and Kraige, L.G., Engineering Mechanics: Dynamics, Wiley, 2011.
- Hibbeler, R.C., Engineering Mechanics: Dynamics, Pearson Prentice Hall, 2010.

2 Kinematics of Rigid Body

UNIT SPECIFIES

This unit discussed the following concepts:

- Definition of a rigid body and motion of a rigid body
- Angular velocity of a rigid body
- Two-Dimensional rotational motion
- Three-Dimensional rotational motion
- Orientation through the integration of angular velocity
- Acceleration formula through motion relation bodies

RATIONALE

The kinematics of a rigid body is a fundamental topic in classical mechanics that deals with the motion of bodies that do not deform during their motion. Understanding the kinematics of rigid bodies is crucial for studying their dynamics, which involves analyzing the forces and torques that cause the motion. Kinematics provides the necessary groundwork for formulating and solving equations of motion. Rigid body kinematics is essential in various fields such as robotics, aerospace engineering, biomechanics, and structural engineering. For instance, in robotics, it is used to design and control the movement of robotic arms and in biomechanics to analyze the motion of limbs. A rigid body in two and three-dimensional rotational motion, and orientation through angular velocity and acceleration study is essential to predict the motion of objects, leading to better designs, improved safety, and enhanced performance in a wide range of applications. The kinematics of rigid bodies adheres to important physical principles such as the conservation of angular momentum and energy.

PRE-REQUISITES

- Nil

UNIT OUTCOMES

List of outcomes of this unit is as follows:

U2-O1: Understanding the rigid body motion

U2-O2: Describe the angular velocity relations of rigid body

U2-O3: Explain the rotation of rigid body

U2-O4: Analyse angular velocity and angular acceleration

U2-O5: Understanding the five-term acceleration equation

Unit - 2 Outcomes	EXPECTED MAPPING WITH COURSE OUTCOMES (*1-Weak Correlation;2-Medium correlation;3-Strong Correlation*)				
	CO-1	CO-2	CO-3	CO-4	CO-5
U2-O1	3	3	3	2	1
U2-O2	3	3	3	2	1
U2-O3	3	3	3	2	1
U2-O4	2	3	3	2	2
U2-O5	2	3	3	2	2

2.1 INTRODUCTION

Rigid body is a body which may not undergo any form of deformation under any type of load or force. The rigid body may fail without any deformation and rigid body cannot hold any amount of stress and strain. Rigid body mainly deals with the subject of "Engineering Mechanics".

Engineering mechanics is the study of rigid bodies under the influence of forces while they are at rest or in motion. The study of rigid bodies which are in state of rest or in motion with constant velocity is called as statics. On the other hand, dynamics is the study of rigid bodies in motion as a result of forces. Kinematics, which simply addresses the motion's geometrical aspects, and kinetics, which examines the forces causing motion, will be presented separately in the two parts of dynamics.

2.1.1 Kinematics

It deals with space, time relationship of a given motion of rigid bodies while ignoring the effect of forces. Also, it is the study of various kinematic parameters without consideration of mass, moment of inertia and turning moment force. Kinematics understands causes of motion, but only at the level of velocity and acceleration. For instance, in the motion study, the accelerations and velocities that the force produces are what cause the motion of a shopping cart to move.

2.2 KINEMATICS OF PARTICLES

Kinematics of particles is the kind of coordinate system we use, its affects, how a particle's motion is described. In general, a coordinate system that is stationery and lack of angular motion is required in order to examine the behaviour of a particle or rigid body. Inertial reference frame is the name given to such a coordinate system. We assume that a particle is travelling across space at a random rate and that, at any given moment in time t, it is at point P with the coordinates (x, y, z) as illustrated in Fig. 2.1 Since the particle changes location while in motion, it is evident that these coordinates are a function of time.

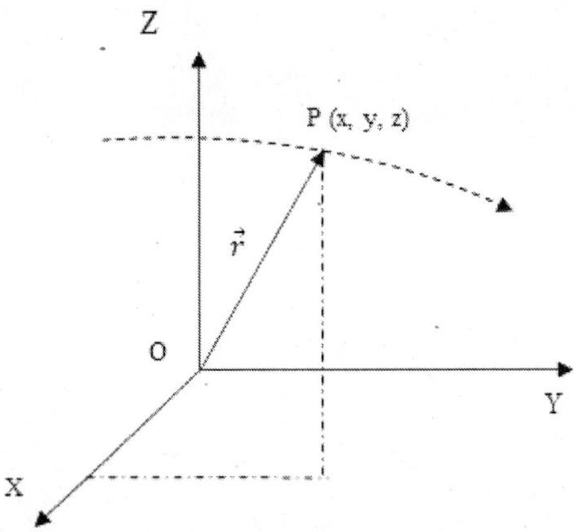

Fig. 2. 1 Position Vector.

The particle's position vector with respect to the origin at time t may be written as the origin, O as $\vec{r} = x\hat{\imath} + y\hat{\jmath} + z\hat{k}$, where $\hat{\imath}, \hat{\jmath}$ and $\hat{k}$ are corresponding vectors along the X, Y, and Z axes.

Please see Fig. 2.2 $\vec{r_2} - \vec{r_1}$ is the displacement vector designated by $\overline{r_1 r_2}$ or $\overrightarrow{r_1 r_2}$ if the particle changes its location from P_1 at time t_1 to P_2 at time t_2 with position vectors and $\vec{r}$ or $\vec{r_2}$. When motion occurs in space, we refer to it as spatial motion (also known as three-dimensional motion), and when it occurs on a plane, we refer to it as plane motion.

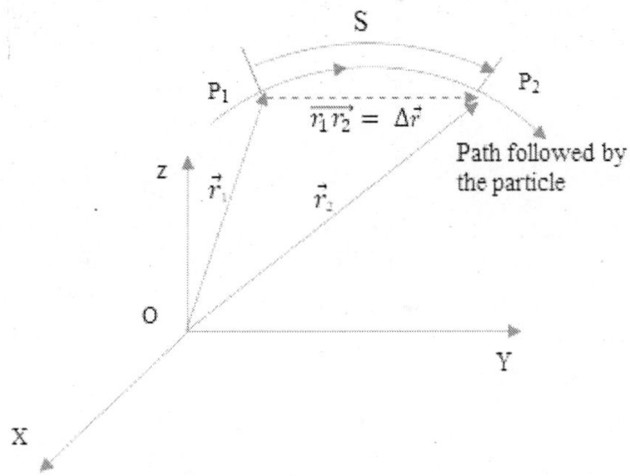

Fig. 2. 2 Displacement Vector.

2.2.1 Speed and Velocity

The rate at which displacement changes is known as velocity, which is a vector quantity. In mathematics, it is described

$$\vec{v} = \lim_{t_2 \to t_1} \frac{\vec{r_2} - \vec{r_1}}{t_2 - t_1} = \lim_{t \to 0} \frac{\Delta\vec{r}}{\Delta t} = \frac{d\vec{r}}{dt} \tag{2.1a}$$

Here, the particle's position vectors at times $t=t_1$ and $t=t_2$, respectively, are denoted by the $\vec{r_1}$ and $\vec{r_2}$. Clearly, $\vec{r_2}$ - $\vec{r_1}$ refers to the particle's movement throughout the time interval (t_2 - t_1). While speed, which is the magnitude of velocity and is represented by the scalar quantity, is

$$v = \frac{ds}{dt} \tag{2.1b}$$

The distance along the motion's route is denoted by the letter s in the equation above.

Instantaneous velocity is the name given to the velocity in Eq. (2.1a). Over a period of time, average velocity is also occasionally utilised. We may define average velocity across the time span (t_2 - t_1) in the formulation of Eq. (2.1c) as

$$\overrightarrow{v_{avg}} = \frac{\vec{r_2}-\vec{r_1}}{t_2-t_1} \tag{2.1c}$$

Similar to that, average speed is determined by dividing the total distance travelled by the particle by the time it took to go that distance.

Therefore, we may say that whereas speed is the rate at which distance changes over time, velocity is the rate at which displacement changes over time.

2.2.2 Acceleration

The rate at which velocity changes is known as acceleration, mathematically,

$$\vec{a} = \frac{d\vec{v}}{dt} \tag{2.2}$$

Since acceleration is a vector variable, discussing it while a particle is moving along a curved path differs greatly from discussing motion along a straight path (rectilinear motion).

2.3 MOTION OF A RIGID BODY

Rigid bodies are solid objects that do not deform when a force is applied. Such bodies experience translational and rotational motion when forces are applied; the system's momentum is altered by these forces. Real life objects are all rigid in nature, and rigid bodies may be found practically everywhere. To be able to forecast their trajectories, it becomes imperative to examine their motion and how various factors affect them.

2.3.1 Rigid Body Dynamics

Rigid body dynamics was developed as a result of the failure of classical mechanics to account for the motion and kinematics of bodies without point masses and prone to rotation under the application of forces. A rigid body has two possible motions when forces are applied to it:

1. Translation Motion (the motion where every point on a moving body moves equally in one direction).
2. Rotational Motion (the motion of an object around a fixed orbital axis).

2.3.1.1 Translational Motion

When every particle on a body moves the same distance during the same period of time, this is referred to as translatory motion, this motion may be rectilinear or curvilinear. A box sliding down on a wedge is one example of translational motion, as is an automobile travelling straight line. Newton's laws and the equations of motion govern this type of motion.

Vector forces F_1, F_2, F_3, and F_N are applied to the body. These forces accelerate motion in the direction of their resultant when they are applied, and Newton's Law governs this motion.

$$\vec{F}_{net} = m\vec{a} \tag{2.3a}$$

$$\vec{F}_1 + \vec{F}_2 + \ldots\ldots \vec{F}_n = m\vec{a} \tag{2.3b}$$

Where, "m" stands for the mass of the system and "a" represents the body's resulting acceleration.

Translation motion or translatory motion can be categorised based on the way the object is moving as

- Rectilinear motion
- Curvilinear motion

Rectilinear motion

Particle in this instance moves in a straight path (say, along X-axis). Basic motion of equation is given here,

$$v = \frac{ds}{dt} = \dot{x} \tag{2.4a}$$

$$\int_{x_1}^{x_2} dx = \int_{t_1}^{t_2} v dt \tag{2.4b}$$

$$a = \frac{dv}{dt} = \frac{d^2x}{dt^2} = \frac{vdv}{dx} \tag{2.4c}$$

$$\int_{v_1}^{v_2} dv = \int_{t_1}^{t_2} a\, dt \tag{2.4d}$$

$$\int_{v_1}^{v_2} v\, dv = \int_{x_1}^{x_2} a\, dx \tag{2.4e}$$

The velocity of a particle can be either uniform (i.e., constant) or non-uniform. It might also have non-uniform acceleration or uniform (i.e., constant) acceleration. We would like to remind you that the block of equations in (2.4) may be integrated in terms of starting time (t_0), initial displacement (x_0), and initial velocity (v_0) as long as the particle travels with uniform acceleration (a_0) as depicted in the equations (2.5)

$$x = x_0 + v_0(t - t_0) + \frac{1}{2}a(t - t_0)^2 \tag{2.5a}$$

$$v = v_0 + a(t - t_0) \tag{2.5b}$$

$$v^2 = {v_0}^2 + 2a(x - x_0) \tag{2.5c}$$

Curvilinear Motion

"When a particle moves along a curved path, curvilinear motion occurs". Vector analysis will be utilised to formulate the particle's position, velocity, and acceleration because this path is generally given in three dimensions.

Position: Look at the particle in Fig. 2.3a, which is situated at a position on a space curve that is defined by the path function s(t). The position vector $\boldsymbol{r} = \boldsymbol{r}(t)$ will be used to indicate the particle's position as determined from a fixed point of O. Be aware that as the particle moves along the curve, this vector's magnitude and direction will change.

Displacement: Assume that after a short time span Δt, the "particle travels a distance Δs along the curve" to a new point, indicated by the equation r' = r + Δr in Fig. 2.3b. By subtracting two vectors, the displacement Δr indicates the "change in the particle's position" and is calculated as follows: Δr = r'- r.

Velocity: The particle's average velocity during the time Δt is

$$\boldsymbol{v}_{avg} = \frac{\Delta r}{\Delta t} \tag{2.6a}$$

The instantaneous velocity, $\boldsymbol{v}$ is determined by Eq. (2.1). The direction of $\boldsymbol{v}$ is tangent to the curve since dr will be tangent to the curve as shown in Fig. 2.3c.

Acceleration: According to Fig. 2.3d, if "a particle has a velocity $\boldsymbol{v}$ at time t" and a velocity $\boldsymbol{v}' = \boldsymbol{v} + \Delta v$ at t + Δt, then its "average acceleration ($\boldsymbol{a}_{avg}$) during the period" of time Δt equals

$$\boldsymbol{a}_{avg} = \frac{\Delta v}{\Delta t} \tag{2.6b}$$

The instantaneous acceleration, *a* is

$$a = \lim_{\Delta t \to 0} \frac{\Delta v}{\Delta t} = \frac{dv}{dt} \tag{2.6c}$$

From Eq. 2.1a and 2.4c, $a = \frac{d^2r}{dt^2}$ (2.6d)

As the particle goes from one location to the next along the curved path, acceleration must take into account the change in both "the magnitude and direction of the velocity $\boldsymbol{v}$". The directional change must always "swing" the velocity vector towards the "inside" or "concave side" of the path for the particle to follow any curved path, hence acceleration cannot stay tangent to the path.

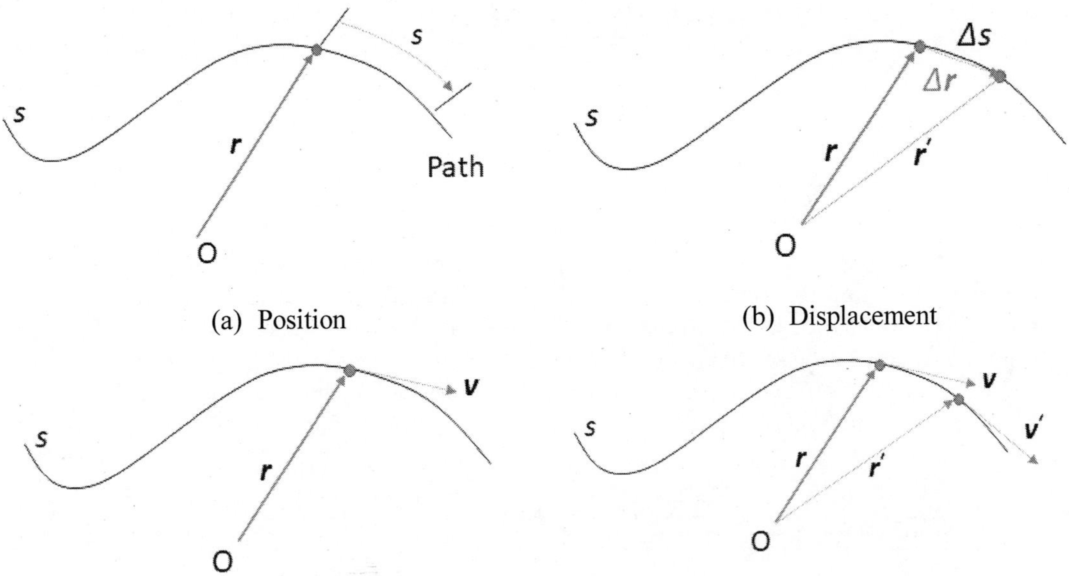

(a) Position (b) Displacement

(c) Velocity (d) Acceleration

Fig. 2. 3 Curvilinear motion of a particle.

2.3.1.2 Rotational Motion

When all the constituent particles of a body are moving in the same direction around a single rotational axis, this is referred to as being in rotational motion. This kind of motion is circular. Examples of circular motions include a ball rolling down a wedge, a car's tyres moving over a road, etc. This motion can also be explained by extending the laws and equations used to describe translational motion. Let T_1, T_2, and T_3 … T_N are the torques acting on the body. The direction in which the resultant of these torques lies experiences angular acceleration.

$$\vec{T}_{net} = I\vec{\alpha} \tag{2.7a}$$

$$\vec{T}_1 + \vec{T}_2 + \ldots \ldots \vec{T}_n = I\vec{\alpha} \tag{2.7b}$$

where "I" denotes the system's moment of inertia and "α" denotes the body's resulting acceleration.

2.3.1.3 Conditions for Equilibrium

The term "mechanical equilibrium" refers to a rigid body whose angular momentum and linear momentum are not changing over time. It can be inferred from this that the body is not experiencing either angular acceleration or linear acceleration. Consider an inertial moment "I" and a mass "m". Assume that the body is being affected by a variety of forces, all of which are producing a torque to be generated on the body in various directions. The body will comply with the aforementioned conditions if it is in mechanical equilibrium:

1. $\vec{F_1} + \vec{F_2} + \ldots \ldots \vec{F_n} = 0$ (2.8)

2. $\vec{T_1} + \vec{T_2} + \ldots \ldots \vec{T_n} = 0$ (2.9)

2.3.2 Rigid Body Coordinate Systems

As shown in Fig. 2.4, which shows a rigid body occupying the domain, the translation and rotation that regulate the rigid body's motion are decoupled using two coordinate systems. This is the first equation of motion for a free rigid body that is described. Both the global coordinate system (XYZ), which is fixed to the ground and the body-fixed coordinate system ($x_1x_2x_3$) have their origins at point O. The relative motion between the body-fixed coordinate system and the global coordinate system can be used to calculate the movement of an arbitrary point, P, within the rigid body. In terms of the global coordinate system, the position vector of point P in the body, $\vec{\rho}$, is defined as

$$\vec{\rho} = \vec{R} + A\vec{r} \tag{2.10}$$

where, $\vec{R}$ denotes the origin of the body-fixed coordinate system's position vector, $\vec{r}$ is the position vector for the body point in terms of the body-fixed coordinate system, and A represents the transformation matrix for the body's angular displacements. According to the proposed model, the degrees of freedom of the rigid body are those that represent the motion of the body-fixed coordinates; i.e., the body-fixed coordinate system's translation, R, and orientation, θ.

$$A = \begin{bmatrix} x_1.X & y_1.X & z_1.X \\ x_1.Y & y_1.Y & z_1.Y \\ x_1.Z & y_1.Z & z_1.Y \end{bmatrix}$$

The projection of one onto the other is produced by the dot product of two unit vectors.

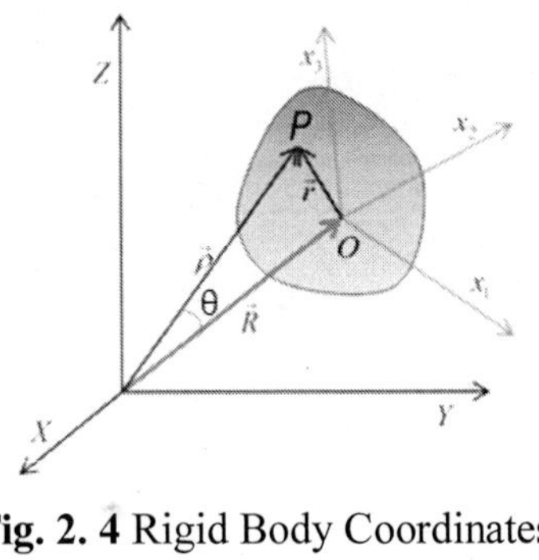

Fig. 2. 4 Rigid Body Coordinates.

2.3.3 Rigid body motion characteristics

The characteristics of rigid body motion is as follows:

- The angular acceleration α and angular velocity ω at every point on a rigid body are same.

- An arbitrary point's translation and subsequent rotation about the point are the two components that make up rigid motion.

2.4 ANGULAR VELOCITY RELATIONS OF RIGID BODY

The rate of change of orientation's first derivative is typically referred to as its angular velocity, while its second derivative, is also known as the rate of change of angular velocity and is known as angular acceleration in the body's plane of motion.

Angular Velocity, $\omega = \frac{d\theta}{dt} = \dot{\theta}$ (2.11)

Angular Acceleration,

$$\alpha = \frac{d\omega}{dt} = \dot{\omega} \quad or \quad \alpha = \frac{d^2\theta}{dt^2} = \ddot{\theta} \tag{2.12}$$

Putting and equating the values of dt from (2.11) & (2.12)

$$dt = \frac{d\theta}{\omega} = \frac{d\omega}{\alpha} \tag{2.13}$$

$$\alpha\ d\theta = \omega\ d\omega \tag{2.14}$$

The positive directions for ω and α, clockwise or anticlockwise, are chosen for each of these relations based on the aforementioned equation by removing the dt from the first two equations. It is mostly used to specify the equations for a particle's rectilinear motion. If the linear parameters *s*, *v*, and *a* are substituted with the angular quantities θ, ω and α, then all relations actually hold for the case of rotation in the plane. The relation between linear and angular motion as follows:

$$s = r\theta; v_T = r\omega;\ a_R = \frac{v_T^2}{r} \tag{2.15}$$

Where r is the radius of rotation, v_T is the tangential velocity and a_R is the radial acceleration.

As we move on to rigid body dynamics, we will notice that the similarities between the equations for linear and angular motion are practically complete. Through kinetics and kinematics, these relations demonstrate how the notions are symmetrical throughout the concepts.

Angular Parameters	Linear Parameters	
$\omega = \omega_o + \alpha t$	$v = u + at$	(2.16)
$\omega^2 = {\omega_o}^2 + 2\alpha(\theta - \theta_o)$	$v^2 - u^2 = 2as$	(2.17)
$\theta = \theta_o + \omega_o t + \frac{1}{2}\alpha t$	$s = ut + \frac{1}{2}at^2$	(2.18)

Here, t is the length of the motion being considered, and θ_o & ω_o are the angular position coordinate's and angular velocity's values, respectively, at t = 0. Given that they are an exact analogue to the corresponding equations for rectilinear motion with constant acceleration, these integrations should be simple to perform. The graphical relationships stated for s, u, and a may be applied to θ, ω and α by simply changing the symbols for t. To represent the rotation of a plane, you should draw these graphical relationships.

Example 2.1: A particle is travelling along a straight line with the following equation

$S = 3t^3 + t + 2$, where S is the displacement (m) and t is the time (s).

Determine all possible kinematic parameters such as velocity, acceleration, average velocity during first 4 seconds and distance travelled initially?

Solution: Initial displacement generally occurs at the initial time (t=0)

$$S_o = 3t^3 + t + 2$$

$$= 3(0)^3 + (0) + 2$$

$$= 2\text{m}$$

Initial velocity, we know that $v = \frac{ds}{dt}$

$$\frac{ds}{dt} = \frac{d(3t^3 + t + 2)}{dt}$$

$$u = 9t^2 + 1$$

$$u = 1\text{m/s} \qquad \text{(at t = 0)}$$

Displacement during first 4 seconds i.e., t = 4s

$$S_4 = 3t^3 + t + 2$$

$$= 3(4)^3 + (4) + 2$$

$$= 198 \text{ m}$$

V_{avg} during first 4 seconds $= \frac{S_4 - S_0}{t_4 - t_o} = \frac{198-2}{4-0} = 49$ m/s

$$V_4 = 9t^2 + 1 = 9(4)^2 + 1 = 145 \text{ m/s}$$

Initial acceleration, we know that $a = \frac{dv}{dt}$

$$\frac{dv}{dt} = \frac{d(9t^2 + 1)}{dt}$$

$$a = 18t$$

$$a = 0 \text{ m/s} \qquad \text{(at t = 0)}$$

Average acceleration, a_{avg} during first 4 seconds $= \frac{V_4 - u}{t_4 - t_o} = \frac{145-1}{4-0} = 36$ m/s^2

2.5 ROTATION OF RIGID BODY

The angular motion of a rigid body can be used to describe rotation. The line on the rigid body will move in a plane with the same angular displacement, angular velocity, and angular acceleration if the angular positions of line is measured from any fixed reference direction. A line's angular motion is determined by its angular position in relation to any chosen fixed reference as well as by the time derivatives of the displacement. It is not necessary for angular motion to have a fixed axis that is normal to the plane motion around which the line and the body revolve. Unlike point masses, rigid bodies allow forces to be applied at many points across the body.

2.5.1 Two-Dimensional Rotational Motion

There are two possible ways to identify the rotation of rigid body in two-dimensional rotational motion.

Fixed axis rotation: The disc rotates about an axis that runs through its centre and is perpendicular to the disc's plane. Most engineered systems include parts that only rotate around a fixed axis. An excellent illustration of rotation on a fixed axis is the flywheel rotation as shown in Fig. 2.5. The movement of a compact disc in a CD player, which is powered by a motor inside the player, is a straightforward another illustration of rotation about a fixed axis. In a streamlined representation of this action, the motor generates angular acceleration, which prompts the disc to spin. The disc starts spinning, but as it moves, resistance prevents it from accelerating in either direction. As a result, the disc spins with a constant angular velocity.

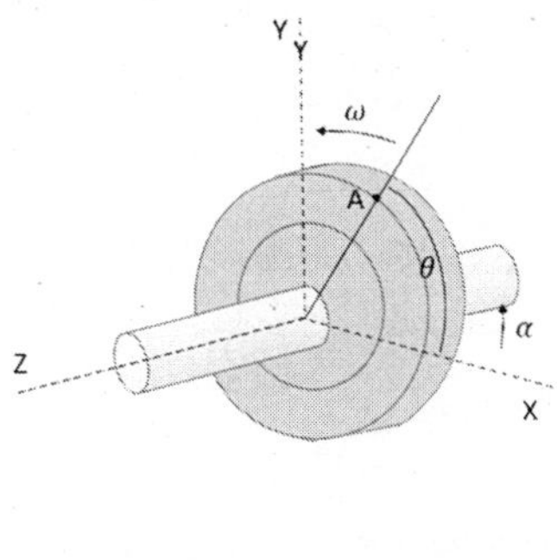

Fig. 2. 5 Fixed axis rotation

Fig. 2. 6 Plane of motion

Plane of motion is fixed: Bicycle wheel revolves in a set direction and mass is translating for motion in a straight line as shown in Fig. 2.6.

2.5.2 Three-Dimensional Rotational Motion

The rotation of rigid bodies about a fixed axis was a topic covered in some of the examples and applications we looked at. But there are several instances of rigid bodies rotating along an axis that is changing. A spinning top, a gyroscope, and a coin rolling across a table are all instances of this kind of motion. These motions can be quite intricate and challenging to analyse. However, we are aware that for each of these motions, the rotational equation of motion requires that the angular momentum around a given point S vary over time if there is a non-zero torque about that point.

We also know that the orbital and spin angular momentums about the mass centre are added to determine the angular momentum about S of a spinning body. The spin angular momentum about the mass centre for fixed axis rotation is

$$\vec{L} = I\,\vec{\omega} \qquad (2.19)$$

Where $\vec{\omega}$ is the angular velocity about the centre of mass of fixed axis of rotation.

The rotating rigid body produces a torque, which then produces an angular acceleration about the mass.

$$\tau = I\,\alpha \qquad (2.20)$$

Where α is the angular acceleration about centre of mass.

2.6 ORIENTATION THROUGH ANGULAR VELOCITY

Angular velocity is the difference between the orientation and the time taken for the change of rate in a rigid body.

- Location of orientation is generally described by the integration of the angular velocity.

$$[\omega dt = d\theta\,] \qquad \theta = \int_0^t \omega\, dt \qquad (2.21)$$

- In Case of same angular acceleration, we get the value of the angular velocity ω ,by integration of α with respect to the time taken for the change of rate of a rigid body.

$$[\alpha dt = d\omega\,] \qquad \omega = \int_0^t \alpha\, dt \qquad (2.22)$$

- Basically, the integration of the angular velocity ω via the angular acceleration α will be used to determine the orientation of the rigid body, which was in radians.

Example 2.2: A variable counter clockwise torque that is first applied at time t = 0 is applied to a flywheel that is freely moving at 2400 rpm clockwise. When a torque is applied, it results in an angular acceleration that is counter clockwise and equal to 4t rad/s^2 (where t is the time in seconds). Calculate (a) how long it will take the flywheel to slow down from its current clockwise angular speed to 1200 rpm, (b) how long it will take it to turn counter clockwise, and (c) how many times the flywheel will turn overall in the first 15 seconds of applying the torque.

Solution: (a) For finding the angular acceleration α, we need to integrate the angular velocity

$[d\omega = \alpha\, dt]$

The initial angular velocity = -2400(2π)/60 = -80π rad/s, we have

$\int_{-80\pi}^{\omega} d\omega = \int_0^t 4t\, dt \quad \Rightarrow \quad \omega = -80\pi + 2t^2$

Replacing it with a 1200 rpm clockwise angular speed, ω = -1200(2π)/60 = -40π rad/s

$-40\pi = -80\pi + 2t^2; \quad t^2 = 20\pi \Rightarrow t = 7.92$ s

(b) When the angular velocity of the flywheel is momentarily zero, it rotates in a different direction. Thus,

$0 = -80\pi + 2t^2 \qquad t^2 = 40\,\pi \qquad t = 11.21$ s

(c) The number of clockwise turns N_1 during the first 11.21 s, plus the number of counter clockwise turns N_2 during the remaining time, equals the total number of spins the flywheel makes in 15 seconds. We may obtain the angular displacement in radians by integrating the expression for in terms of t. Thus, for the first interval

$[d\theta = \omega\, dt] \qquad \int_0^{\theta_1} d\theta = \int_0^{11.21} (-80\pi + 2t^2)\, dt$

$$\theta_1 = \left[-80\pi\, t + \tfrac{2}{3}t^3\right]_0^{11.21} = -1878.25 \text{ rad}$$

Or N_1 = 1878.25/2π = 299 rpm clockwise.

For the second interval

$$\int_0^{\theta_2} d\theta = \int_{11.21}^{15} (-80\pi + 2t^2)\, dt$$

$$\theta_2 = \left[-80\pi\, t + \tfrac{2}{3}t^3\right]_{11.21}^{15} = 358.34 \text{ rad}$$

Or N_2 = 358.34/2π = 57 rpm counter clockwise. As a result, there were a total of 15 revolutions throughout that time, $N = N_1 + N_2 = 299 + 57 = 356$ rpm

2.7 MOTIONS OF ROTATING RIGID BODY

The rectilinear motion of the rigid body comes under the translation form of the motion and while coming to the rotation paths we generally go into the curvilinear motion. Plane curvilinear motion and space curvilinear motion are the two different types of curvilinear motion that exist. In a plane curved motion, there are three different sorts of components: rectangular components, tangential and normal components, and radial and transverse components. If the particle moves in three dimensions along a space curve, that motion is known as space curvilinear motion. Rectilinear, cylindrical, and spherical coordinate systems can all be used to explain the space's curvilinear motion.

2.7.1 Rectangular Components of Curvilinear Motion

When describing a particle's motion with respect to a fixed frame of reference, it is often helpful to refer to the particle's x, y, and z components, or rectangular components. The position vector can define the particle's position at any time (Fig. 2.7).

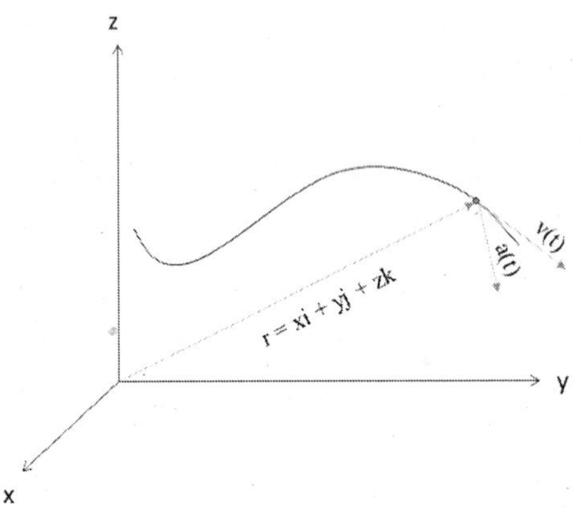

Fig. 2. 7 Position, velocity and acceleration vectors.

$$\mathbf{r} = \mathbf{x(t)\ i + y(t)\ j + z(t)\ k} \quad (2.23)$$

Position vector's magnitude, $r = \sqrt{x^2 + y^2 + z^2}$

The position vector's time derivative is the velocity vector, $v(t) = \frac{dr}{dt} = \frac{dx(t)}{dt}i + \frac{dy(t)}{dt}j + \frac{dz(t)}{dt}k$

The velocity vector can be written as $v(t) = v_x i + v_y j + v_z k,$ (2.24)

Where $v_x = \frac{dx(t)}{dt}, v_y = \frac{dy(t)}{dt},$ and $v_z = \frac{dz(t)}{dt}$

Velocity vector's magnitude, $v = \sqrt{(v_x)^2 + (v_y)^2 + (v_z)^2}$ (2.25)

The velocity vector, v, is always tangent to the motion's path.

The second derivative of the position vector, or the velocity vector's time derivative, is the acceleration vector (a). $a(t) = \frac{d^2r}{dt^2} = \frac{dv(t)}{dt} = \frac{dv_x(t)}{dt}i + \frac{dv_y(t)}{dt}j + \frac{dv_z(t)}{dt}k = a_x i + a_y j + a_z k$ (2.26)

Acceleration vector's magnitude, $a = \sqrt{(a_x)^2 + (a_y)^2 + (a_z)^2}$ (2.27)

In most cases, acceleration vector direction is not tangent to the particle's path.

Example 2.3: $x = 2(t+2)^2$ & $y = 2(t+1)^{-2}$ when t = 1 second

Determine (a) Path traced by particle

(b) Cartesian displacement and resultant displacement

(c) Cartesian velocity and resultant velocity

(d) Cartesian acceleration and resultant acceleration

Solution: (a) Path traced by the particle: xy= 4

Equation shows the main equation as Rectangular Hyperbola.

(b) Cartesian displacement and results displacement

$x = 2(t+2)^2$ = 2(2)2 = 8m

$y = 2(t+1)^{-2}$ = 2(2)-2 = 0.5m

$r = \sqrt{x^2 + y^2} = \sqrt{8^2 + 0.5^2} = 8.01$

$\tan\varphi = \frac{y}{x}$ $\varphi = \tan^{-1}\left(\frac{0.5}{8}\right) = 3.57^{o}$

(c) Cartesian velocities and resultant velocities

$V_x = \frac{dx}{dt} = 2(2(t+1)).1 = 4(t+1) = 8m/s$

$V_y = \frac{dy}{dt} = \frac{d}{dt}\left(\frac{2}{t+1}\right) = = 0.5\ m/s$

$V_r = \sqrt{8^2 + 0.5^2} = 8.01m/s$

(d) Cartesian acceleration and resultant acceleration

$a_x = \frac{dV_x}{dt} = 4m/s^2$

$a_y = \frac{d}{dt}(4(t+1)^{-3}) = (-4(-3(t+1)^{-4}) = (12(1+1)^{-4}) = 0.75m/s^2$

2.7.2 Curvilinear Motion: Tangential and Normal components

Consider there is a fixed bar of length (r) and it is moving with an angular velocity ω and angular acceleration α as shown in Fig. 2.8. While the bar displaces from the original position and it's oriented with an angle δθ, the angular motion can be resolving it into trigonometric functions (*cos* and *sin* functions). Obtain two modes of acceleration one is centrifugal acceleration and another is the tangential acceleration, the centrifugal acceleration always directed towards the centre of rotation, the tangential acceleration is always perpendicular to the rotating body.

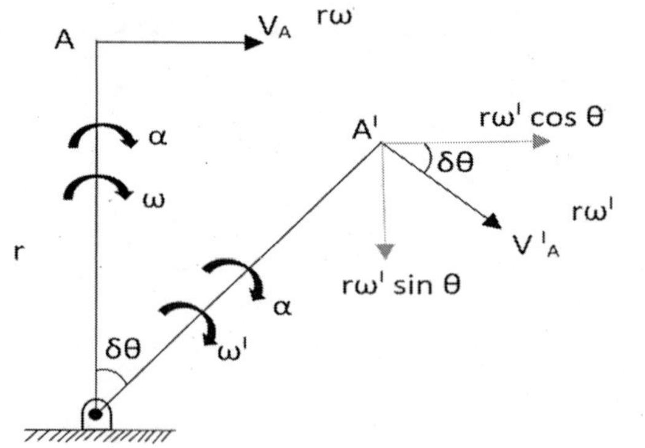

Fig. 2. 8 Movement of body with angular velocity and angular acceleration.

and the terms obtained here is $\mathbf{v}_t$ = Tangential velocity

$\mathbf{v}_r$ = Centrifugal / Normal Velocity

Some other terms, $\mathbf{a}_t$ = Tangential acceleration

$\mathbf{a}_c$ = Centrifugal acceleration

But, $\mathbf{a}_t = \dfrac{r\omega'\cos\delta\theta - r\omega}{\delta t}$

$$\mathbf{a}_t = \frac{r(\omega'\cos\delta\theta - \omega)}{\delta t}$$

$$\mathbf{a}_t = \frac{r(d\omega)}{\delta t} = r\alpha$$

Centrifugal acceleration, $\mathbf{a}_c = r\omega^2$ (2.28)

$$\mathbf{a}_c = r\left(\frac{V}{r}\right)^2 = \frac{V^2}{r} \tag{2.29}$$

Thus, the total acceleration is the square root, over the sum of the squares of tangential acceleration $\mathbf{a}_t$ and centrifugal acceleration $\mathbf{a}_c$,

$$a = \sqrt{a_t^2 + a_c^2} \qquad (2.30)$$

Case 1: If only rotation with constant tangential velocity is present, then $\mathbf{a_t}$ = 0, then the total acceleration, $\mathbf{a} = \mathbf{a_c}$

Case 2: If only Translation velocity is present without any centrifugal i.e., no rotation is occurring in the body then the centrifugal acceleration becomes zero $a_c = 0$, then the total acceleration, $\mathbf{a} = \mathbf{a_t}$

Example 2.4: A motorcycle ahead to turn a left turn (Fig. E2.1) while the radius of the turning road is about 200m and the initial velocity is about 72 kmph and the second case is while the time increased to 10sec the velocity is half of the initial velocity. Determine the total acceleration of the first and second case with the jerk too.

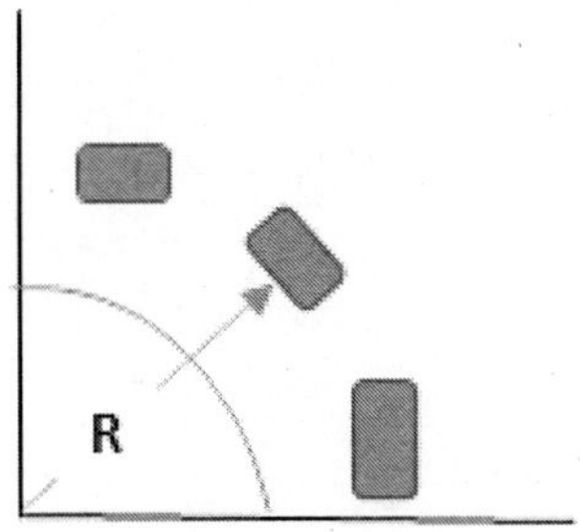

Fig. E2.1 Motorcycle taking left turn

Solution: Radius (R) = 200 m; initial velocity (v) = 72 × $\frac{5}{18}$ = 20 m/s

(a) $a_t = \frac{dv}{dt} = 0$

$$a_c = \frac{v^2}{r} = \frac{20^2}{200} = 2 \text{ m/s}^2$$

$$a = \sqrt{a_t^2 + a_c^2} = 2 \text{ m/s}^2$$

(b)Velocity, v = 36 kmph × $\frac{5}{18}$ = 10 m/s at time t=10 s

$$a_t = \frac{10-20}{10-0} = -1 \text{ m/s}^2$$

$$ac = \frac{v^2}{r} = \frac{10^2}{200} = 0.5 \text{ m/s}^2$$

$$a = \sqrt{a_t^2 + a_c^2} = 0.5 \text{ m/s}^2$$

$$\text{Jerk, } J_N = J_c = \frac{da_c}{dt} = \frac{0.5-2}{10} = -0.15 \text{ m/s}^3$$

'Drift' is result of centrifugal jerk / Normal jerk, the first and second case with the jerk too.

2.7.3 Curvilinear Motion: Radial and Transverse components

When a particle moves along a 3-D curve, the cylindrical coordinate system is used. An illustration of the cylindrical coordinate system in two dimensions is called a polar coordinate system. The particle's path of motion can be expressed using the polar coordinate system when it moves in a plane (2-D) and its radial distance, r, is not constant. The location of P in polar coordinates (Fig.

2.9) as **r** = r u_r. Keep in mind that the radial direction, r, radiates outward from the fixed origin, O, and that the transverse coordinate, θ, is measured counter clockwise from the horizontal.

It is known that instantaneous velocity as, $v = \frac{dr}{dt} = \frac{d(ru_r)}{dt} = \frac{dr}{dt}u_r + r\frac{du_r}{dt}$

Apply the chain rule on the second term, $\frac{du_r}{dt} = \frac{du_r}{d\theta}\frac{d\theta}{dt}$; But $\frac{du_r}{d\theta} = u_\theta$;

$$\text{Hence, } v = \frac{dr}{dt}u_r + r\frac{d\theta}{dt}u_\theta \tag{2.31}$$

In the equation (2.27), the first part is radial component ($v_r = \frac{dr}{dt}$) and second part is transverse component ($v_\theta = r\frac{d\theta}{dt}$) for the velocity vector.

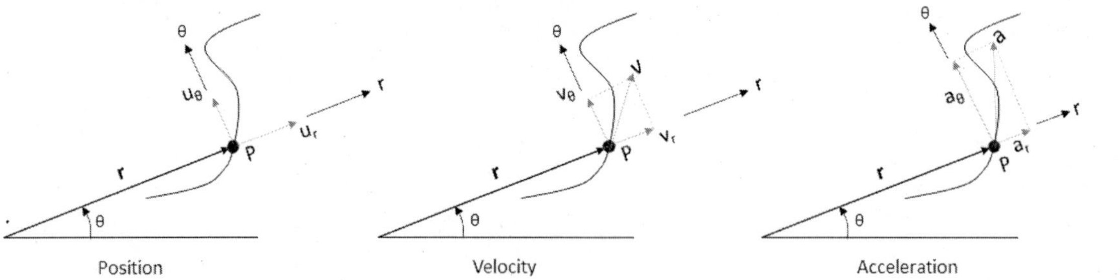

Fig. 2. 9 Position, Velocity and Acceleration vectors in Polar Coordinates.

$$\text{The magnitude of velocity of the particle, } v = \sqrt{\left(\frac{dr}{dt}\right)^2 + \left(r\frac{d\theta}{dt}\right)^2} \tag{2.32}$$

The acceleration at any instant, $a = \frac{dv}{dt} = \frac{d\left(\frac{dr}{dt}u_r + r\frac{d\theta}{dt}u_\theta\right)}{dt} = \frac{d^2r}{dt^2}u_r + \frac{dr}{dt}\frac{du_r}{dt} + \frac{dr}{dt}\frac{d\theta}{dt}u_\theta + r\frac{d\left(\frac{d\theta}{dt}u_\theta\right)}{dt}$

Apply the chain, $\frac{du_r}{dt} = \frac{du_r}{d\theta}\frac{d\theta}{dt} = u_\theta\frac{d\theta}{dt}$ and $\frac{du_\theta}{dt} = \frac{du_\theta}{d\theta}\frac{d\theta}{dt} = -u_r\frac{d\theta}{dt}$

$$a = \frac{d^2r}{dt^2}u_r + \frac{dr}{dt}\left(u_\theta\frac{d\theta}{dt}\right) + \frac{dr}{dt}\frac{d\theta}{dt}u_\theta + r\left[\frac{d^2\theta}{dt^2}u_\theta - -u_r\left(\frac{d\theta}{dt}\right)^2\right]$$

$$a = \left[\frac{d^2r}{dt^2} - r\left(\frac{d\theta}{dt}\right)^2\right]u_r + \left[2\frac{dr}{dt}\frac{d\theta}{dt} + r\frac{d^2\theta}{dt^2}\right]u_\theta = a_r u_r + a_\theta u_\theta \tag{2.33}$$

Radial acceleration component, $a_r = \frac{d^2r}{dt^2} - r\left(\frac{d\theta}{dt}\right)^2$ and transverse component, $a_\theta = 2\frac{dr}{dt}\frac{d\theta}{dt} + r\frac{d^2\theta}{dt^2}$

$$\text{The magnitude of acceleration, } a = \sqrt{\left[\left(\frac{d^2r}{dt^2} - r\left(\frac{d\theta}{dt}\right)^2\right)^2 + \left(2\frac{dr}{dt}\frac{d\theta}{dt} + r\frac{d^2\theta}{dt^2}\right)^2\right]} \tag{2.34}$$

If the particle P moves along a space curve, its position can be written in cylindrical coordinate system as,

$$r_p = r\,u_r + z\,u_\theta \tag{2.35}$$

Velocity, $v_p = \frac{dr_p}{dt} = \frac{dr}{dt} u_r + r\, \frac{d\theta}{dt}\, u_\theta + \frac{dz}{dt} u_z$ (2.36)

Acceleration, $a_p = \left[\frac{d^2r}{dt^2} - r\left(\frac{d\theta}{dt}\right)^2\right] u_r + \left[2\frac{dr}{dt}\frac{d\theta}{dt} + r\,\frac{d^2\theta}{dt^2}\right] u_\theta + \frac{d^2z}{dt^2}\, u_z$ (2.37)

Example 2.5: A wheel and a connected blower are housed inside the housing and are turned by the motor shown in the image. In Fig. E2.2 (a), the design's specifics are depicted. If the motor-connected pulley A starts rotating at rest with a constant angular acceleration of α_A = 4 rad/s^2, calculate the magnitudes of the velocity and acceleration at point P on the wheel after the pulley has completed two revolutions. Assume the wheel and pulley will not experience any transmission belt slippage.

Solution: $\theta_A = 2\, rev\left(\frac{4\pi\, rad}{1\, rev}\right) = 25.14\, rad$

The pulley's A angular velocity, because α_A is constant,

$$(\circlearrowleft +)\ \omega^2 = \omega_0^2 + 2\alpha_c(\theta - \theta_0)$$

$$\omega^2 = 0 + 2(4\ \text{rad/s}^2)(25.14\, rad - 0)$$

$$\omega_A = 14.18\, rad/s$$

The belt passes over the pulley and wheel at the same speed and with the same tangential component of acceleration, thus

$$v = \omega_A r_A = \omega_B r_B = 14.18\, rad/s\, (0.1\, m) = \omega_B\, (0.5\, m)$$

$$\omega_B = 2.836\, rad/s$$

$$\alpha_t = \alpha_A r_A = \alpha_B r_B = 4\, rad/s^2\, (0.1\, m) = \alpha_B\, (0.5\, m)$$

$$\alpha_B = 0.8\, rad/s^2$$

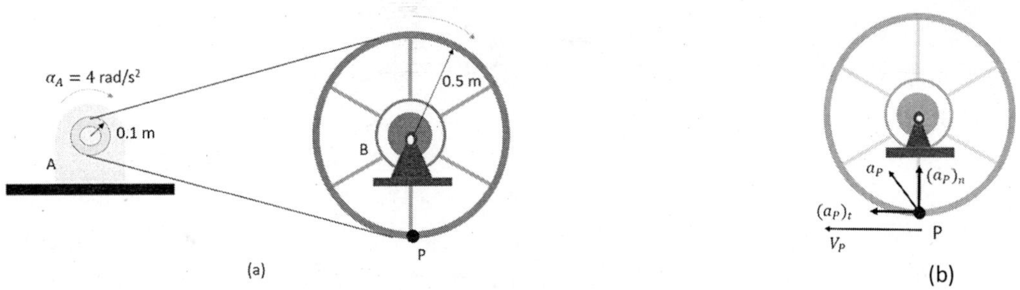

Fig. E2.2

The kinematic diagram Fig. E2.2 (b) depicted the motion of P as

$$v_P = \omega_B r_B = 2.836\, rad/s\, (0.5\, m) = 1.418\, m$$

$$(a_P)_t = \alpha_B r_B = 0.8\, rad/s^2\ (0.5\, m) = 0.4\, m/s^2$$

$$(a_P)_n = \omega_B^2 r_B = (2.836\, rad/s^2)^{\,2}(0.5\, m) = 4.02\, m/s^2$$

The magnitude of acceleration at P is,

$$a_P = \sqrt{(0.4\, m/s^2)^2 + (4.02\, m/s^2)^2} = 4.04\, m/s^2$$

Example 2.6: A string that is wrapped around the drum is attached to the plate as shown in Fig. E2.3. Calculate the drum's angular acceleration and velocity after 20 revolutions if it accelerates from rest at a rate of 10 feet per second. After the drum has made 20 revolutions and the hook has moved downward for 5 seconds, how many more revolutions will the drum make?

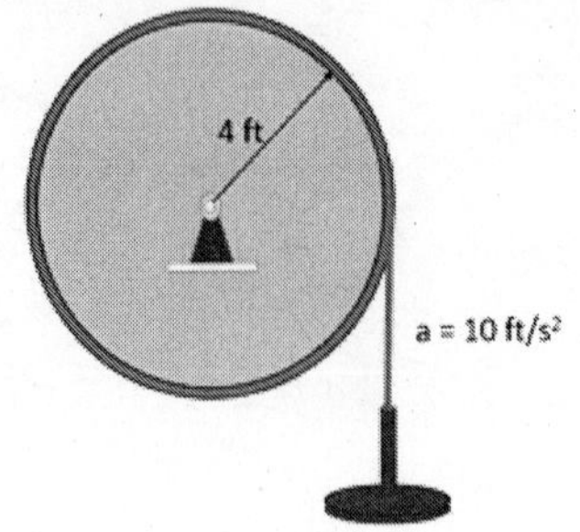

Fig. E2.3

Solution: Angular acceleration of the drum, $a_t = \alpha r$; $10 = \alpha(4)$;

$$\alpha = 2.5\ rad/s^2$$

$$a_C = \alpha = 2.5\,rad/s^2 \text{ and } \theta = (20\,rev) \times \left(\frac{2\pi\,rad}{1rev}\right) = 40\pi\,rad,$$

$$\omega^2 = \omega_0^2 + 2\alpha_c(\theta - \theta_0)$$

$$\omega^2 = 0 + 2(2.5\,rad/s^2)(40\pi - 0)$$

$$\omega = 25.07\,rad/s^2$$

The drum's angular displacement, measured in 5 sec after it has made 20 revolutions,

$\omega_0 = 25.07\,rad/s^2$

$$\theta = \theta_0 + \omega_0 t + \frac{1}{2}\alpha_c t^2$$

$$= 0 + 25.07(5) + \frac{1}{2}(20)(5^2)$$

$$= (375.35\,rad) \times \left(\frac{1\,rev}{2\pi\,rad}\right) = 59.71\,rev$$

2.8 FIVE TERM ACCELERATION EQUATION

The movement of the slider on long bar as illustrated in Fig. 2.10, five components of accelerations are acting on this mechanism. These accelerations equation may be termed as five term acceleration formula. They are tangential acceleration, Coriolis acceleration, radial acceleration, centrifugal acceleration and normal acceleration. The tangential acceleration and Coriolis acceleration will be summative and subtractive according to the direction of slider movement in horizontal direction as shown in Fig. 2.11 (a) and (b).

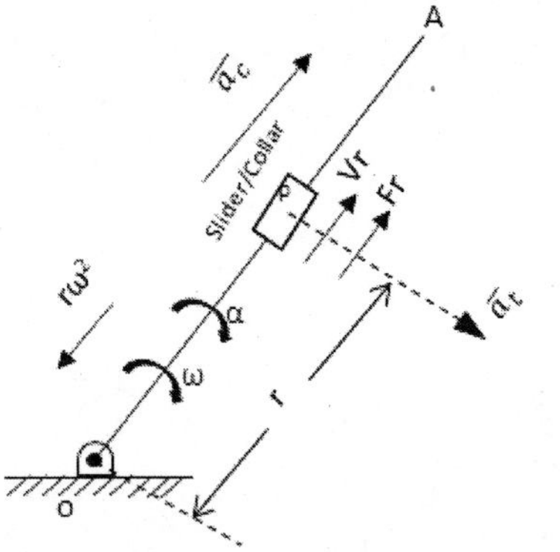

Fig. 2. 10 Slider on Long Bar.

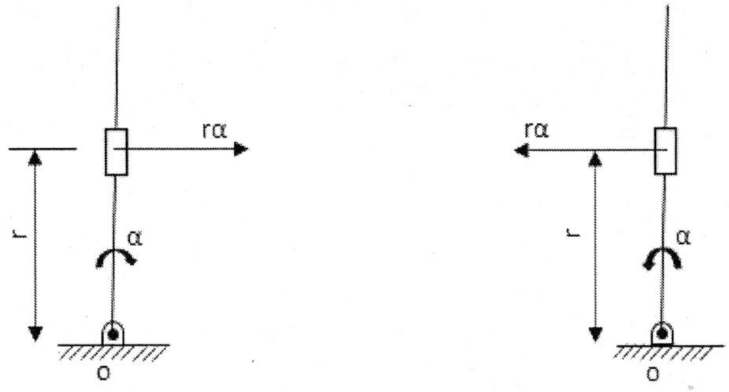

(a) **(b)**

Fig. 2. 11 (a) Right Tangential Acceleration, **(b)** Left Tangential Acceleration.

In Fig. 2.11 (a), the direction of the angular acceleration i.e., α is in the direction of the rightward or positive axis and it generally denotes the motion of the body is in the positive direction. It also says that the tangential acceleration is positive. α is positive and the Coriolis must be in the positive direction then the tangential acceleration will be positive a_t = positive.

In Fig. 2.11 (b), the direction of the angular acceleration i.e., α is in the direction of the leftward or negative axis and it generally denotes the motion of the body is in the negative direction. It also says that the tangential acceleration is negative. α is negative and the Coriolis must be also in the negative direction then the tangential acceleration will be negative a_t = negative.

Tangential acceleration is the summation of the normal tangential acceleration (rα) and the Coriolis acceleration (2ωVr) because it is in the direction of the tangential velocity,

The total tangential acceleration is, $\bar{a_t} = r\alpha + 2\omega Vr$ (2.38)

It is implemented when the rate of change of angular acceleration is in the positive direction.

$$\bar{a_t} = r\alpha\text{-}2\omega Vr \quad 2.39)$$

It is implemented when the rate of change of angular acceleration is in the negative direction.

Now, centrifugal Acceleration is basically the combination of the radial acceleration and centripetal acceleration. Fig. 2.12 (a) shows the direction of the centrifugal acceleration $\boldsymbol{a_c}$ is always in the direction of the downward and it generally denotes the motion of the body is in the negative direction. It also says that the radial acceleration is negative. The centripetal acceleration must be in the downward direction then the centrifugal acceleration will be positive.

The direction of the centrifugal acceleration (**a_c**) is always in the direction of the downward and it generally denotes the motion of the body is in the positive direction as shown in Fig. 2.12 (b). It also says that the radial acceleration is negative. The centripetal acceleration must be in the downward direction then the centrifugal acceleration (**a_t**) will be positive.

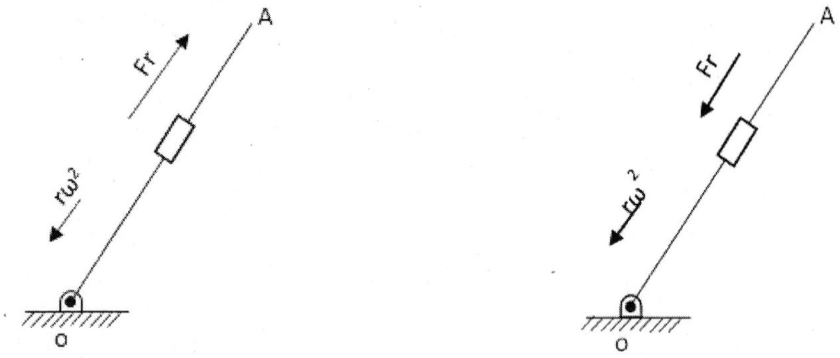

(a) **(b)**

Fig. 2. 12 (a) Up Radial Acceleration, **(b)** Down Radial Acceleration.

Centrifugal acceleration is the summation of the radial acceleration (fr) and the centripetal acceleration ($r\omega^2$) because it is in the direction of the linear velocity. The summation of the total centrifugal acceleration is, $\overline{a_c} = fr + r\omega^2$, if the rate of change of radial acceleration is in the positive direction.

If the rate of change of radial acceleration is in the negative direction, $\overline{a_c} = fr - r\omega^2$ (2.40)

Thus, the total acceleration is the square root, over the sum of the squares of tangential acceleration $\bar{a_t}$ and centrifugal acceleration $\overline{a_c}$

$$\bar{a} = \sqrt{\bar{a_t}^{\,2} + \overline{a_c}^{\,2}} \qquad (2.41)$$

Dynamics uses a coordinate system to describe motion that moves and rotates, allowing analysis of the motion of a vehicle's parts even while the vehicle is moving in a fixed frame. In order to make the so-called five term acceleration equation easier to see, build the same in two dimensions. When two frames-one stationary, the other moving and rotating are utilised to represent the motion. It is describing the motion to be investigated, which includes every motion to be analysed and the Fig. 2.13 depicts the situation.

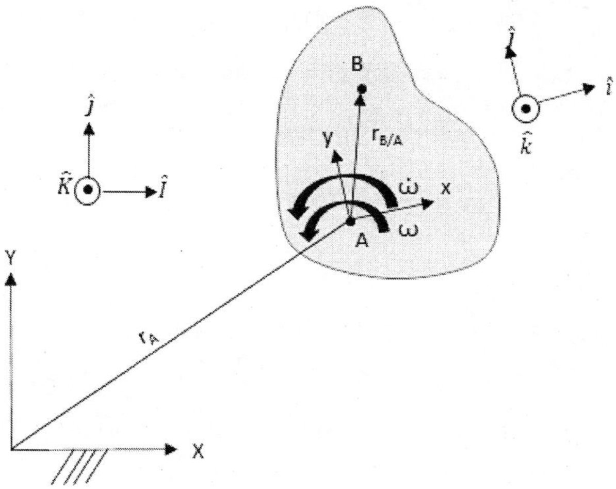

Fig. 2. 13 Co-ordinate of moving and rotating on X-Y, x-y axis.

The position vector for motion B in XY-frame can be represented as:

$$\overrightarrow{r_B} = \overrightarrow{r_A} + \overrightarrow{r_{B/A}} \tag{2.42}$$

$$\overrightarrow{r_B} = r_{AX}\hat{I} + r_{AY}\hat{J} + r_{B/A-X}\hat{\imath} + r_{B/A-X}\hat{\jmath} \tag{2.43}$$

Velocity: The velocity vector, $\overrightarrow{v_B} = \dot{\overrightarrow{r_B}}$

To differentiate $\dot{\overrightarrow{r_B}}$. Use the product rule and the velocity vector as

$$\overrightarrow{v_B} = \dot{\overrightarrow{r_B}} = \dot{r}_{AX}\hat{I} + r_{AX}\dot{\hat{I}} + \dot{r}_{AY}\hat{J} + r_{AY}\dot{\hat{J}} + \dot{r}_{B/A-X}\hat{\imath} + r_{B/A-X}\dot{\hat{\imath}} + \dot{r}_{B/A-X}\hat{\jmath} + r_{B/A-X}\dot{\hat{\jmath}} \tag{2.44}$$

The magnitudes of the corresponding r vectors for every $\dot{r}$ are altered, either speeding up or slowing down. The equation (2.44) becomes,

$$\overrightarrow{v_B} = v_{AX}\hat{I} + v_{AY}\hat{J} + v_{B/A-X}\hat{\imath} + v_{B/A-X}\hat{\jmath} + r_{B/A-X}\dot{\hat{\imath}} + r_{B/A-X}\dot{\hat{\jmath}} \tag{2.45}$$

The slab has made a little angle, Δθ turn in the counter-clockwise after a brief period, Δt. The unit vectors $\hat{\imath}$ and $\hat{\jmath}$ changes are represented as $\Delta\hat{\imath}$ and $\Delta\hat{\jmath}$ respectively. As Δt gets shorter and shorter, $\Delta\hat{\imath}$ rotates 90 degrees counterclockwise to $\hat{\imath}$ becoming perpendicular to $\hat{\imath}$ and moving in the direction of $\hat{\jmath}$ and $\Delta\hat{\jmath}$ is directed in the $-\hat{\imath}$-direction. The magnitude of $\Delta\hat{\imath}$ approaches the arc length in the limit. Since $\hat{\imath}$ is a unit vector, the arc length becomes 1*Δθ . So $\Delta\hat{\imath} = \Delta\,\hat{\jmath}$. From this, in the limit,

$$\dot{\hat{\imath}} = \frac{\Delta\hat{\imath}}{\Delta t} = \frac{\Delta\theta}{\Delta t}\hat{\jmath} = \omega\,\hat{\jmath} \tag{2.46}$$

$$\dot{\hat{\jmath}} = \frac{\Delta\hat{\jmath}}{\Delta t} = \frac{\Delta\theta}{\Delta t}(-i) = -\,\omega\,\hat{\imath} \tag{2.47}$$

Now we can substitute these two equations in $\overrightarrow{v_B}$.

$$\overrightarrow{v_B} = v_{AX}\hat{I} + v_{AY}\hat{J} + v_{B/A-X}\hat{\imath} + v_{B/A-X}\hat{\jmath} + r_{B/A-X}\omega - r_{B/A-y}\omega\hat{\imath} \tag{2.48}$$

$$\overrightarrow{v_B} = v_{AX}\hat{I} + v_{AY}\hat{J} + (v_{B/A-X} - r_{B/A-y}\omega)\hat{\imath} + (v_{B/A-X} + r_{B/A-X}\omega)\hat{\jmath} \quad (2.49)$$

Because B's position and velocity are given in the x-y coordinate system, we should eliminate the B/A term. Thus the equation (2.45) becomes

$$\overrightarrow{v_B} = v_{AX}\hat{I} + v_{AY}\hat{J} + (v_X - r_y\omega)\hat{\imath} + (v_y + r_X\omega)\hat{\jmath} \quad (2.50)$$

Acceleration:

Acceleration is the time derivative of velocity, $\vec{a}_B = \dot{\vec{v}}_B$ (2.51)

The time derivative of the equation (2.45) and substitute $\dot{r}$ is v, $\dot{v}$ is a with $\hat{I}$ and $\hat{J}$ fixed frame,

$$\overrightarrow{a_B} = a_{AX}\hat{I} + a_{AY}\hat{J} + (a_X - 2v_y\omega - r_y\dot{\omega} - r_X\omega^2)\hat{\imath} + (a_y - 2v_x\omega - r_x\dot{\omega} - r_y\omega^2)\hat{\jmath} \quad (2.52)$$

The five-term acceleration equation is given by the expression (2.52). Nevertheless, the five accelerations were just the result of applying the product rule to vector differentiation.

1. $\overrightarrow{a_A} = a_{AX}\hat{I} + a_{AY}\hat{J}$, the acceleration of point A in the fixed XY-frame
2. $\overrightarrow{a_{xy}} = a_X\hat{\imath} + a_Y\hat{\jmath}$, the acceleration of B within the moving, rotating xy-frame
3. $\overrightarrow{a_B} = -2v_y\omega\hat{\imath} + 2v_x\omega\hat{\jmath}$, the Coriolis acceleration of B
4. $\overrightarrow{a_t} = -r_y\dot{\omega}\hat{\imath} + r_x\dot{\omega}\hat{\jmath}$, the tangential acceleration of B
5. $\overrightarrow{a_n} = -r_X\omega^2\hat{\imath} - r_y\omega^2\hat{\jmath}$, the normal acceleration of B

The acceleration, $\overrightarrow{a_B} = \overrightarrow{a_A} + \overrightarrow{a_{xy}} + \overrightarrow{a_c} + \overrightarrow{a_t} + \overrightarrow{a_n}$

(2.53)

UNIT SUMMARY

This unit, discussed various basic concepts of kinematic parameters, the range of the motion in both rotation and tangential, described the velocities present in various motions either in tangential, centrifugal, radial, centripetal and Coriolis acceleration i.e. five terms of acceleration which is included in the total acceleration in resolving of radial and transverse components. The knowledge gained in this unit will be definitely helping to deal with various problems in engineering mechanics mainly in the kinematics of the rigid body.

EXERCISES

Multiple Choice Questions

1) The axis of rotation of a purely rotating body

a) must pass through the centre of mass and a particle of the body

b) may pass through the centre of mass and a particle of the body

c) must pass through a particle of the body only

d) may pass through a particle of the body only

2) A circular object with radius r rolls with a centre velocity of V and a velocity at the point of contact with the floor of being on a horizontal surface without sliding.

a) V in the direction of motion.

b) V vertically upward from the floor

c) V opposite to the direction of motion.

d) Zero

3) A cylinder of radius 'R' rolling on ground with an angular velocity ω velocity of particle on the surface of cylinder having maximum velocity is

a) ωr/2

b) ωr

c) 3ωr/2

d) 2ωr

4) When a solid disc of radius r is rolling on a horizontal surface with an angular velocity (ω) and acceleration (α), the magnitude of the acceleration at the point of contact is ----

a) $\sqrt{r\alpha^2 + (r\omega^2)^2}$

b) Zero

c) rα

d) $r\omega^2$

5) The centre of a circular disc moves at a constant velocity v while it rolls on a horizontal surface without slipping. Which of the following can apply to the velocity at a particular location on the disc's rim?

a) v

b) 2v

c) zero

d) All of the above

6) A circular disc with radius R moves with constant velocity V_0 along a horizontal surface without slipping. The acceleration of point P which is located on the disc's surface is

a) Constant magnitude

b) Constant in direction

c) Constant in magnitude as well as direction

d) None of the above

7) The following types of planar motion occur when a rigid body

a) Translation

b) Rotation about a fixed axis

c) General plane motion

d) All of the above

8) All of the particles in a body that is translating rectilinearly follow ———————paths.

a) along parallel straight-line

b) along curved

c) along horizontal linc

d) along vertical line

9) When the particle rotates along a fixed axis while moving in a circular motion. Every line segment in the body goes through the

a) same angular displacement

b) same angular displacement and angular velocity

c) same angular velocity and angular acceleration

d) same angular displacement, angular velocity and angular acceleration

10) The sliding velocity is preceded by the Coriolis component of acceleration-----

a) 45^0

b) 90^0

c) 180^0

d) 135^0

Answers of Multiple-Choice Questions

1. (b) 2. (d) 3. (d) 4. (d) 5. (d) 6. (c) 7. (d) 8. (a) 9. (d) 10. (b)

Short and Long Answer Type Questions

1) What types of rigid body planar motion are there?
2) What is the rigid body's equation of motion?

3) What are the three fundamental rigid motions?
4) What distinguishes rigid bodies from rigid bodies in two dimensions?
5) How can a rigid body be rotated?
6) What is the most notable distinction between rigid body motion and particle motion?
7) What modifies the rigid body's rotational motion?
8) What do rigid body translational motion and rotational motion mean?
9) Describe rotational motion with an example.
10) What are the five kinematics variables?

Numerical Problems

1) When travelling at a uniform acceleration, a car will travel 450 m in 5 sec and 700 m in the following 5 sec. Find the acceleration of the car.
2) The gear (Fig. P2.1) begins at rest and after 25 rotations has reached an angular velocity of $\omega = 30$ rad/s. Find the needed time and constant angular acceleration.

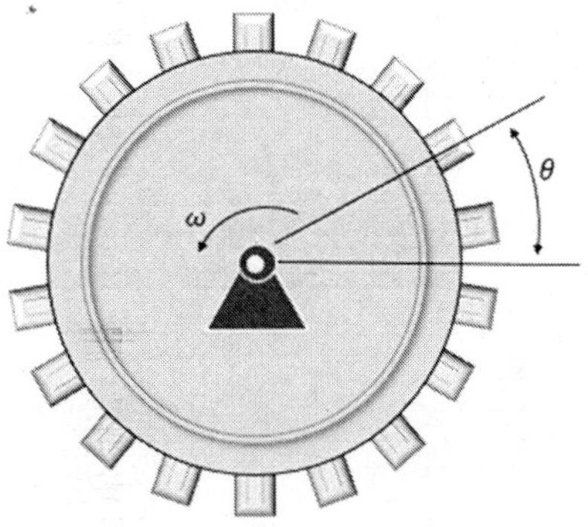

Fig. P2.1

3) As seen in Fig. P2.2, Person A is positioned in the centre of a revolving platform and is facing Person B, who is travelling east on a bicycle. The given Fig. P2.2 displays the pertinent distances and speeds. What is the apparent velocity of B as observed by A at the moment in question?

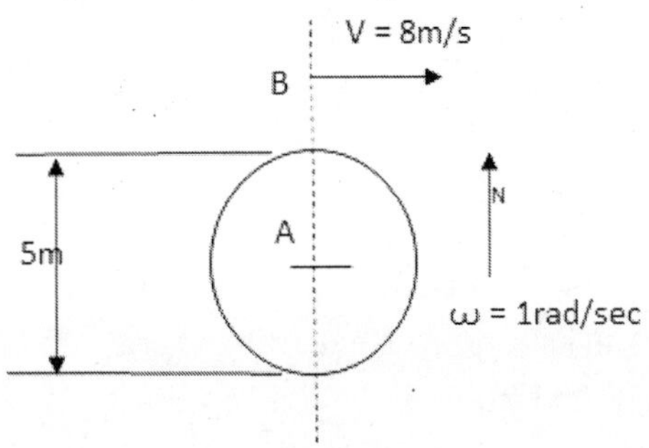

Fig. P2.2

4) Equation, $x = 3t^4 + t^3 + 5t$ describes how a particle's position changes over time in rectilinear motion. Find the velocity and acceleration in consistent units when the motion began with.
5) An object's starting speed is 50 m/s. The following expression gives the object's acceleration, a = -0.5v where v denotes the object's instantaneous speed. Find the object's speed after three seconds.
6) An angular velocity of w = (2t + 3) rad/s, where t is measured in seconds, is used to describe the rotation of a disc with a radius of 6 in around a fixed axis. When the angular displacement is 40 rad, calculate the tangential and normal components of the acceleration at a location on the disk's rim.
7) A wheel revolves around O in the illustration on Fig. P2.3. The speeds at points A and B, which are situated along the radius of the wheel, are 80 m/s and 140 m/s, respectively. Between points A and B, there is a 300mm gap. Find the wheel's diameter (in mm).

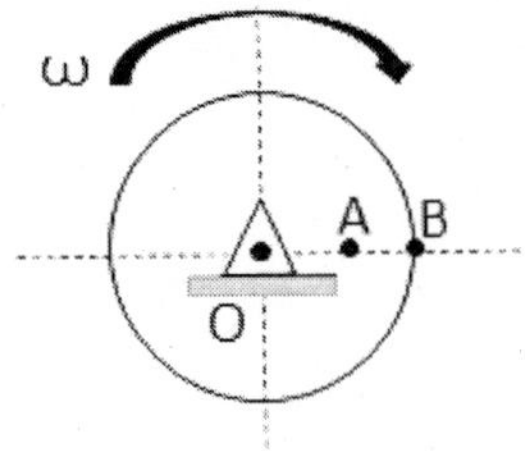

Fig. P2.3

8) As depicted in the Fig, P2.4, a uniformly thin rod of 10 m length and 3 kg mass rotates in a vertical plane around a horizontal axis 2 m from its end. Find the rod's angular acceleration at the free ends in rad/s^2.

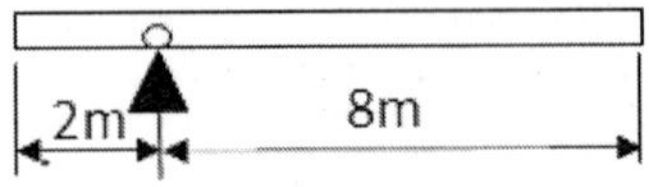

Fig. P2.4

9) As seen in the Fig. P2.5, a rod that is 2 m long is sliding around a corner. The speed of point A is 1 m/s at the instant when the rod forms an angle of 60^0 with the horizontal plane. Find the rod's angular velocity at this instant.

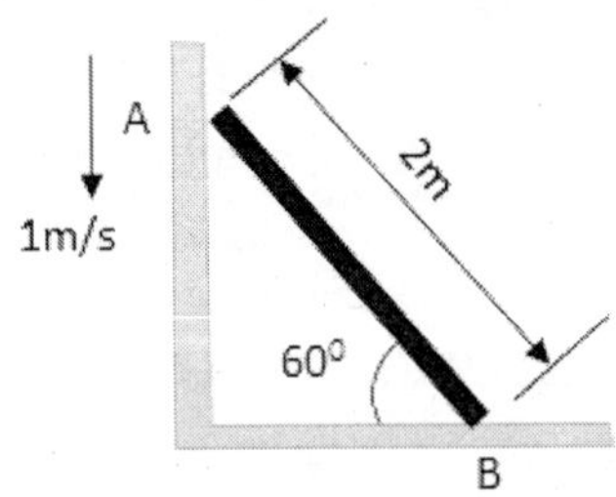

Fig. P2.5

10) A slender rod 10m long moves with its ends in contact with a horizontal floor and an inclined surface as shown in Fig. P2.6. At the given position, end A has a velocity of 36m/s towards left. Find the velocity of the end B.

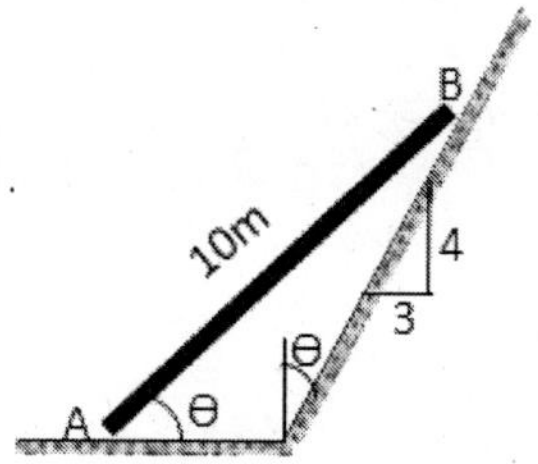

Fig. P2.6

11) A wheel is seen spinning around "O" in Fig. P2.7. Two points, A and B, are situated along the radius of the wheel and are moving at speeds of 80 m/s and 140 m/s, respectively. Find the diameter of the wheel given that A and B are separated by 30 cm.

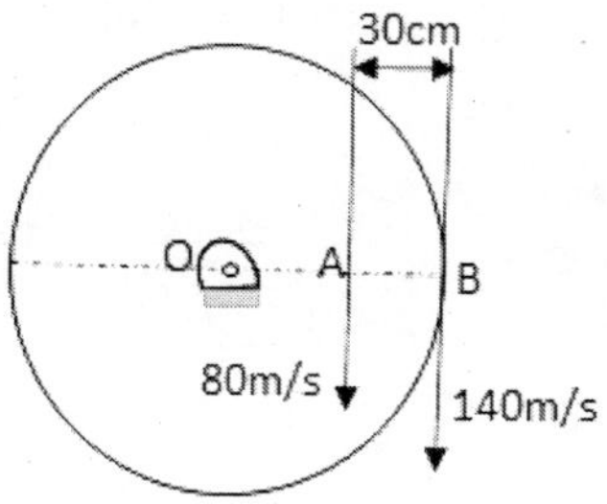

Fig. P2.7

12) A circular roller of radius, $r = 200$mm is contacted at the top and bottom points of its circumference by two conveyer belts AA and BB as shown in Fig. P2.8. If the belts run with uniform speeds, $v_1 = 1.8$ m/s and $v_2 = 1.2$ m/s, find the linear velocity v_{ac} of the roller and also the angular velocity of the roller.

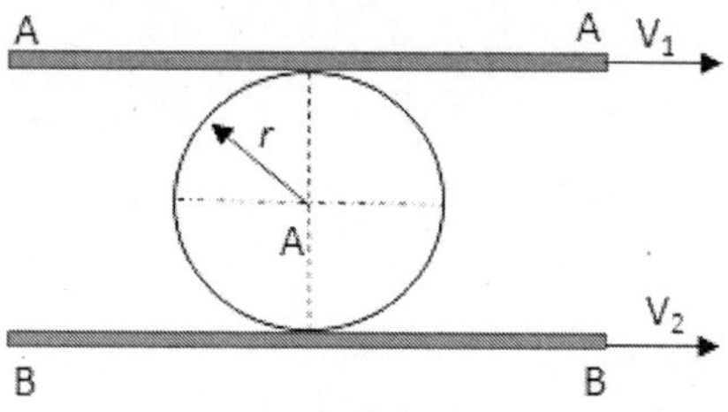

Fig. P2.8

13) The disc shown in Fig. P2.9 rolls without slip on two level rails with constant speed $V_C = 1$m/s. Find the velocities V_D and V_E of points D and E at the top and bottom of the rim.

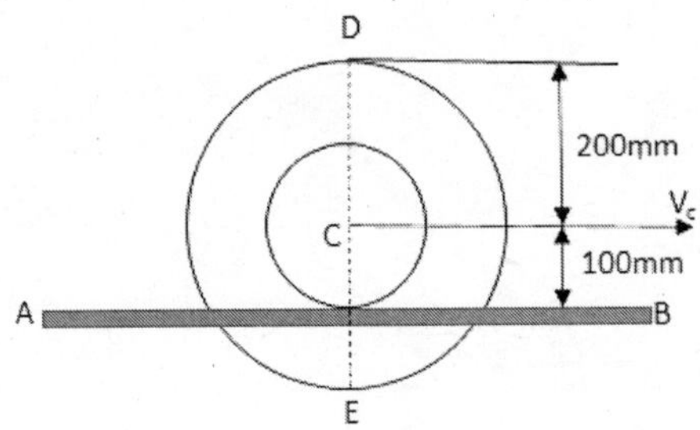

Fig. P2.9

14) The hoop is positioned on the uneven surface as shown in Fig. P2.10 with a $\omega = 8$ rad/s angular velocity and an $\alpha = 10$ rad/s^2 angular acceleration. Additionally, the center's velocity, v_0 is 10 m/s and its acceleration, a_0 is 4 m/s^2. Calculate the current acceleration of point A and B.

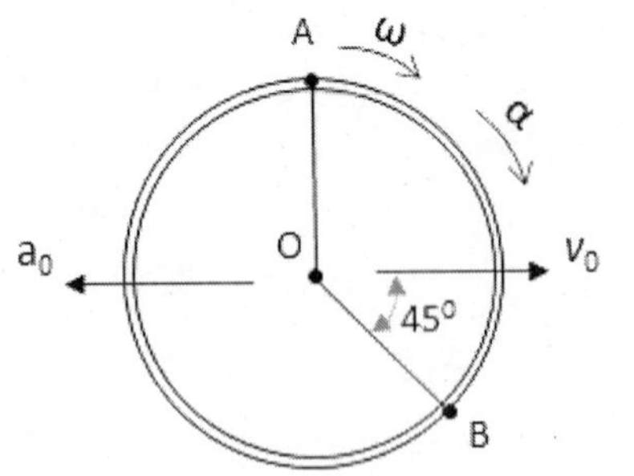

Fig. P2.10

15) With a constant angular velocity of $\omega = 20i + 10j + 15k$ rad/sec, a circular disc revolves along a fixed axis. A point P on its rim has a velocity of 100 m/s for its x component and 60 m/s for its y component at a given instant. Find the radial distance R from P to the rotation axis and the magnitude v of P's velocity. Additionally, determine P's acceleration.

PRACTICAL

Aim: To calculate the velocity of a crawler moving in a circular motion

Materials required: string, stopwatch, measure stick, ring stand

Procedure:

1. Calculate the cart's mass in kg
2. A 1 m length of string has been calculated.
3. Create loops on both ends, spacing the knots 0.9 m apart.
4. Overlap the string's ring end with the pole. With the push pin, fasten the other loop of the string to the cart.

5. With a measuring tape, mark the car's starting position on the ground.
6. Start the car and the stopwatch and observe the car for 3 revolutions
7. Turn off the car. Note the time taken for 3 revolutions.
8. Steps 6 and 7 should be repeated three times at this radius.
9. Repeat these steps for other string lengths (may be a string length of 0.70 m and 0.50 m).

Observation:

Mass of the cart, *m*: __________ kg

String length (m)	Trail-1, Time (sec)	Trail-2, Time (sec)	Trail-3, Time (sec)	t_{avg} (sec)

Calculations:

Time for one revolution = $t_{avg}/3$

String Length (m)	t_{avg} (sec)	Time for 1 revolution (sec)

Linear Velocity:

The circumference of your circles is the distance travelled, measured in metres. You may determine the linear velocity by multiplying the circumference of each circle by the amount of time it took the car to complete that round.

String length, *L* (m)	Circumference of circle, *d* (m)	Time to complete the circle, *t* (sec)	Velocity, $v = d/t$ (m/s)

Angular Velocity and Tangential Velocity:

String length, L (m)	Time for one revolution, t_{avg} (sec)	Angular velocity, $\omega = \frac{\Delta\theta}{t_{avg}} = \frac{2\pi}{t_{avg}}$ (rad/s)	Tangential velocity, $v_t = r\omega$ (m/s)

Centripetal Acceleration and Centripetal Force:

String length, L (m)	Tangential velocity, $v_t = r\omega$ (m/s)	Centripetal acceleration, $a_c = \frac{{v_t}^2}{r}$ (m/s^2)	Centripetal force, $F_c = ma_c$ (N)

Conclusion:

1. What connection exists between radius (string length) and angular velocity?
2. What is the relationship between angular and linear velocity?
3. What changed with the centripetal force as the radius shrunk?

KNOW MORE

André-Marie Ampère

The word "kinematic" is the English translation of "cinematique" by **A.M. Ampère**.

André-Marie Ampère: (20 January 1775 – 10 June 1836) was a French physicist, mathematician, and lay Catholic. He is credited with helping to develop the field of classical electromagnetism, which he called "electrodynamics," and was also a lay Catholic. In addition, he is the creator of many applications, including the electrical telegraph and the solenoid (a name he coined).

Ampère was an autodidact who taught at the École polytechnique and the Collège de France in addition to being a member of the French Academy of Sciences. He is the subject of the ampere, the SI unit used to measure electric current. One of the 72 names engraved on the Eiffel Tower includes his name.

The members of the French Academy of Sciences were shown Ampère's friend and eventual eulogist François Arago's startling discovery—that a magnetic needle is deflected by an adjacent electric current—in September 1820. Ampère started creating a mathematical and physical theory to comprehend how electricity and magnetism interact. Ampère built the theoretical framework for electrodynamics by extending experimental work by demonstrating that two parallel wires carrying electric currents will either attract or repel one another depending on whether the currents are flowing in the same or the opposite directions. He used mathematics to extrapolate general physical rules from these experimental findings.

Ampère's law, which asserts that the mutual action of two lengths of current-carrying wire is proportional to their lengths and current intensities, is the most significant of these principles. Ampère demonstrated the compatibility of his law with Charles Augustin de Coulomb's law of electric action by applying the same idea to magnetism. Ampère's science was firmly established in the developing domains of experimental physics thanks to his attention to and mastery of experimental methods.

A physical explanation of the electromagnetic relationship was also supplied by Ampère, who proposed the presence of an "electrodynamic molecule" (the ancestor of the concept of the electron) that acted as the constituent element of both electricity and magnetism. Ampère created a physical description of electromagnetic phenomena that was both theoretically and practically provable using this physical explanation of electromagnetic motion.

Ampère's seminal work, Memoir on the Mathematical Theory of Electrodynamic Phenomena, Uniquely Deduced from Experience, was published in 1827 and is regarded as the work that gave rise to the field of electrodynamics. Ampère earned the moniker "Newton of electricity" for his "Memoir on the Mathematical Theory of Electrodynamic phenomena inferred exclusively from experiment."

(Source: https://en.wikipedia.org/wiki/André-Marie Ampère)

REFERENCES AND SUGGESTED READINGS

- Meriam, J.L. and Kraige, L.G., Engineering Mechanics: Dynamics, Wiley, 2011.
- Hibbeler, R.C., Engineering Mechanics: Dynamics, Pearson Prentice Hall, 2010.

3 Kinetics of Rigid Body

UNIT SPECIFIES

This unit discussed the following concepts:

1. Angular momentum acting on a point
2. Inertia tensor: Definition and computation
3. Principal moments and axis of inertia
4. Parallel and perpendicular axis theorems
5. Mass, area, and polar moment of inertias
6. Newton – Euler's laws of Rigid body motion

RATIONALE

The kinetics of a rigid body is the study of the relationships between the forces and torques acting on a body and the resulting motion of the body. This field is crucial for understanding and predicting the behavior of physical systems under various conditions. Engineers use kinetics to design mechanical systems, such as vehicles, machinery, and structures, ensuring they can withstand expected loads and operate efficiently. The foundation of rigid body kinetics lies in Newton's and Euler's laws, which describe the relationship between the forces acting on a body and its acceleration. These laws are applied to both translational and rotational motion. By studying the kinetics of rigid bodies, engineers and scientists gain a comprehensive understanding of how forces and torques produce motion, allowing them to design, analyze, and optimize mechanical systems for a wide range of applications. This knowledge is essential for ensuring the safety, reliability, and efficiency of engineering solutions in various fields.

PRE-REQUISITES

1. Unit 1
2. Unit 2

UNIT OUTCOMES

List of outcomes of this unit is as follows:

U3-O1: Understanding the concepts of kinetic motion and the angular momentum

U3-O2: Describe Inertia tensor, principal moments and axes of inertia

U3-O3: Explain the parallel and perpendicular axis theorems

U3-O4: Demonstrate moment of inertia

U3-O5: Applying Newton-Euler's law of rigid body motion

Unit - 3 Outcomes	EXPECTED MAPPING WITH COURSE OUTCOMES (1-Weak Correlation; 2-Medium Correlation; 3-Strong Correlation)				
	CO-1	CO-2	CO-3	CO-4	CO-5
U3-O1	3	3	3	2	1
U3-O2	3	3	3	2	1
U3-O3	3	3	3	2	1
U3-O4	2	3	3	2	2
U3-O5	2	3	3	2	2

3.1 INTRODUCTION

Kinetics is the inspection of the relationship associated between forces and the motion that results from rigid bodies due to those forces. The foundational principle of kinetics, second law of motion by Newton, can be utilised to show how forces and motion interact in the case of rigid bodies. According to Newton's second law, a particle that is subjected to an unbalanced force will accelerate in the direction of the force with an amount that is proportionate to the force. Since the relationship between the force and the acceleration is directly proportional, it is possible to calculate the constant of proportionality, m, using the formula $m = F / a$. The particle's mass is denoted by the positive scalar m. The constant value of m during any acceleration serves as a quantitative indicator of a particle's inertia, or resistance to a change in velocity. The momentum equation can be used to explain particle motion in three dimensions. In comparison to motion in a plane, the change of angular momentum includes additional elements for motion in three dimensions.

3.2 KINETICS OF PARTICLES

Consider a particle of mass m moving randomly (i.e., it might be moving in a rectilinear or curvilinear direction) as shown in Fig. 3.1 while being influenced by various external forces $F_1, F_2, \ldots\ldots, F_n$, as shown by the particle's Free Body Diagram (FBD). Assume that $\grave{a}$ represents the particle's acceleration vector as calculated from an inertial reference system, (a reference frame that has no acceleration and is neither stationary nor moving with uniform speed nor spinning). This results from Newton's second law.

$$\vec{F}_1 + \vec{F}_2 + \dots\dots.. + \vec{F}_n = \frac{d\vec{g}}{dt} = \dot{\vec{G}} \tag{3.1}$$

where G stands for the particle's linear momentum vector, which is $\vec{G} = m\vec{v}$, and $\vec{v}$ represents the observed particle's velocity.

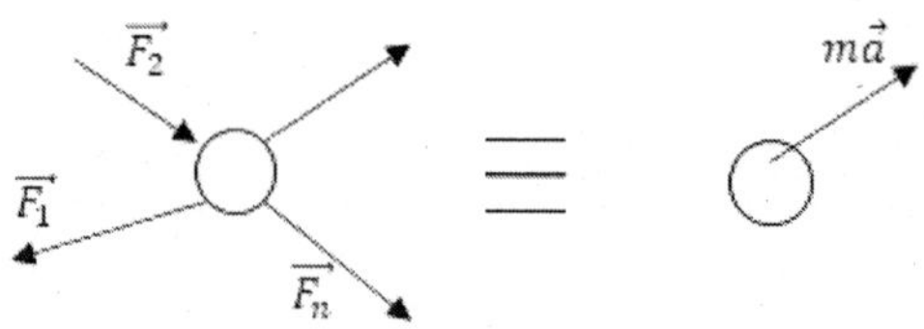

Fig. 3. 1 FBD of Kinetics of Particle

Therefore,

$$\dot{\vec{G}} = m\vec{a} \qquad \text{(m being constant)}$$

Or,

$$\Sigma\vec{F} = m\vec{a}$$

In the x-y plane, the aforementioned vector equation is equal to two separate scaler motion for motion in a plane or three independent scaler motion for motion in space.

Thus, we get,

$$\left.\begin{aligned} \Sigma F_x &= ma_x \\ \Sigma F_y &= ma_y \\ \Sigma F_z &= ma_z \end{aligned}\right\} \text{for space motion}$$

Or

$$\left.\begin{aligned} \Sigma F_x &= ma_x \\ \Sigma F_y &= ma_y \end{aligned}\right\} \text{for plane motion in x-y plane}$$

Where ΣF_x, ΣF_y and ΣF_z represent sum of forces in x, y and z direction respectively, and a_x, a_y and a_z are the components of acceleration of the particle in X, Y, Z direction.

3.2.1 D'Alembert Principle

If we take into account the speed of the particles and Newton's second law of motion, we may state that,

$$\Sigma\vec{F} + (-m\vec{a}) = \bar{0} \tag{3.2}$$

The aforementioned assumption is basically one of static equilibrium made mathematically, yet it is incorrect since the particle is not in static equilibrium. But we have purposefully created such a circumstance by adding a false or fake force $-m\vec{a}$ to the genuine collection of forces acting on m. The outcome is the FBD shown in Fig. 3.2. The pseudo force is shown by the dotted vector. This situation is known as dynamic equilibrium, sometimes known as pseudo-static equilibrium. An inertia force is the $-m\vec{a}$ force, which is represented by the dotted vector in the FBD.

As shown in Fig. 3.3, we may use a rigid body to demonstrate D'Alembert's concept. The inertia couple, $-\bar{I}\vec{\alpha}$ and the inertia force, $-m\vec{a}_G$ are considered for this instance.

Therefore, in order to apply D'Alembert's concept to a rigid body, inertia force and moment must be added to the existing set of external forces and moments, as demonstrated by its FBD. Use the fundamental equations of equilibrium ($\Sigma\vec{F} = \vec{0}$ and $\Sigma\vec{M} = \vec{0}$) to solve a dynamics problem.

$$\Sigma F = maG\ \Sigma\vec{F} = m\vec{a}_G\ ;\ \Sigma\vec{M} = \bar{I}\vec{\alpha} \quad (3.3)$$

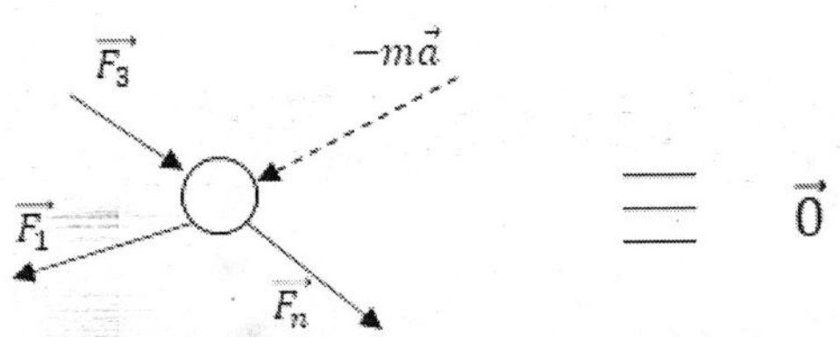

Fig. 3. 2 FBD of particles showing inertia force (D'Alembert's principle).

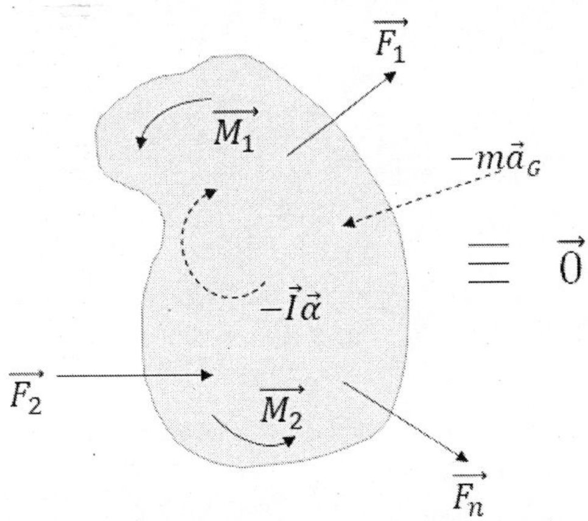

Fig. 3. 3 FBD of rigid body showing inertia force and inertia couple.

3.3 MASS MOMENT OF INERTIA

The equation of rotational motion for a rigid body in planar motion has an integral that depends on how the mass is distributed in relation to the moment axis. When a rigid body experiences an angular acceleration α with reference to its axis of rotation, this integral take place. Consider the case of a body with mass *m* rotating at an angular acceleration α about an axis Z. All the body's particles travel in parallel planes that are normal to the Z axis of rotation.

The resultant tangential force acting on this element as, $F_t = r\,\alpha\,dm$.

Moment resulting from tangential force about axis Z for this element as, $M_z = r^2\,\alpha\,dm$

Sum of these forces' moments across all elements, $\sum M_i = \int r^2\,\alpha\,dm$ (3.4a)

Equation (3.4a) can be expressed as, if α is the same for all the radial lines in the rigid body

$$M = \alpha \int r^2\,dm = \alpha\,I \tag{3.4b}$$

Where *I* is the rigid body's mass moment of inertia, $I = \int r^2\,dm$ (3.4c)

Equation (3.4c) can be expressed alternatively as, $I = \sum r_i^2 m_i$ (3.4d)

Where r_i is the radial distance between the inertia axis and the particle having mass m_i.

Elemental mass *dm* in the equation (3.4c) can be conveyed in regard of volume and density of an element as $dm = \rho\,dV$,

$$I = \int r^2 \rho\,dV \tag{3.4e}$$

If the density ρ is constant all through the body, the equation (3.4e) can be expressed as

$$I = \rho \int r^2 dV \tag{3.4f}$$

The integral in the equation (3.4f) indicates purely body's geometrical property.

3.3.1 Radius of Gyration

Mass moment of inertia also stated in regard of *radius of gyration*, *k* (which is radial distance between the axis of rotation to the body's center of mass (Fig. 3.4)) as

$$I = k^2 m \text{ or } k = \sqrt{\frac{I}{m}} \tag{3.5}$$

Radius of gyration, k is measured in length units and it is represented as the geometrical property

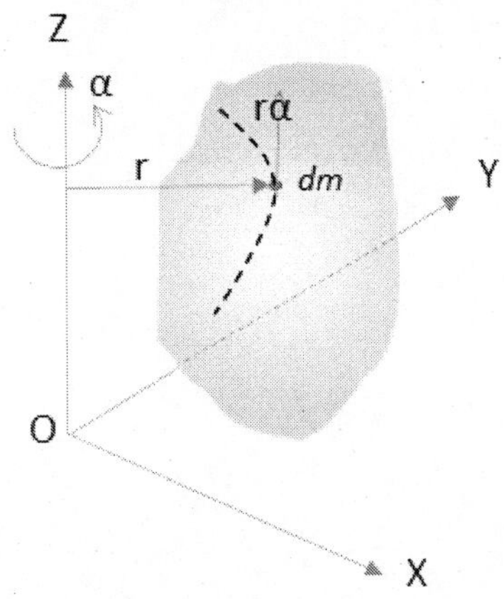

Fig. 3. 4 Moment of Rigid Body.

3.3.2 Parallel Axis Theorem

When moment of inertia of a body about an axis that moves through its mass centre is known, the parallel-axis theorem can be made use of to determine the moment of inertia about any other parallel axis. Consider the two parallel axes in Fig. 3.5, one of which an axis going through the mass centre G and the other via another point B that is spaced l away, to demonstrate this claim. According to Fig. 3.5,

$$r_b{}^2 = (l+y)^2 + z^2 = l^2 + y^2 + z^2 + 2\,l\,y \tag{3.6a}$$

Moment of inertia about axis P with reference to equation (3.4c) is,

$$I_P = \int_m r_b{}^2\,dm = \int_m (y^2+z^2)\,dm + l^2 \int_m dm + 2l \int_m y\,dm \tag{3.6b}$$

Since X-axis moving through mass center, G of the body then $\int_m y\,dm = \bar{y}\,m = 0$, therefore, the equation (3.6b) becomes, $I_P = I_G + ml^2$ (3.6c)

Where $I_G = \int_m (y^2+z^2)\,dm$ is the moment of inertia about axis X moving through the mass center G, l is the perpendicular distance between the parallel axes of X and P, the total mass of the body is denoted by m.

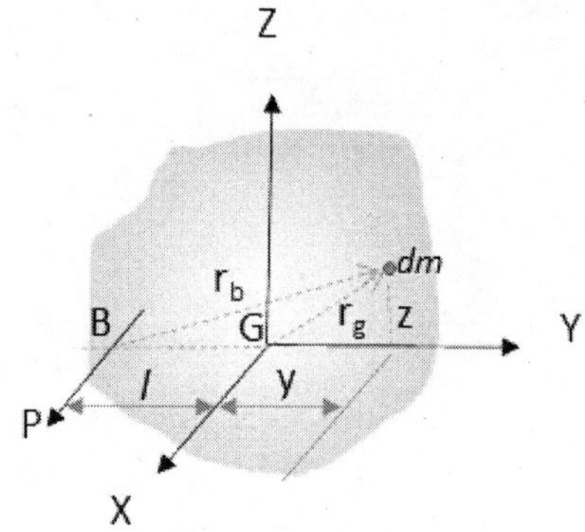

Fig. 3. 5 Parallel axis theorem.

3.3.3 Perpendicular Axis Theorem

The perpendicular axis theorem, also known as the plane figure theorem, states that the moment of inertia of a planar lamina, or two-dimensional body, about an axis perpendicular to its plane is "equal to the sum of the moments of inertia of the lamina about the two axes at right angles to each other, in its own plane, intersecting each other at the location where the perpendicular axis passes through it".

The shape of the body is symmetric about two of the three axes (Fig. 3.6), the perpendicular axis theorem is applied.

The moment of inertia about X-axis for the symmetric shape (Fig. 3.6) is

$$I_X = \int_m (y^2 + z^2)\, dm = \int_m y^2 dm + \int_m z^2 dm \quad (3.7)$$

$$I_X = I_Y + I_Z$$

From above equation, the moment of inertia about the third axis can be determined if the moments of inertia about the two axes are known.

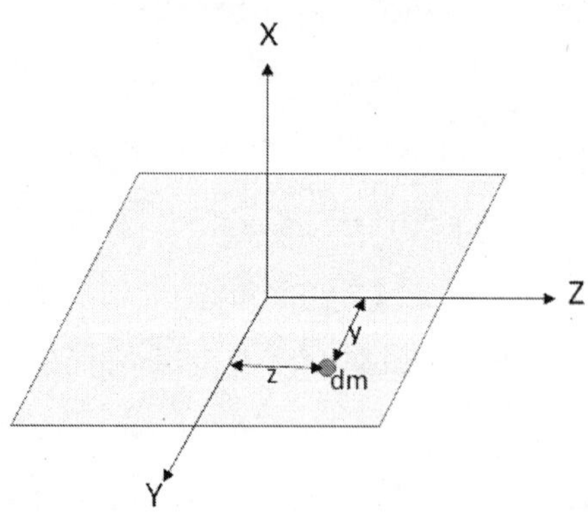

Fig. 3. 6 Perpendicular axis theorem.

The perpendicular axes theorem gives the following useful relationship if a planar object has rotational symmetry such that I_Y and I_Z are equal:

$$I_X = 2I_Y = 2I_Z \quad (3.8)$$

3.3.4 Mass moment of inertia of symmetrical bodies

3.3.4.1 Slender thin rod

Fig. 3.7 (a) depicts the midpoint O of a slender rod PQ of length l.

Consider about a uniform rod PQ of length l, with its midpoint O as depicted in Fig. 3.7 (a).

Let ρ = Mass of the rod per unit length

m = Total mass of the rod = ρl

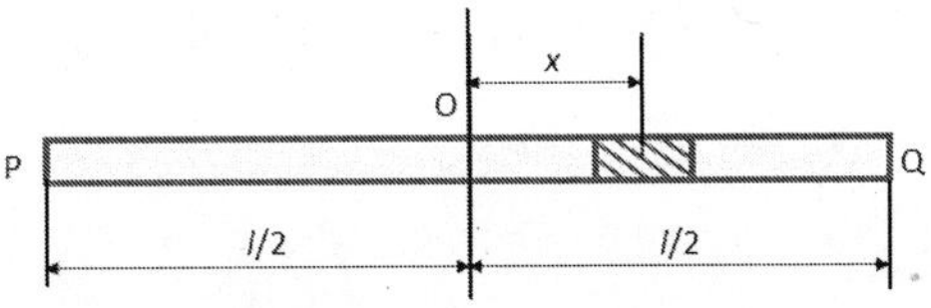

Fig. 3. 7 (a) Slender thin rod

Consider a simple strip of length dx, located x units away from the axis done the centre of O.

Mass moment of inertia of strip length dx about O = (mass of strip length) X (square of distance from centroid of strip to midpoint O, i.e., x)

$$dI_o = \rho dx \times x^2 = \rho x^2 dx$$

Mass moment of inertia of slender rod,

$$I_0 = \int_{-\frac{l}{2}}^{\frac{l}{2}} \rho x^2\, dx = \rho \left[\frac{x^3}{3}\right]_{-\frac{l}{2}}^{\frac{l}{2}} = \frac{\rho l^3}{12} = \frac{(\rho l)l^2}{12} = \frac{ml^2}{12} \qquad (3.9a)$$

Mass moment of inertia of a uniform thin rod with an axis moving through the end P and perpendicular to its length, inferred from the parallel axis theorem

$$I_P = I_o + m\left(\frac{l^2}{2}\right) = \frac{ml^2}{12} + \frac{ml^2}{4} = \frac{ml^2}{3} \qquad (3.9b)$$

3.3.4.2 Rectangular lamina

Take into consideration a lamina with the dimensions l, h, and centre O as illustrated in Fig. 3.7 (b).

Let ρ = Mass of the lamina per unit area and m = Total mass of the lamina = ρlh

Consider a basic strip with dimensions of h and dx that is located x distance from the centre O, as illustrated in the Fig.3.7 (b) and moment of inertia about Y-axis is

$$dI_Y = (\rho)(hdx)x^2$$

Mass moment of inertia of total rectangle lamina

$$I_Y = \rho h \int_{-\frac{l}{2}}^{\frac{l}{2}} x^2\, dx = \rho h \frac{l^3}{12} = \frac{ml^2}{12} \quad (3.10a)$$

Similarly, about X-axis, moment of inertia is,

$$I_X = \frac{mh^2}{12} \quad (3.10b)$$

The perpendicular axis theorem provides us with, $I_Z = I_X + I_Y$

$$\therefore I_Z = \frac{m(h^2+l^2)}{12} \quad (3.10c)$$

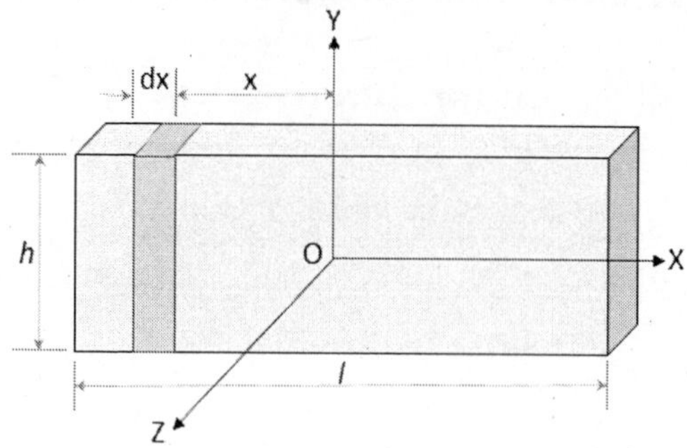

Fig. 3. 7 (b) Rectangular lamina

3.3.4.3 Thin circular ring

Considering the narrow circular ring with radius r and centre O in the Fig. 3.7 (c). The whole mass of the ring, m = 2πrρ, where ρ is the ring's mass per unit circumferential length. Take into consideration a ring element of length dx as illustrated in the Fig. 3.7 (c).

Mass moment of inertia of the narrow strip,

$$dI_Z = (\rho dx) \times r^2; dx = rd\theta$$

Mass moment of inertia of total ring,

$$I_Z = \int_0^{2\pi} (\rho r d\theta) \times r^2$$

$$I_Z = (2\pi r \rho) \times r^2 = mr^2 \quad (3.11a)$$

The perpendicular axis theorem provides us with, $I_Z = I_X + I_Y$

$$I_X = I_Y = \frac{mr^2}{2} \quad (3.11b)$$

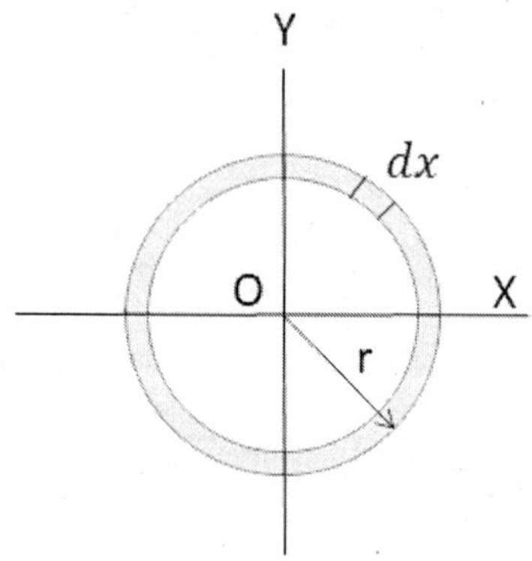

Fig. 3. 7 (c) Thin circular ring

3.3.4.4 Circular cylinder

The perpendicular axis theorem provides us with,

$$I_Z = I_X + I_Y$$

$$I_X = I_Y = \frac{mr^2}{4} \qquad (3.12a)$$

Let m = Lamina's total mass = $\pi r^2 \rho$;

where ρ = lamina's mass per unit area

r = radius of circular lamina

According to Fig. 3.7 (d), an elementary ring with a centre at O, a radius of x, and a thickness of dx has a mass moment of inertia, $dI_Z = (\rho 2\pi x\, dx)\, x^2$

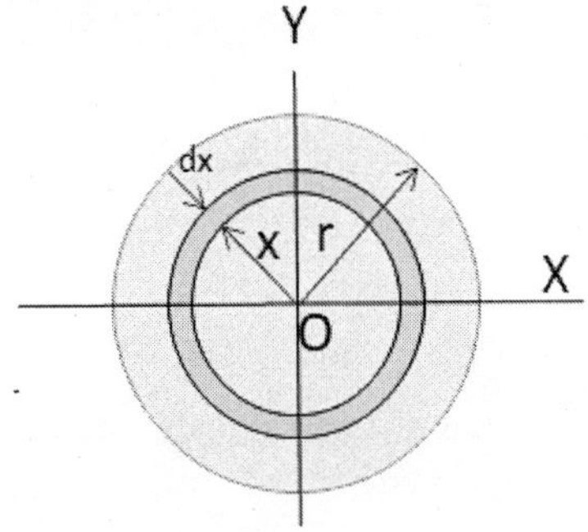

Fig. 3. 7 (d) Circular lamina

Mass moment of inertia of total circular lamina

$$I_z = \int_0^r \rho 2\pi x^3 dx = \frac{\rho\pi r^4}{2} = \frac{(\pi r^2 \rho) r^2}{2}$$

$$I_z = \frac{mr^2}{2} \qquad (3.12b)$$

Similarly, mass moment of inertia of the cylinder (Fig. 3.7 (e)) will be $I_z = \frac{mr^2}{2}$ (3.12c)

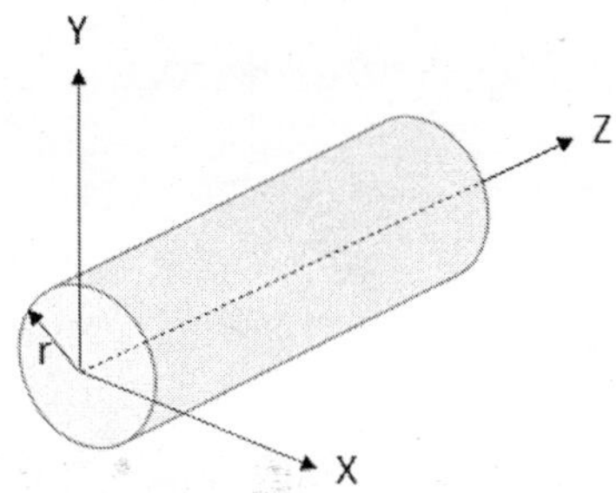

Fig. 3. 7 (e) Circular cylinder

3.3.4.5 Circular sphere

Think about a radius of *R* solid sphere and with a centre O as shown in fig. 3.7 (f)

Let ρ = Sphere's mass per unit volume

m = Total mass of sphere = $\frac{4\pi (R)^3}{3} * \rho$

Consider a simple thin disc of thickness dx, located at x distance from the sphere's centre O.

Radius of thin disc = $\sqrt{r^2 - x^2}$

Mass of thin disc = $\rho\, \pi\, (\sqrt{r^2 - x^2}\,)^2\, dx$

$= \rho\, \pi\, (r^2 - x^2)\, dx$

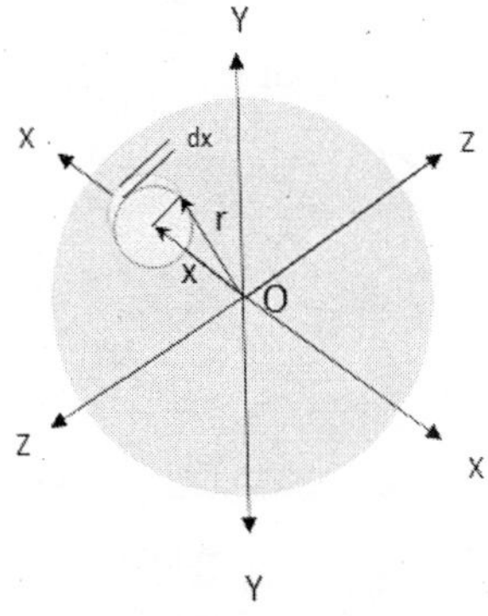

Fig. 3. 7 (f) Circular sphere

The mass moment of inertia of thin disc about an axis that is perpendicular to its plane and passes through its centre O.

$$dI_X = \frac{\text{mass} * (\text{radius})^2}{2} = \frac{\rho\,\pi\,(r^2 - x^2)\,dx * (\sqrt{r^2 - x^2}\,)2}{2}$$

$$I_X = \frac{\rho\pi}{2}(r^4 + x^4 - 2r^2x^2)dx$$

$$I_X = \int_{-r}^{r} \frac{\rho\pi}{2}(r^4 + x^4 - 2r^2x^2)dx = \frac{8\rho\pi r^5}{15}$$

$$I_X = \frac{2mr^2}{5} \qquad (3.13a)$$

Due to complete symmetry, the mass moment of a sphere around an axis that passes through its centre O is the same.

$$I_{XX} = I_{YY} = I_{ZZ} = \frac{2mr^2}{5} \qquad (3.13b)$$

3.3.4.6 Right circular cone

Right circular cone as depicted in Fig. 3.7 (g) with base radius R, height h, and mass density ρ. Consider a thin, circular disc with a radius of r and a thickness of dz that is situated z distance from the apex O. Let dm represent the disc's mass.

$$dm = \rho\,\pi r^2 dz$$

From Fig. 3.7 (g), $\frac{R}{h} = \frac{r}{z}; r = \frac{R\,z}{h}$

$$dm = \frac{\rho\pi R^2 z^2}{h^2} dz$$

Mass moment of inertia about Z-axis, $dI_Z = \frac{1}{2} dm\, r^2$

$$dI_Z = \frac{1}{2}\left(\frac{\rho\pi R^2 z^2}{h^2} dz\right)\left(\frac{R^2 z^2}{h^2}\right) = \frac{1}{2}\frac{\pi\rho R^4 z^4}{h^4} dz$$

Mass moment of inertia of total cone,

$$I_Z = \int_0^h \frac{1}{2}\frac{\pi\rho R^4 z^4}{h^4} dz = \frac{1}{2}\frac{\pi\rho R^4 h^5}{5h^4} = \frac{\pi\rho R^4 h}{10}$$

$$I_Z = \frac{3mR^2}{10}; where\ m = \frac{1}{3}\pi R^2 h\rho \qquad (3.14a)$$

The centroid is situated h/4 of a distance above the base.

$$I_X = I_Y = \frac{3}{5} m\left(\frac{h^2}{4^2} + r^2\right) = \frac{3}{80} m(4r^2 + h^2) \qquad (3.14b)$$

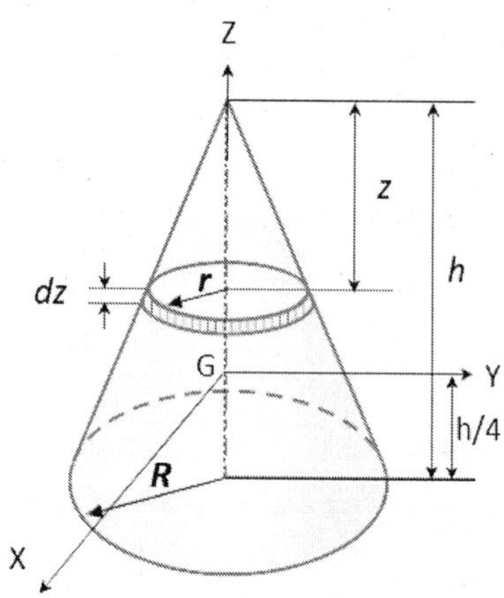

Fig. 3. 7 (g) Right circular cone

Example 3.1: As shown in Fig. E3.1, the ends of a uniform rod (mass m_3 kg) are welded together by a sphere of mass m_1 kg at the left end and a circular disc of mass m_2 kg at the right end. Compute the mass moment of inertia about the pinned axis.

Solution: $I_O = I_{Sphere} + I_{Disc} + I_{Rod}$

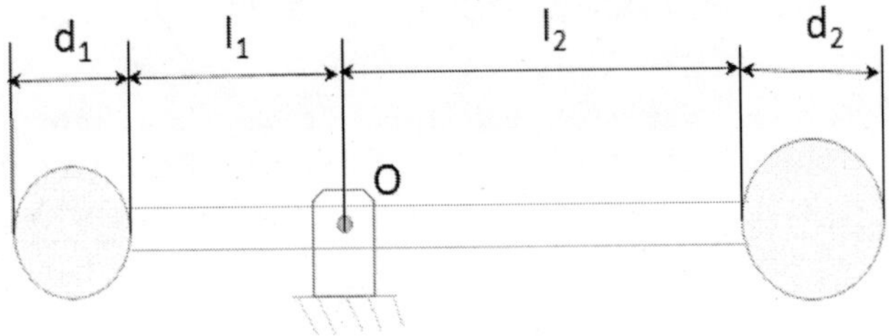

Fig. E3.1

$$I_O = \left[\frac{2m_1{r_1}^2}{5} + m_1{x_1}^2\right] + \left[\frac{m_2{r_2}^2}{2} + m_2{x_2}^2\right] + \left[\frac{m_3l^2}{12} + m_3{x_3}^2\right]$$

Where $r_1 = \frac{d_1}{2}, r_2 = \frac{d_2}{2}, l = l_1 + l_2, x_1 = r_1 + l_1, x_2 = r_2 + l_2, and\ x_3 = \frac{l}{2} - l_1$

Example 3.2: Determine the paraboloid's moment of inertia about the x axis, which is created by the equation presented in Fig. E3.2, and direct your result in respect of the total paraboloid's mass *m*. The density of the paraboloid, *ρ*, is constant.

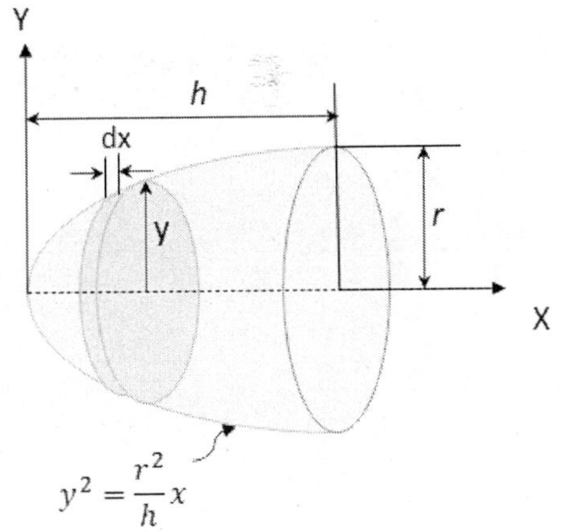

Fig. E3.2

Solution: mass of the strip, $dm = \rho\, dV = \rho\,(\pi y^2 dx)$

Mass moment of inertia of elementary strip,

$$dI_X = \frac{1}{2}dm\, y^2 = \frac{1}{2}\rho\,(\pi y^2 dx)y^2 = \frac{1}{2}\rho\pi\left(\frac{r^4}{h^2}x^2\right)dx$$

$$I_X = \int_0^h \frac{1}{2}\rho\pi\left(\frac{r^4}{h^2}x^2\right)dx = \frac{1}{6}\rho\pi r^4 h$$

$$m = \int_0^h \rho\,(\pi y^2 dx) = \int_0^h \rho\pi\frac{r^2}{h}x\, dx = \frac{1}{2}\rho\pi r^2 h$$

$$I_x = \frac{1}{3}mr^2$$

3.4 AREA MOMENT OF INERTIA AND POLAR MOMENT OF INERTIA

A body's area and mass are distributed relatively according to its reference axis, and this is known as area moment of inertia. The area moment of inertia for the plane area can be conveyed as about the Z-axis normal to its plane and about the X and Y axes in its plane, as illustrated in Fig. 3.8,

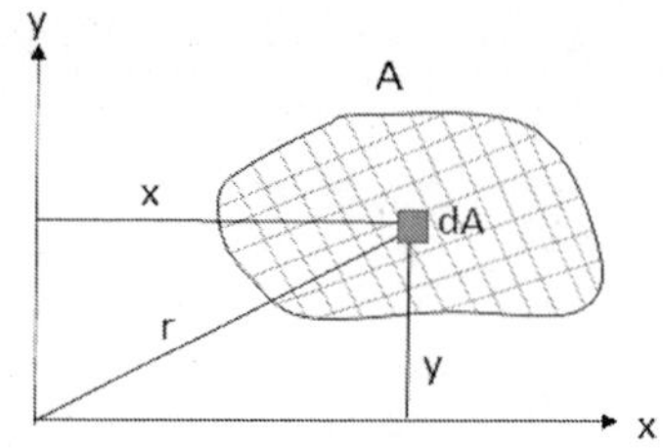

Fig. 3. 8 Rigid body under area moment of inertia.

$$I_x = \int y^2\, dA \tag{3.14a}$$

$$I_y = \int x^2\, dA \tag{3.14b}$$

$$I_z = \int r^2\, dA \tag{3.14c}$$

In the equation (3.14c), r^2 can be replaced with $(x^2 + y^2)$,

$$I_z = \int r^2\, dA = \int x^2\, dA + \int y^2\, dA \tag{3.14d}$$

$$I_z = I_x + I_y \tag{3.14e}$$

Where I_z is called *polar moment of inertia* which is represents the total moment of inertia about two symmetrical axes about the third axis.

When the moment of inertia about two of the three axes is known and the body is symmetric about two of the three axes as shown in Fig. 3.9, the moment of inertia about the third axis can be calculated using the *perpendicular axis theorem*:

$$I_c = I_a + I_b \tag{3.14f}$$

Where I_a, I_b, and I_c are the area moment of inertia about the axes a, b, and c respectively.

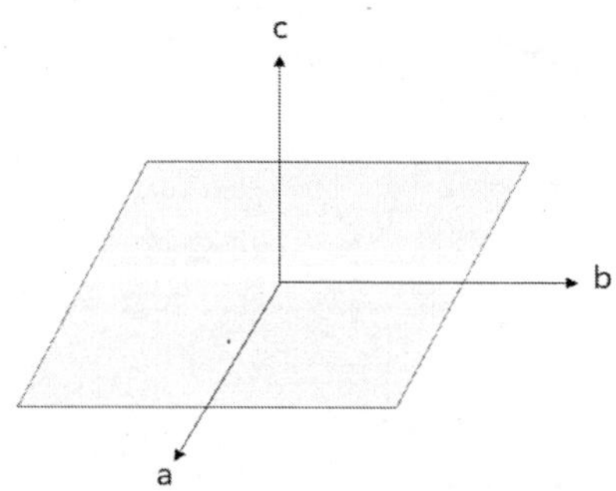

Fig. 3. 9 Symmetric body for perpendicular axis theorem.

3.4.1 Moment of inertia of geometrical shapes

3.4.1.1 Rectangle

Consider a rectangle (Fig. 3.10 (a)) with a base of *b* and a height of *h*. The reference axis across base is AB. Consider a strip of elements with width *dy* that is separated from the reference axis AB by *y*.

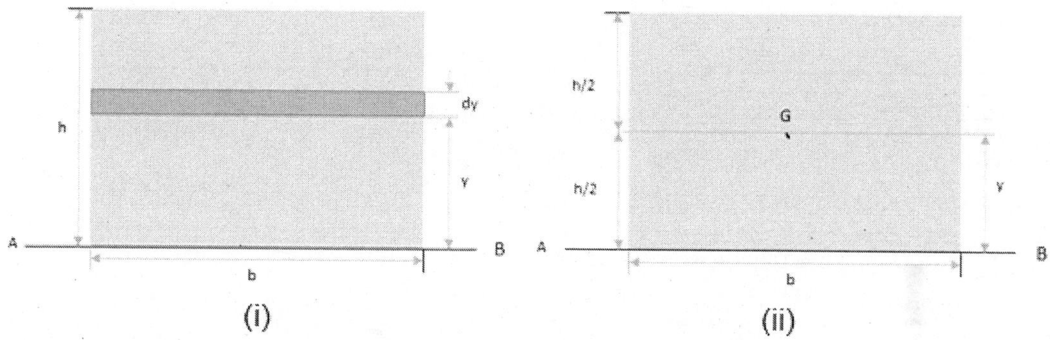

Fig. 3. 10 (a) Rectangle

Using basic principle of moment of inertia.

$I = \int r^2 dA$

Area of elemental strip dA = b.dy

$$I_{AB} = \int_0^h y^2 b\, dy = b\int_0^h y^2\, dy = \text{b}\left[\frac{y^3}{3}\right]_0^h = \frac{\text{b}h^3}{3} \tag{3.15a}$$

Moment of inertia of rectangle with regard to a parallel centroidal axis as determined by the parallel axis theorem, $I_{AB} = I_G + Ad^2$

Where d = distance between the parallel centroidal axis and the reference axis.

$$I_G = I_{AB} - Ad^2 = \frac{\text{b}h^3}{3} - bh\left(\frac{h}{2}\right)^2 = \frac{\text{b}h^3}{3} - \frac{\text{b}h^3}{4} = \frac{\text{b}h^3}{12} \tag{3.15b}$$

$$\text{I}_\text{G} = \frac{\text{b}h^3}{12} \text{ [Fig. 3.10 (a-ii)]} \tag{3.15c}$$

3.4.1.2 Triangle

Consider about a triangle illustrated in Fig. 3.10 (b) with base (b) and height (h). Take into consideration an elemental strip with width *dy* and placed at a distance of *y* from the AB as shown in Fig. 3.10 (b-i).

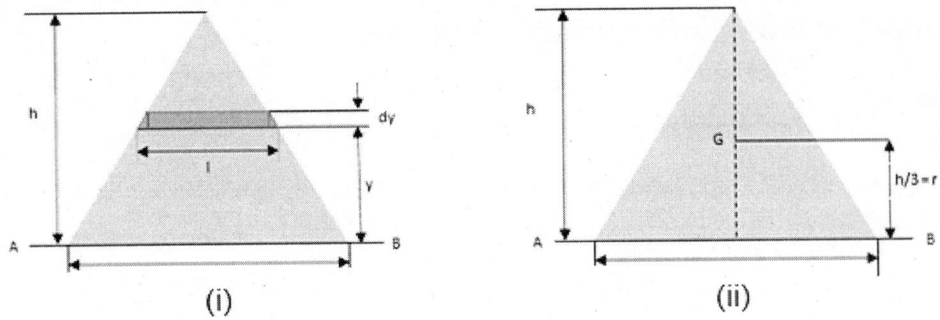

Fig. 3. 10(b) Triangle

Moment of inertia of triangle with reference to base, $dI = \int r^2 dA$

With reference to Fig. 3.10 (b-i), $r = y, dA = l\,dy$

From geometry property of triangle, $l = \frac{b}{h}(h - y)$

$$I_{AB} = \int_0^h y^2 \frac{b}{h}(h-y)dy = \frac{b}{h}\int_0^h (hy^2 - y^3)\,dy = \frac{b}{h}\left[\frac{y^3 h}{3} - \frac{y^4}{4}\right]_0^h = \frac{b}{h}\left[\frac{h^4}{3} - \frac{h^4}{4}\right] = \frac{bh^3}{12} \quad (3.16a)$$

Moment of inertia using the parallel axis theorem, a triangle's base is centred on a parallel centroidal axis [Fig. 3.10(b-ii)], $I_{AB} = I_G + Ad^2$.

Here $d = \frac{h}{3}$ is the distance the between base AB and centroid G of triangle.

$$I_G = I_{AB} - Ad^2 = \frac{bh^3}{12} - \frac{1}{2}bh\left(\frac{h}{3}\right)^2 = \frac{bh^3}{12} - \frac{bh^3}{4} = \frac{bh^3}{18} = \frac{bh^3}{36}$$

$$I_G = \frac{bh^3}{36} \quad (3.16b)$$

3.4.1.3 Circle

Consider a very thin circular ring as shown in Fig. 3.10(c) with a radius of *r* and a width of *dr* for circle radius *R* centroid is located at O. The thin ring's area $dA = 2\pi r dr$. About Z-axis, moment of inertia which is perpendicular to X and Y-Axes,

$$I_{zz} = \int r^2\,dA = \int_0^R r^2\,2\pi dr = 2\pi\int_0^R r^3\,dr$$

$$= 2\pi\left[\frac{r^4}{4}\right]_0^R$$

$$I_{zz} = \frac{\pi R^4}{2} \quad (3.17a)$$

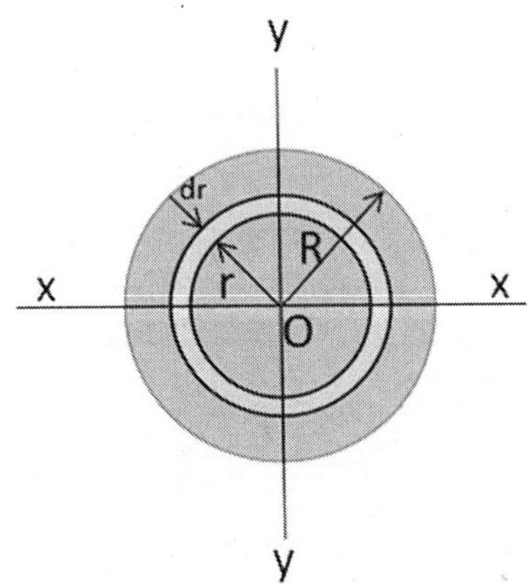

Fig. 3. 10(c) Circle

Due to symmetry of circle with regard to the X- and Y-axes, $I_{xx} = I_{yy}$

The perpendicular axis theorem provides us with, $I_{zz} = I_{xx} + I_{yy}$

$$I_{xx} = I_{yy} = \frac{I_{zz}}{2} = \frac{\pi R^4}{4} \tag{3.17b}$$

3.4.1.4 Semicircle

Consider semicircle as shown in Fig. 3.10(d) of radius R and procedure as like the above section.

$$Semicircle = \frac{1}{2}\, circle$$

$$I_{AB} = \frac{\pi R^4}{8};\ I_{OC} = \frac{\pi R^4}{8} \tag{3.18a}$$

$$I_{zz} = \frac{\pi R^4}{4} \tag{3.18b}$$

Use parallel axis theorem, $I_{AB} = I_G + Ay^2$

$y = \frac{4R}{3\pi}$ is the distance between base AB and G

$$I_G = \frac{\pi R^4}{8} - \frac{\pi R^2}{2}\left(\frac{4R}{3\pi}\right)^2 = 0.11R^4 \tag{3.18c}$$

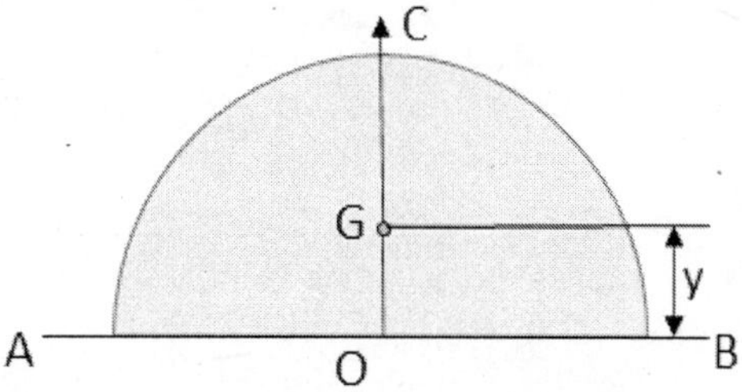

Fig. 3. 10(d) Semicircle

3.4.1.5 Quarter circle

Consider a quarter circle as shown in Fig. 3.10(e) of radius R and procedure for moment of inertia determination is as like the semi- circle.

$$I_{OB} = I_{OC} = \frac{\pi R^4}{16} \tag{3.19a}$$

Use parallel axis theorem, $I_{OB} = I_G + Ay^2$

$y = \frac{4R}{3\pi}$ is the distance between base OB and G

$$I_G = \frac{\pi R^4}{16} - \frac{\pi R^2}{4}\left(\frac{4R}{3\pi}\right)^2 = 0.055R^4 \tag{3.19b}$$

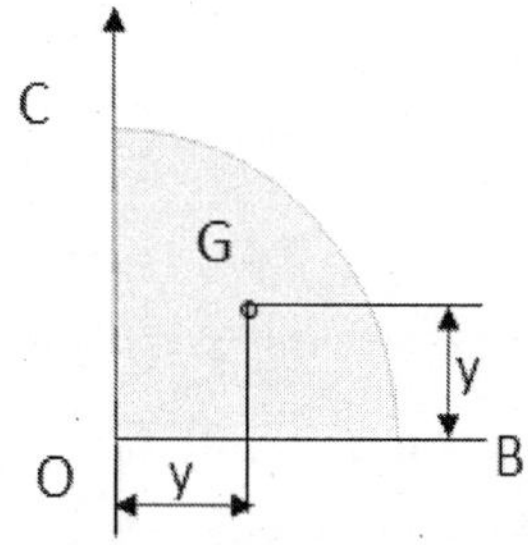

Fig. 3. 10(e) Quarter circle

Example 3.5: Establish the moment of inertia of shaded region about OX and G as shown in Fig. E3.5

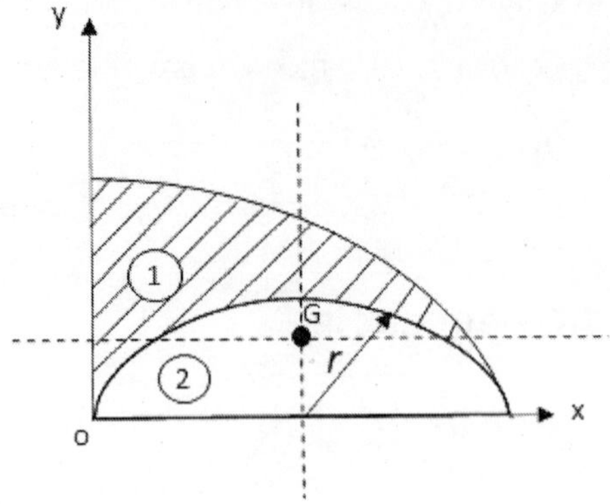

Fig. E3.5

Solution: Shaded area's moment of inertia with respect axis, I = moment of inertia of quarter circle with respect to same axis (I_1) – moment of inertia of semicircle with respect to same axis (I_2).

$$I_{OX} = \frac{\pi(2r)^4}{16} - \frac{\pi r^4}{8} = \frac{7\pi r^4}{8}$$

$$I_G = 0.055(2r)^4 - 0.11r^4$$

$$I_G = 0.77r^4$$

Example 3.6: An isosceles triangle ABC with 90 mm base and 60 mm height. As seen in Fig. E3.6, a circle with radius *r* is taken out of the triangle's lamina at centroidal location, causing a 20% reduction in the moment of inertia of the triangle about its base. Determine the circle's size.

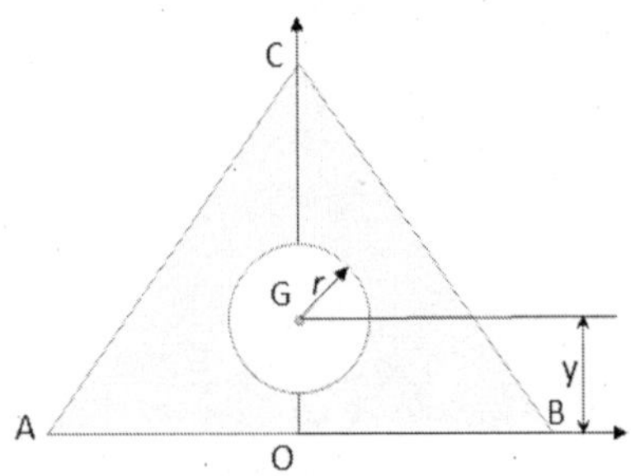

Fig. E3.6

Solution: Moment of inertia of triangle with respect base, AB is, $I_{AB}^{\Delta} = \frac{90 \times 60^3}{12} = 1620000\ mm^4$

Moment of inertia of circle with regard to AB is

$I_{AB}^{O} = I_G^O + A^O \times y^2$, where y is the distance between base AB to centroid of triangle

y = 20 mm.

$$I_{AB}^{O} = \frac{\pi r^4}{4} + \pi r^2 \times 20^2$$

$$I_{AB} = I_{AB}^{\Delta} - I_{AB}^{O} \qquad \text{(i)}$$

But $I_{AB} = I_{AB}^{\Delta} \times 0.2 = 324000\ mm^4$

From equation (i), $324000 = 1620000 - \frac{\pi r^4}{4} + \pi r^2 \times 20^2$

$$1296000 = \frac{\pi r^4}{4} + \pi r^2 \times 20^2$$

$$\pi r^4 + 1600\pi r^2 = 5184000$$

After solving above quadratic equation, *r* = 26.7 mm (circle radius)

3.5 ANGULAR MOMENTUM

An object's angular momentum is divided into orbital and spin angular momentums of two different categories. The orbital angular momentum is the amount of angular momentum about a certain centre of rotation, whereas the spin angular momentum is the amount of angular momentum about the object's centre of mass. Orbital and spin angular momenta are combined to form total angular momentum. A point particle's orbital angular momentum vector and orbital angular velocity vector, ω are always parallel and exactly proportional, where the proportionality constant depends on the particle's mass and distance from the source. The spin angular momentum vector and the spin angular velocity vector, Ω of a rigid body are proportional but not always parallel, therefore the constant of proportionality is a second-rank tensor rather than a scalar.

A particle with mass *m* and linear momentum (*p=mv)* at position **r** in relation to the origin O as depicted in Fig. 3.11. The term "angular momentum" refers to the moment of linear momentum with respect to O that is the origin. The velocity's perpendicular component ($v_\perp$) or, in other words, the perpendicular distance ($r_\perp$) from the origin are both directly proportional to the angular momentum (m).

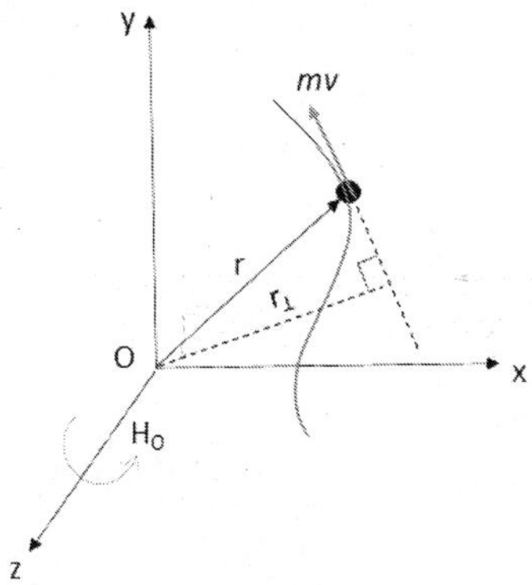

Fig. 3. 11 Angular Momentum of a space curve.

Angular momentum in scalar formulation, $(H_O)_z = p\, r_\perp = (mv)\,(r_\perp)$ (3.20)

Angular momentum can be expressed in vector formulation as,

$H_O = \boldsymbol{r} \times mv$ (3.21)

Where **r** is symbolizing position vector

r and *mv* must be penned in regard of their Cartesian components in order to calculate the cross product, which will then allow the determinant to be used to calculate the angular momentum:

$$H_O = \begin{vmatrix} e_1 & e_2 & e_3 \\ r_x & r_y & r_z \\ mv_x & mv_y & mv_z \end{vmatrix} \tag{3.22a}$$

Where e_1, e_2, and e_3 are unit vectors in the x, y, and z axis directions correspondingly.

The linear velocity can be expressed as, $v = \omega \times r$

Angular momentum for each particle, $H_O = \boldsymbol{r} \times m\,(\omega \times r)$ (3.22b)

The angular momentum about O for rigid body, $H_O = \int [r \times (\omega \times r)] dm$ (3.22c)

3.6 INERTIA TENSOR

Equation (3.22c) can be expanded with triple vector product, and the ω components are constant about the integrals over the body and it becomes constant multiplier.

$$H = e_1 \int \left[(y^2 + z^2)\omega_x - xy\omega_y - xz\omega_z\right] dm$$

$$+ e_2 \int \left[-yx\omega_x + (z^2 + x^2)\omega_y - yz\omega_z\right] dm$$

$$+ e_3 \int \left[-zx\omega_x - zy\omega_y + (x^2 + y^2)\omega_z\right] dm \tag{3.23}$$

Let

$$I_{xx} = \int (y^2 + z^2) dm;\ I_{xy} = \int xy\, dm$$

$$I_{yy} = \int (z^2 + x^2) dm;\ I_{xz} = \int xz\, dm \tag{3.24}$$

$$I_{zz} = \int (x^2 + y^2) dm;\ I_{yz} = \int yz\, dm$$

Where I_{xx}, I_{yy} and I_{zz} are the body's moment of inertia about the respective axes, and I_{xy}, I_{yz} and I_{xz} are the product of inertia about the coordinate axes. These inertia terms explained in the subsequent sections.

Further simplifies the equation (3.23) with substituting the above inertia terms described in the equation (3.24),

$$H = (I_{xx}\omega_x - I_{xy}\omega_y - I_{xz}\omega_z)e_1 + (-I_{yx}\omega_x + I_{yy}\omega_y - I_{yz}\omega_z)e_2 + (-I_{zx}\omega_x - I_{zy}\omega_y + I_{zz}\omega_z)e_3 \tag{3.25}$$

The above equation (3.25) can be written as, $H = H_x\, e_1 + H_y\, e_2 + H_z\, e_3$

Where H_x, H_y and H_z are the H's components and can be expressed in the matrix form,

$$\begin{bmatrix} H_x \\ H_y \\ H_z \end{bmatrix} = \begin{bmatrix} I_{xx} & -I_{xy} & -I_{xz} \\ -I_{yx} & I_{yy} & -I_{yz} \\ -I_{zx} & -I_{zy} & I_{zz} \end{bmatrix} \begin{bmatrix} \omega_x \\ \omega_y \\ \omega_z \end{bmatrix} \tag{3.26}$$

Equation (3.26) could be penned in short form as, {H} = [I]{ω}

Where $[I] = \begin{bmatrix} I_{xx} & -I_{xy} & -I_{xz} \\ -I_{yx} & I_{yy} & -I_{yz} \\ -I_{zx} & -I_{zy} & I_{zz} \end{bmatrix}$ is the *inertia tensor* or inertia matrix. The orientation of the axis with relation to the rigid body influences the inertia magnitude (moment and product of inertia) terms. Additional benefits of studying the inertia tensor include providing approaches that may be used to examine the deformable solids and fluids mechanics, where deformation rate tensors, stress and strain are essential.

3.7 PRINCIPAL MOMENTS AND AXIS OF INERTIA

At any orientation of the axes x-y-z for a specific origin where the moment of inertia terms ((I_{xx}, I_{yy} and I_{zz})) take on unvarying values, the product of inertia terms vanishes. The inertia tensor is transformed to this orientation as

$$\begin{bmatrix} I_{xx} & 0 & 0 \\ 0 & I_{yy} & 0 \\ 0 & 0 & I_{zz} \end{bmatrix}$$

The *principal axes of inertia* are the x-y-z for which the product of inertia vanishes, whereas the *principal moments of inertia* are I_{xx}, I_{yy} and I_{zz}.

For the angular momentum about the mass centre or a stationary point, if the coordinate axes match with the principal axes of inertia the equation (3.26) becomes,

$$H = I_{xx}\omega_x e_1 + I_{yy}\omega_y e_2 + I_{zz}\omega_z e_3 \tag{3.27}$$

For an extensive three-dimensional rigid body, the principal axes of inertia can always be identified. Except in cases where around one of the principal axes of inertia or when $I_{xx} = I_{yy} = I_{zz}$, the body rotates, where the vectors H and ω have unlike axes, the equation (3.27) remains.

Transfer principle for Angular Momentum

The final resulting angular momentum vector H_G about the mass centre and outcome linear momentum vector $G = m\bar{v}$ across the mass centre, as shown in Fig.3.12, may be used to depict the momentum characteristics of a rigid body. Although H_G has free vector characteristics, it is conveniently represented as G. These vectors have characteristics that are similar to those of a couple and a force. As a result, around any point P the angular momentum equals the summation of the free vector H_G and the linear momentum vector G about P. Therefore,

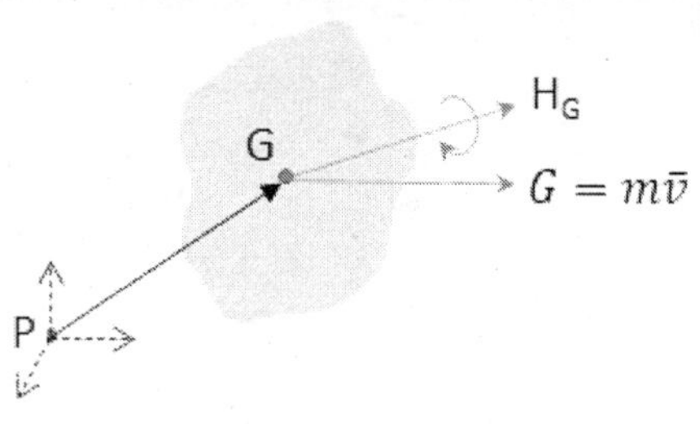

Fig. 3. 12 Transfer of angular momentum

$$H_p = H_G + \bar{r} \times G \tag{3.28}$$

Example 3.3: Ball-and-socket joints are used to attach the disc and collar to the 2 kg rod CD. If the disc rotates at a constant angular velocity of 2 rad/s, find the rod CD's angular momentum about its mass centre at the point depicted in Fig. E3.3. Suppose that the rod's angular velocity is perpendicular to that of rod's axis.

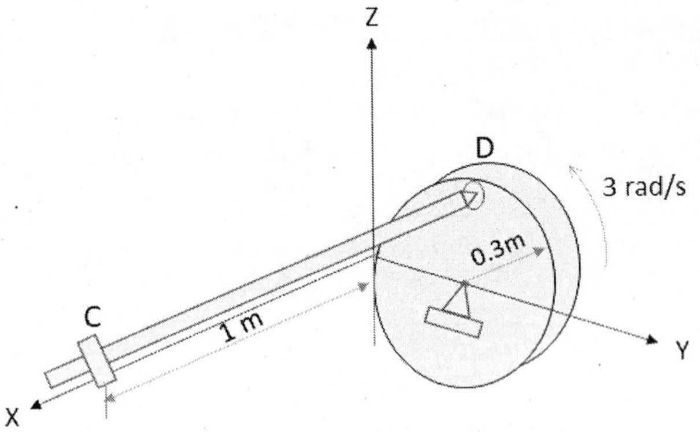

Fig. E3.3

Solution: m = 2 kg, ω = 3 rad/s

Linear velocity at C, $v_C = v_D + \omega \times r_{C/D}$

$$v_C e_1 = -(0.3)(3)e_2 + \begin{vmatrix} e_1 & e_2 & e_3 \\ \omega_x & \omega_y & \omega_z \\ 1 & -0.3 & -0.3 \end{vmatrix}$$

Equating the components:

$v_C = 0.3(-\omega_y + \omega_z)$ (i)

$0 = -0.9 + (0.3\omega_x + \omega_z)$ (ii)

$0 = -0.3\ \omega_x - \omega_y$ (iii)

$\omega . r_{C/D} = 0 \Rightarrow \omega_x - 0.3\ \omega_y - 0.3\ \omega_z = 0$ (iv)

Solving the above equations (i)-(iv), $\omega_x = 0.41 \frac{rad}{s}$; $\omega_y = -0.123 \frac{rad}{s}$; $\omega_z = 2.59 \frac{rad}{s}$;

$$v_C = 0.8139\ m/s$$

$$r_{C/D} = \sqrt{(1^2 + (-0.3)^2 + (-0.3)^2} = 1.6733\ m$$

Mass moment of inertia of the rod CD about mass center G, $I_G = \frac{1}{12} m r_{\frac{C}{D}}{}^2 = \frac{1}{12}(2)(1.6733)^2$

$$I_G = 0.467\ kg.m^2$$

Angular momentum rod CD about mass center, $H_G = I_G(\omega_x e_1 + \omega_y e_2 + \omega_z e_3)$

$H_G = (0.467)(0.41e_1 - 0.123e_2 + 2.59e_3) = 0.1915e_1 - 0.0574e_2 + 1.2095e_3$ kg.m²/s

$$H_G = \sqrt{(0.1915)^2 + (-0.0574)^2 + (1.2095)^2} = 1.226\ kg.m^2/s$$

Example 3.4: The uniformly thin plate of mass 10 kg and it is falling with v_c = 4 m/s velocity vertically and is not rotating just before its corner P hits the hook (Fig. E3.4). In the instant after corner P striking the hook without rebounding, ascertain its angular velocity.

Solution: m = 10 kg, v_c = {-4 e₃} m/s

Let us consider the X, Y, and Z axes parallel to x, y and z axes at centroid C.

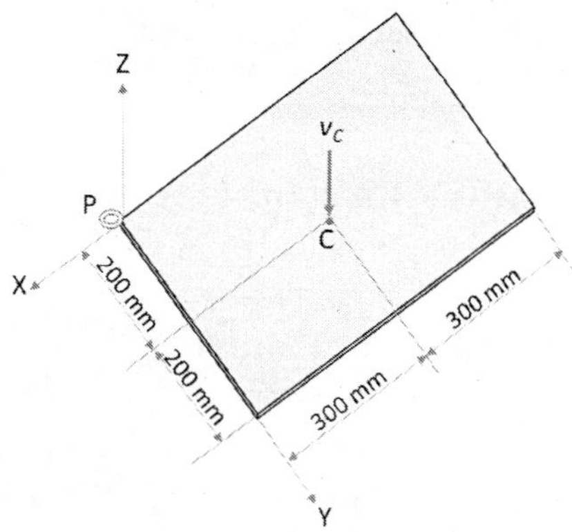

Fig. E3.4

With reference to the equation (3.6c) and x_C=-0.3m, y_C=0.2m, and z_C = 0 are coordinates with respect to X-Y-Z axes.

$$I_X = I_x + m({y_C}^2 + {z_C}^2) = \frac{1}{12}(10)(0.4)^2 + 10(0.2^2 + 0^2) = 0.533\ kg.m^2$$

$$I_Y = I_y + m({x_C}^2 + {z_C}^2) = \frac{1}{12}(10)(0.6)^2 + 10((-0.3)^2 + 0^2) = 1.2\ kg.m^2$$

$$I_Z = I_z + m({x_C}^2 + {y_C}^2) = \frac{1}{12}(10)[(0.4)^2 + (0.6)^2] + 10((-0.3)^2 + 0.2^2) = 1.433\ kg.m^2$$

Since the thin plate is symmetry about C, $I_{xy} = I_{yz} = I_{zx} = 0$

$$I_{XY} = I_{xy} + mx_Cy_C = 0 + 10(-0.3)(0.2) = -0.6\ kg.m^2$$

$$I_{YZ} = I_{yz} + my_Cz_C = 0 + 10(0.2)(0) = 0\ kg.m^2$$

$$I_{ZX} = I_{zx} + mz_Cx_C = 0 + 10(0)(-0.3) = 0\ kg.m^2$$

Prior to impact, the plate falls without rotating, therefore its angular momentum about point P is

$$H_P = r_{C/A} \times mv_C = (-0.3e_1 + 0.2e_2) \times 10(-4e_3) = [-8e_1 - 12e_2]kg.m^2/s \qquad \text{(a)}$$

The plate rotates about point P immediately following impact; hence it is possible to compute the elements of its angular momentum at this instant.

$$(H_P)_X = I_X\omega_X - I_{XY}\omega_Y - I_{XZ}\omega_Z = 0.533\omega_X - (-0.6)\omega_Y = 0.533\omega_X + 0.6\omega_Y \qquad \text{(b1)}$$

$$(H_P)_Y = -I_{XY}\omega_X + I_Y\omega_Y - I_{YZ}\omega_Z = -(-0.6)\omega_X + 1.2\omega_Y = 0.6\omega_X + 1.2\omega_Y \qquad \text{(b2)}$$

$$(H_P)_Z = -I_{XZ}\omega_X - I_{YZ}\omega_Y + I_Z\omega_Z = 1.433\omega_Z \qquad \text{(b3)}$$

Using principle of angular momentum, angular momentum before impact and after impact; equation the equations (a) and (b1, b2, b3),

$$-8e_1 - 12e_2 = (0.533\omega_X + 0.6\omega_Y)e_1 + (0.6\omega_X + 1.2\omega_Y)e_2 + (1.433\omega_Z)e_3$$

Equating the components of vectors, $-8 = 0.533\omega_X + 0.6\omega_Y$ (c1)

$$-12 = 0.6\omega_X + 1.2\omega_Y \qquad \text{(c2)}$$

$0 = 1.433\omega_Z$ (c3)

From equations c1, c2, and c3, $\omega_X = -11.42\frac{rad}{s}$; $\omega_Y = -4.29\frac{rad}{s}$; $\omega_Z = 0$

Angular velocity, $\omega = -1142e_1 - 4.29e_2\ rad/s$

3.8 NEWTON-EULER'S LAWS OF RIGID BODY MOTION

If a rigid body undergoing a combination of translation and rotation motion (i.e., general motion), "rate of change of both linear and angular momentum" equations will be considered for equilibrium conditions (i.e., $\Sigma\vec{F} = m\vec{a}_G$; $\Sigma\vec{M}=\dot{H}_G = \bar{I}\vec{a}$).

A rigid body's combined translational and rotational motions are described by the Newton-Euler equations. The body's mass centre acceleration, which is determined from an inertial X, Y, and Z reference, is used to characterise the body's translational motion. The vector form of the body's translational motion equation is as follows:

$$\Sigma\vec{F} = m\vec{a}_G \quad (3.29)$$

In the equation (3.29), the force F is expressed as

$$\Sigma\vec{F} = \sum F_x e_1 + \sum F_y e_2 + \sum F_z e_3 \quad (3.30)$$

Where, $F_x = ma_x$, $F_y = ma_y$, and $F_z = ma_z$

The equation (3.30) indicates, summation of all the external forces acting on the body.

The equations given below are intended for the rotational motions,

$$M_x = I_{xx}\alpha_x,\ M_y = I_{yy}\alpha_y,\ \text{and } M_z = I_{zz}\alpha_z \quad (3.31)$$

If the principal axes of inertia become the reference coordinate system, equation (3.31) remains valid. The general moment relation expressed by $\Sigma\vec{M}=\dot{H}_G = \bar{I}\vec{a}$ (3.32)

According to equation (3.32), about a fixed-point O, the rate of change in the total angular momentum of the body equals the sum of the moments of all external forces that are acting on a system of particles contained in a rigid body over time.

When the components of angular momentum H are measured in relation to a moving coordinate system, as in Fig. 3.13, which has angular velocity ${}^A\omega^B$, then the equation (3.32) becomes

$$\sum \boldsymbol{M}_c = \frac{dH_c}{dt} + {}^A\omega^B \times H_C \quad (3.33)$$

Where $\frac{dH_c}{dt}$ is the time rate of change of angular momentum (H_C), and ${}^A\omega^B$ is the angular velocity of B with reference to another axis frame A. ${}^A\omega^B$ is different from the body's angular velocity. The first term of the equation (3.33) represents the body's rate of change of angular momentum and the second term indicates the moment with reference to the rotation of B with respective to A (i.e., product of angular velocity of B with respective to A and angular momentum).

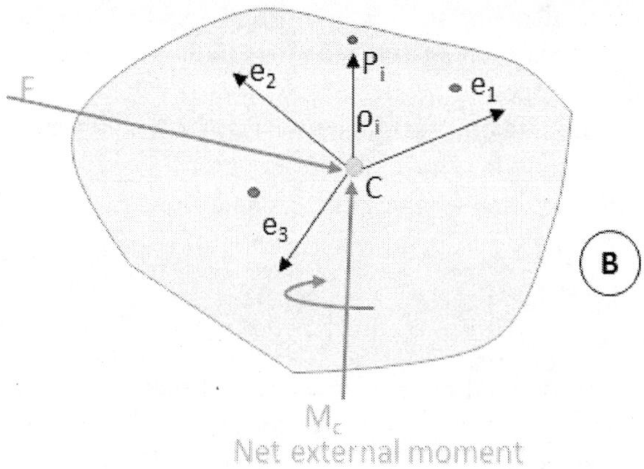

Fig. 3. 13 Body under Principal axis

The physical interpretation of H_C and ${}^{A}\omega^{B}$ are depicted in Fig. 3.14.

Let e_1, e_2, e_3, be along principal axes,

$${}^{A}\omega^{B} = \omega_1 e_1 + \omega_2 e_2 + \omega_3 e_3 \tag{3.34}$$

and with reference to equation (3.18), $H_C = I_{11}\omega_1 e_1 + I_{22}\omega_2 e_2 + I_{33}\omega_3 e_3$ and

$$\frac{dH_c}{dt} = I_{11}\dot{\omega}_1 e_1 + I_{22}\dot{\omega}_2 e_2 + I_{33}\dot{\omega}_3 e_3 \tag{3.35}$$

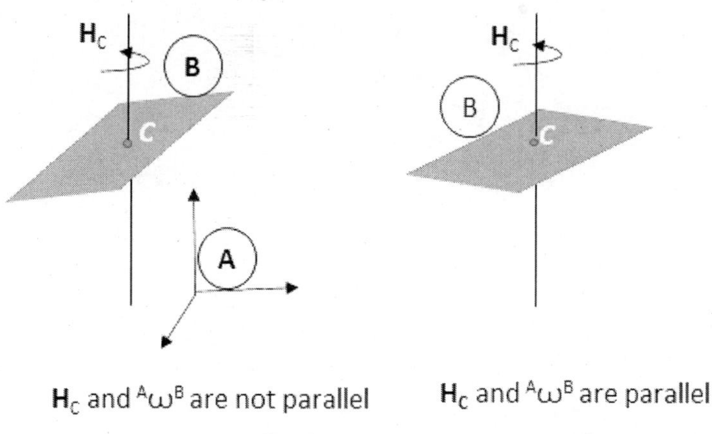

Fig. 3. 14 Physical interpretation of H_C and ${}^{A}\omega^{B}$

Perform the cross-product and simplify the equations (3.33), (3.34) and (3.35),

$$\begin{bmatrix} M_{C1} \\ M_{C2} \\ M_{C3} \end{bmatrix} = \begin{bmatrix} I_{11} & 0 & 0 \\ 0 & I_{22} & 0 \\ 0 & 0 & I_{33} \end{bmatrix} \begin{bmatrix} \dot{\omega}_1 \\ \dot{\omega}_2 \\ \dot{\omega}_3 \end{bmatrix} + \begin{bmatrix} 0 & -\omega_3 & \omega_2 \\ \omega_3 & 0 & -\omega_1 \\ -\omega_2 & \omega_1 & 0 \end{bmatrix} \begin{bmatrix} I_{11} & 0 & 0 \\ 0 & I_{22} & 0 \\ 0 & 0 & I_{33} \end{bmatrix} \begin{bmatrix} \omega_1 \\ \omega_2 \\ \omega_3 \end{bmatrix} \tag{3.36}$$

Leonhard Euler, a Swiss mathematician, is credited with creating this system of equations, which are known as the Euler equations of motion.

The Newton-Euler equations, as they are known, are formed by combining Euler's two laws of motion into a single equation with six components utilising column vectors and matrices for a rigid body. These rules establish a connection between a rigid body's centre of gravity movement and the total force and moment applied on the rigid body. They can be represented as a matrix form with regard to a coordinate frame whose origin is the centre of mass of the body for the moment (M) and an inertial frame of reference for the force (F) as follows:

$$\begin{bmatrix} F \\ M \end{bmatrix} = \begin{bmatrix} mI_3 & 0 \\ 0 & I_C \end{bmatrix} \begin{bmatrix} a_C \\ \alpha \end{bmatrix} + \begin{bmatrix} 0 \\ \omega \times I_C \omega \end{bmatrix} \tag{3.37}$$

Where I_3 = the identity matrix 3 X 3

The more complex "multi-body" formulations (screw theory) that explain the dynamics of systems of rigid bodies joined by joints and other constraints are built on the Newton-Euler equations. Numerous numerical algorithms can solve multi-body problems.

Example 3.7: The disc, which radius 100 mm and weighs 6 kg, is eccentrically attached on shaft PQ as shown in Fig. E3.7 (a). If ω_S = 10 rad/s is the constant speed at which the shaft is rotating and at the same time, an angular velocity of ω_P = 5 rad/s the shaft rotating (precising) about the thrust bearing at P, determine the reactions at P (the thrust bearing) and Q (journal bearing) supports while the disc is in the position displayed.

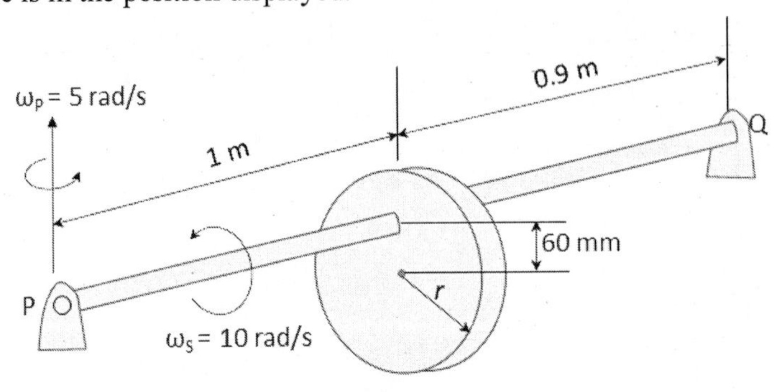

(a)

Solution: Free-body diagram as illustrated in Fig. E3.7 (b), the disc's centroid G is offset from O, the x, y, and z coordinate system's origin.

The moment of inertia of disc, $I_y = \frac{1}{2}(6)(0.1)^2 + (6)(0.06)^2 = 0.0516$ *kg-m²*

$I_x = I_z = \frac{1}{2} I_y = 0.0258$ *kg-m²*

Angular velocity of disc, $\omega_x = 0, \omega_z = 5\, rad/s, \omega_y = 10\, rad/s$

It is essential to determine the time derivative of ω in relation to the x, y, and Z axes. In this instance, ω_p and ω_s do not alter in magnitude or direction, and as a result

$$\dot{\omega_x} = \dot{\omega_y} = \dot{\omega_z} = 0$$

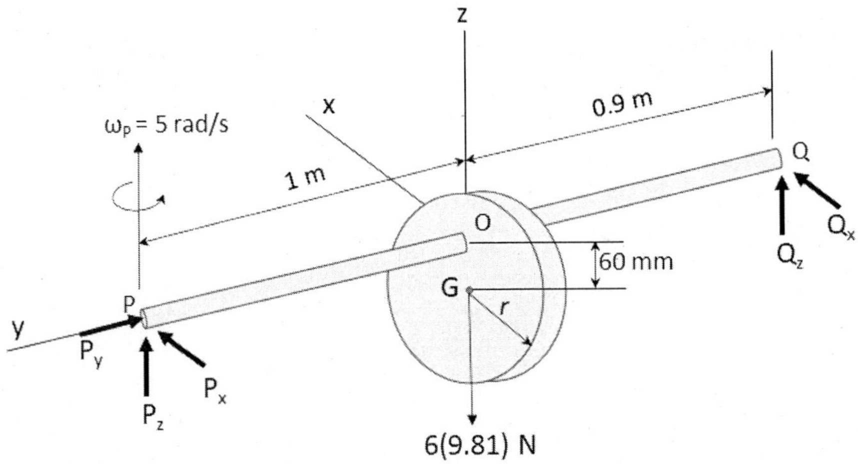

(b)

Fig. E3.7

From equation (3.27),

$$\begin{bmatrix} M_x \\ M_y \\ M_z \end{bmatrix} = \begin{bmatrix} 0 & -\omega_z & \omega_y \\ \omega_z & 0 & -\omega_x \\ -\omega_y & \omega_x & 0 \end{bmatrix} \begin{bmatrix} I_x & 0 & 0 \\ 0 & I_y & 0 \\ 0 & 0 & I_z \end{bmatrix} \begin{bmatrix} \omega_x \\ \omega_y \\ \omega_z \end{bmatrix}$$

$$\begin{bmatrix} P_z(1) - Q_z(0.9) \\ 0 \\ -P_x(1) + Q_x(0.9) \end{bmatrix} = \begin{bmatrix} 0 & -5 & 10 \\ 5 & 0 & 0 \\ -10 & 0 & 0 \end{bmatrix} \begin{bmatrix} 0.0258 & 0 & 0 \\ 0 & 0.0516 & 0 \\ 0 & 0 & 0.0258 \end{bmatrix} \begin{bmatrix} 0 \\ 10 \\ 5 \end{bmatrix}$$

$P_z(1) - Q_z(0.9) = -1.29$ (i)

$-P_x(1) + Q_x(0.9) = 0$ (ii)

From equation (3.20),

$\sum F_x = ma_x;\ P_x + Q_x = 0$ (iii)

$\sum F_z = ma_z;\ P_z + Q_z - 6(9.81) = 6(10)^2(0.06);\ P_z + Q_z = 70.86$ (iv)

Solving the above equations (i)-(iv), $P_x = Q_x = 0;\ P_z = 32.89\ N;\ Q_x = 37.97\ N$

UNIT SUMMARY

This unit covered a variety of fundamental kinetics concepts, including Newton's Euler laws of motion, inertia and non-inertia frames, inertia tensor, angular momentum, polar moment of inertia, area moment of inertia, and mass moment of inertia. Thus, the skills learned in this unit will undoubtedly be able to handle a variety of engineering mechanics issues, particularly those involving the dynamics of rigid bodies.

EXERCISES

Multiple Choice Questions

1. The Fig. P3.1 below depicts a reel has a radius of gyration "k" is gradually rolling downward from rest with one end of the thread wound on it being kept in the ceiling. The symbol "g" stands for the acceleration caused by gravity because the thread's thickness and mass are negligible in comparison to the hub's radius "r" and the reel's mass "m."

   i) The Linear acceleration of the reel is:

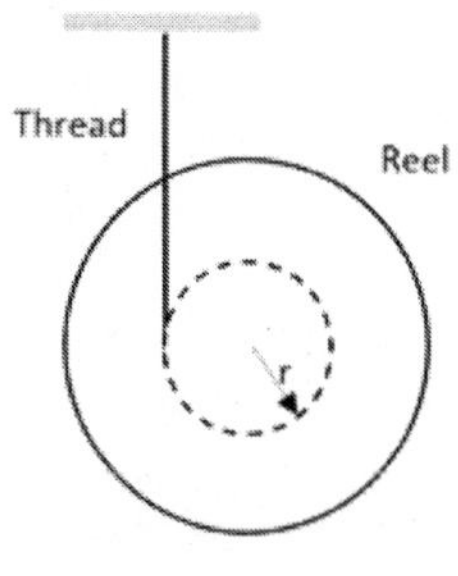

Fig. P3.1

a) $\frac{gr^2}{(r^2+k^2)}$ b) $\frac{gk^2}{(r^2+k^2)}$ c) $\frac{grk}{(r^2+k^2)}$ d) $\frac{mgr^2}{(r^2+k^2)}$

   ii) The tension in thread is:

a) $\frac{mgr^2}{(r^2+k^2)}$ b) $\frac{mgk^2}{(r^2+k^2)}$ c) $\frac{mgrk}{(r^2+k^2)}$ d) $\frac{mg}{(r^2+k^2)}$

2. A wheel with a mass m and radius r is rolling more quickly and without sliding under an axle torque T that is constant. If the friction coefficient is μ, the friction force of ground acting on the wheel is.

   a) zero b) μmg c) $\frac{T}{r}$ d) $\frac{2T}{3r}$

3. A point mass is vertically propelled upward from the earth at 3 m/s speed and at time t=0. Its impact velocity decreases by 20% each time it strikes the ground. The mass stops bouncing and comes to a complete rest on the ground after ----- seconds, ignoring the air resistance and assuming g = 10 m/s^2.

   a)1 b)1.5 c)2 d)3

4. A thin cylinder ring of mass "M" and radius "r" spinning along its axis at an angular velocity of "ω" Two particles that each have a mass of "m" are now joined at an angle. The angular speed of the ring will increase.

   a) $\frac{2M\omega}{M+m}$ b) $\frac{2M\omega+m}{M}$ c) $\frac{M\omega}{M+2m}$ d) $\frac{M\omega}{2M+m}$

5. Assuming perfect elastic impact, when a ball A of mass m collides (drops from a height h due to gravity) with a ball B of mass m held at rest by a spring of stiffness k, the velocity of -------------

   a) ball B is $\frac{1}{2}\sqrt{2gh}$.

   b) ball B is zero

   c) both balls are $\frac{1}{2}\sqrt{2gh}$.

d) ball A is zero.

6. On an inclined plane with an angle of θ, a solid cylinder has mass *m* rolls without sliding. Find the cylinder's linear acceleration.

   a) $\frac{1}{3}gsin\theta$ b) $\frac{1}{3}gcos\theta$ c) $\frac{2}{3}gsin\theta$ d) $\frac{2}{3}gcos\theta$

7. Moment of inertia theory is based on ________ distribution.

   a) Both area and mass b) only area c) only mass d) either mass nor area

8. Using the perpendicular axis theorem, the moment of inertia of a ____________ is calculated.

   a) Triangle b) square c) rectangle d) circle

9. If there is no axis of symmetry in a figure, locating the ______________ is necessary to determine the moment of inertia about the centroidal axis.

   a) X-axis b) Y-axis c) centroid d) origin

10. In mathematics, the term "area moment of inertia" also refers to ______________ moment of inertia.

    a) First b) second c) third d) fourth

Answers of Multiple-Choice Questions

1. i (a) 1.ii (b) 2 (d) 3 (b) 4 (c) 5 (d) 6 (c) 7 (a) 8 (d) 9 (c) 10 (b)

Short and Long Answer Type Questions

1) Describe two physical aspects of the moment of inertia?
2) What are mass and force's rotational counterparts in terms of physical quantities?
3) Define condition required for pure rolling?
4) How can the moment of inertia of composite regions be determined?
5) State and substantiate the perpendicular axis and parallel axis theorems.
6) What does Radius of Gyration mean?
7) How is angular momentum calculated and what does it mean?
8) At 4 m/s of linear speed, a 180 g, 25 cm in diameter frisbee is launched. The frisbee travels at an angle of 5 rad/s. What is the total kinetic energy of Frisbee? Suppose the frisbee is a thin, uniformly dense disc.
9) A bicycle is riding at 15 km/h downhill. The frame weighs 10 kg, and each wheel has 0.35 m radius and 3 kg mass. What is the bicycle's total kinetic energy? Assume that the tires can be compared to thin rings and that they don't move relative to the surface.
10) What is the difference in kinetic energy between spinning around the point P as depicted in Fig. P3.2, which is located a distance of 7 m away, and rotating about the rectangular plate whose dimensions are a=5 and b=4? The plate is the same mass and has 4 rad/s of angular velocity in both scenarios.

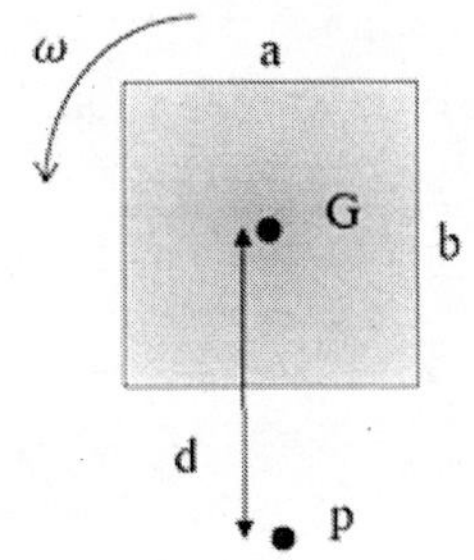

Fig. P3.2

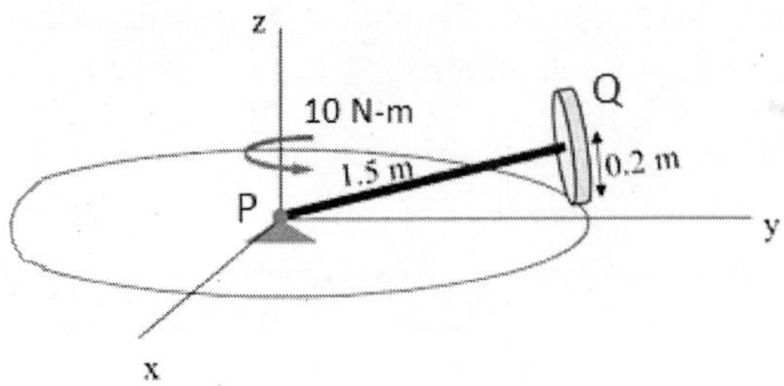

Fig. P3.3

Numerical Problems

1) Thin rod of mass 6 kg is linked to the 10-kg disc as shown in the Fig. P3.3. If at P the assembly is linked to a ball-and-socket joint and experiences a 10 N-m couple moment, it will move with an angular velocity of ω_z =4 rad/s about the z axis. Quantify the angular momentum of the assembly with respect to P.

2) Each of the two discs weighs 5 kg and radius r = 0.5 m. The weight of axle AB is 1.5 kg. If an assembly as shown in Fig. P3.4 rotates around the z axis at ω_z =6 rad/s, about z axis, determine its angular momentum. The discs roll without slipping.

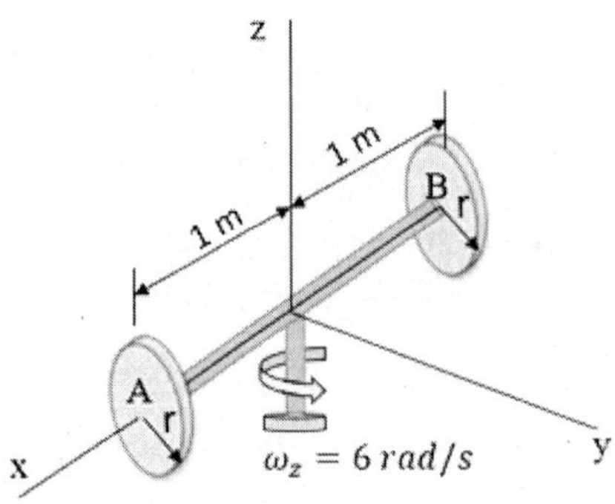

Fig. P3.4

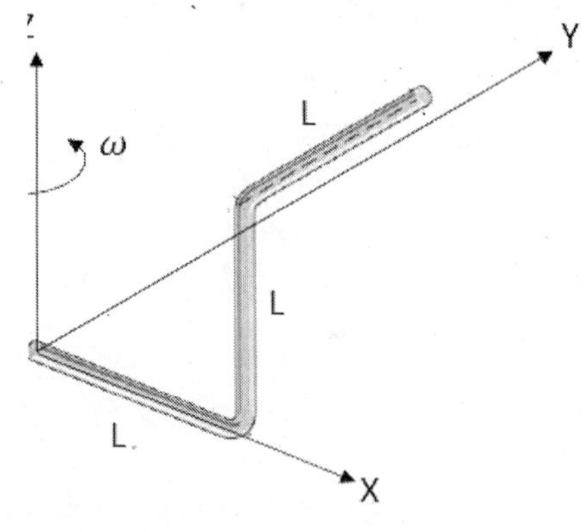

Fig. P3.5

3) As depicted in Fig. P3.5, about the Z-axis, a bent rod rotates with an angular velocity ω and has a mass ρ per unit length. Calculate the angular momentum of the rod with regard to the fixed origin O.

4) The equation shown in Fig. P3.6 creates the moment of inertia of the paraboloid about Y axis. Calculate this moment and translate your answer in relation with the paraboloid's total mass m. The density of paraboloid, ρ, remains constant.

5) As illustrated in Fig.3.7, a T-shaped assembly is created by welding together two uniform rods, each of which has mass m and length a. Determine the assembly's mass moment of inertia around the pin axis at O.

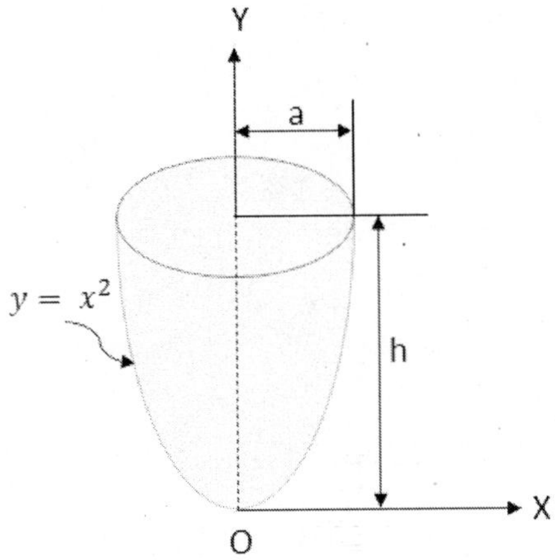

Fig. P3.6

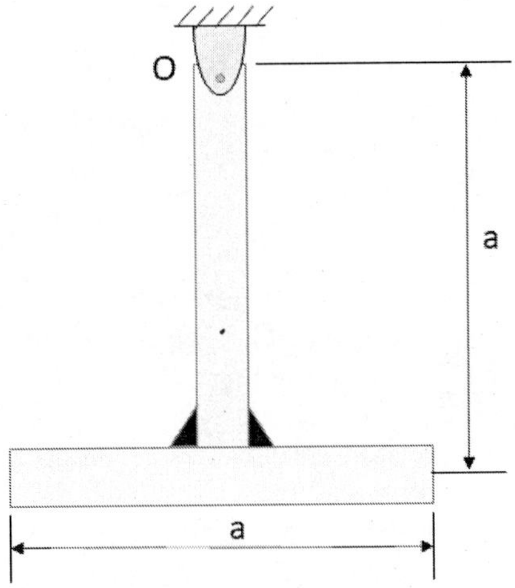

Fig. P3.7

6) As depicted in the Fig. P3.8, about the y-axis, the thin shaft rotates with 10 rad/s constant angular velocity while carrying two offset (1 m) particles, each weighing 1 kg. Determine how the thrust bearing at P and the journal bearing at Q will respond to the shaft's dynamic imbalance at the position indicated.

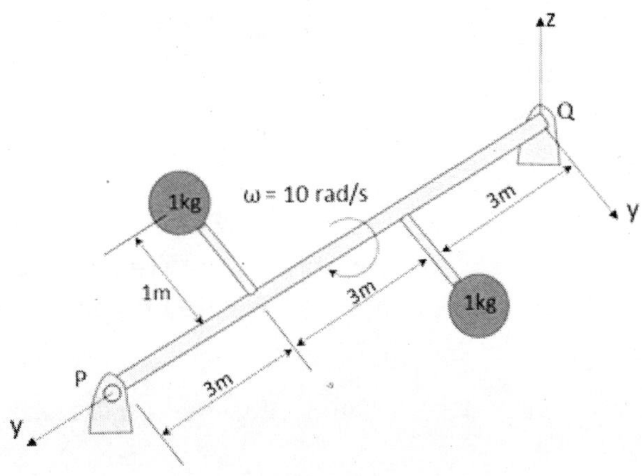

Fig. P3.8

7) As seen in Fig. P3.9, a 1 m long rod is slipping around a corner. The velocity of point A on the rod is 1 m/s at the instant when an angle of 60^0 is formed between rod and the horizontal plane. Calculate the rod's angular velocity at this instant.

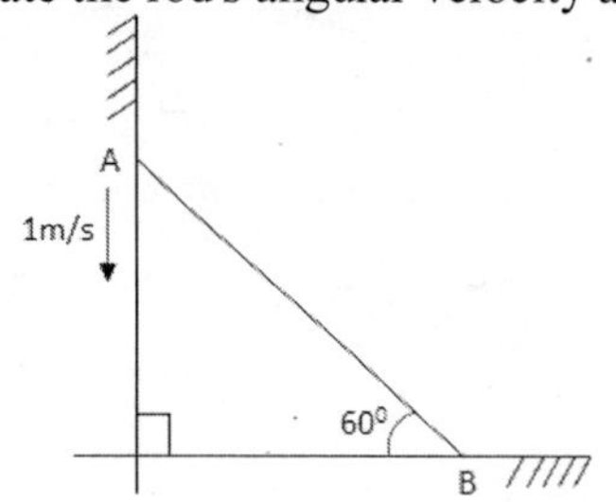

Fig. P3.9

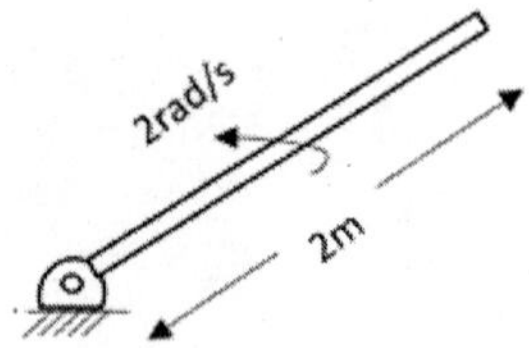

Fig. P3.10

8) The cannon fires a shell as displayed in Fig. P3.10. The velocity of the shell in relation to the barrel at this precise moment is 3 m/s, while the barrel is swinging upward at 2 rad/s of constant angular velocity. Establish the shell's absolute velocity's magnitude.

9) When a gunshot is fired, a bullet spins. Two helical slots are carved into the barrel for this purpose, as seen in the image P3.11. The bullet's projections A and B enter each of the slots. One helix turn is performed over 0.5 m in helical slots. When a bullet leaves the barrel at speed of 20 m/s, 20 rad/s is its spinning speed.

10) As seen in the Fig. P3.12, on the horizontal surface, a wheel with radius r rolls without slipping. Evaluate the point Q's velocity magnitude (in m/s) is if the horizontal velocity of point P is 10 m/s.

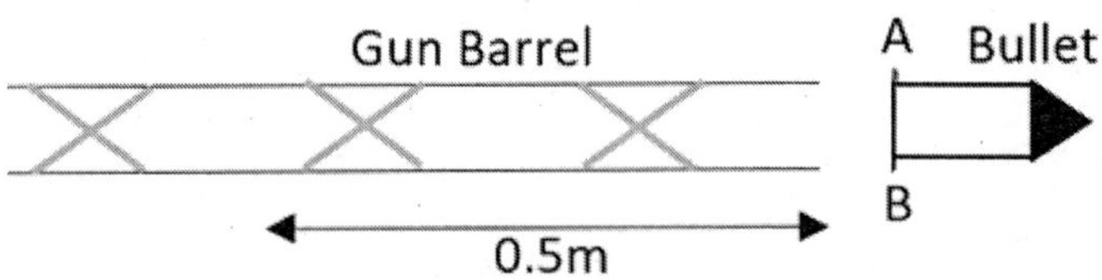

Fig. P3.11

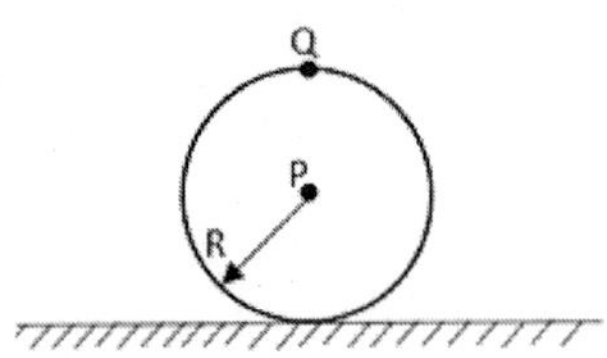

Fig. P3.12

11) A frictionless fixed axis rotates through the centre of a disc with a rotational inertia of 5 kg-m^2 and a radius of 0.25 m. An 8N tangential force is subjected to the rim. If the disc begins at rest, find the angular velocity after completing half a rotation.

12) A circular disc of velocity V, radius "R" rolls without slipping as depicted in Fig. P3.13. Determine the velocity at point "P".

13) As seen in the Fig. P3.14, the one end of a uniform stiff rod with mass denoted by M and length denoted by L is hinged. The rod swings to the right by applying a force at a distance of 2L/3 from the hinge. Determine the hinge's horizontal response.

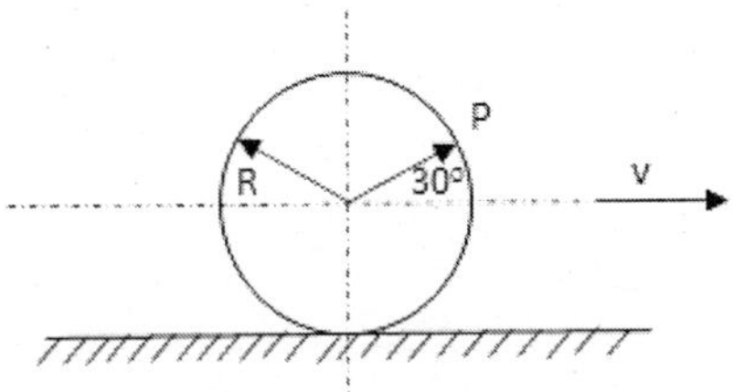

Fig. P3.13

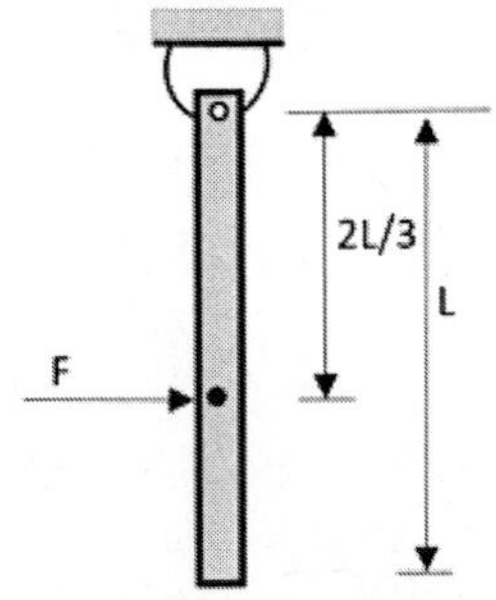

Fig. P3.14

14) At the position depicted in the Fig. P3.15, the rod PQ, which has a length of L = $\sqrt{2}m$ and a uniformly dispersing M (mass 10 kg), is released from rest. The ends go effortlessly along the faces OP and OQ. Assume that g = 10m/s^2 for the gravitational acceleration. The rod's mass moment of inertia, measured along the axis parallel to the figure's plane and around its centre of mass, is ($ML^2/12$). At this moment, Determine the rod's angular acceleration magnitude.

15) A circular disc with a 100 mm radius and 1 kg mass rolls down a curved path without slipping as indicated in the illustration Fig. P3.16, initially in rest at position a. Find the speed of disc that is moving when it reaches location B. (Assume 10 m/s^2 as the acceleration due to gravity).

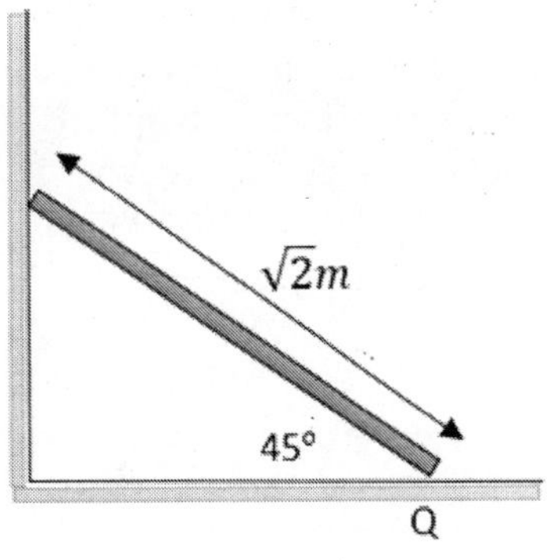

Fig. P3.15

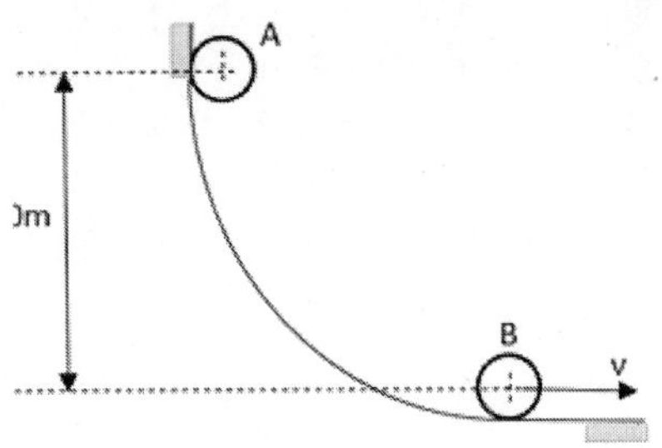

Fig. P3.16

PRACTICAL

Aim: Determining the flywheel's moment of inertia.

Apparatus: The flywheel, the weight hanger with slots for the weights, the stopwatch, the metre scale, etc.

Theory: A flywheel is a form of inertial energy storage. When the supply of energy exceeds the demand, it stores energy and acts as a reservoir; when the supply is surpassed by demand, it releases energy. It absorbs mechanical energy. A fly wheel's primary job is to reduce speed variations in a shaft that result from sudden changes in torque. Numerous machine load patterns result in cycle-to-cycle variations in torque. Systems with flywheels include one- or two-cylinder internal combustion engines, punch presses, rock crushers, and piston compressors.

A large wheel with a sturdy axle extending on either side is called a flywheel. The axle is mounted on ball bearings on two permanent supports, as shown in Fig. P3.17. On the axle, there is a little peg that is loosely placed. A string that has been wrapped around a peg has a weight hanging connected to one end. Near the rim of the flywheel is a pointer.

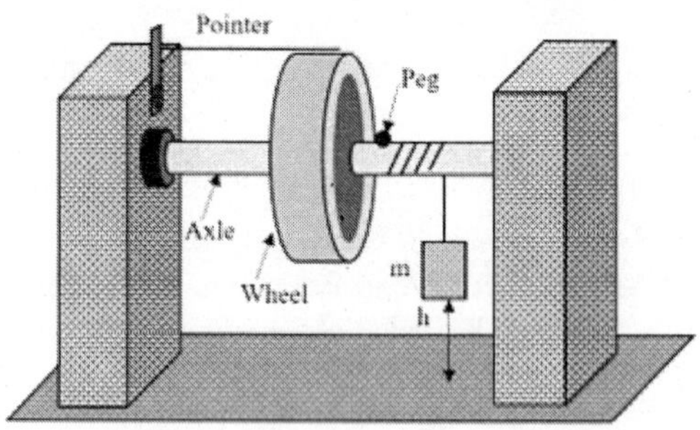

Fig. P3.17

Procedure: To conduct the experiment, the string's length is sets o that the peg's axle must come loose when the falling mass barely reaches the ground. Now, a line is marked with chalk on the rim beneath the pointer. After that, the string is attached to the peg, and the wheel is rotated n times until the string coils n times without contacting the axle. The mass m is currently 'h' above the floor. The mass is then let to drop. It causes the flywheel's axle to experience a torque. Because of this torque, the flywheel rotates with an angular acceleration. Let W denote the work against friction performed during each revolution and denote the wheel's angular velocity while the peg is just separating from the axle in order to apply the law of conservation of energy.

$$mgh = \frac{1}{2}I\omega^2 + \frac{1}{2}mv^2 + nW \qquad \text{(P1)}$$

Assume, N represent how many revolutions the wheel completes before coming to a stop. When the flywheel comes to rest, its kinetic energy has fully dissipated, thus we may say,

$$NW = \frac{1}{2}I\omega^2 \text{ Or, } W = \frac{I\omega^2}{2N} \qquad \text{(P2)}$$

by using eq. (P1) and eq. (P2),

$$mgh = \frac{1}{2}I\omega^2 + \frac{1}{2}mv^2 + n\frac{I\omega^2}{2N} = \frac{1}{2}I\omega^2 + \left(1 + \frac{n}{N}\right) + \frac{1}{2}mr^2\omega^2$$

$$\text{Therefore, } I = \frac{Nm}{N+M}\left(\frac{2gh}{\omega^2} + r^2\right) \qquad \text{(P3)}$$

where "r" stands for the axle's radius. In order to calculate ω, we use the assumption that the angular retardation of flywheel is uniform when the mass separates as of the axle. Then,

$$Average\ angular\ velocity = \frac{Total\ angular\ displacement}{Time\ taken}$$

$$\frac{\omega+0}{2} = \frac{2\pi N}{t};\ \omega = \frac{4\pi N}{t} \qquad \text{(P4)}$$

One end of the rope is wrapped around the peg to begin the experiment, and an appropriate weight is added to the weight hanger. The string is looped around the axle n times without overlapping by rotating the fly wheel n times. In this state, the flywheel is kept motionless. The distance 'h' between the ground and the weight hanger's bottom is calculated. After that, the flywheel is let go. While the flywheel turns, the mass is lowered. Just as the peg separates from the axle, start a stopwatch. The number of rotations the wheel does between the moment the peg separates as of the axle and the moment it approaches to rest, or 'N' is counted. Additionally, the time span "t" is indicated. The experiment is run again with the same number "n" and mass "m". It is established what "N" and "t" average out to. Equations (P3) and (P2) are used to compute the moment of inertia "I" (P4). For various values of "n" and "m," the entire experiment is repeated, and the average value of I is computed.

- To ensure that the peg separates from the axle when the mass just hits the ground, make sure the rope is long enough.
- The peg is tightly fastened to the axle in certain wheels. In this scenario, the string's one end is loosely wrapped around the peg so that it slips off the peg when the mass barely hits the ground.
- The weight hanger's mass and the extra mass resting on it are added to form the value m'.

Observation and tabulation:

Using vernier callipers, calculate the axle's radius

One main scale division's value (1 m s d) = ---------- cm.

The vernier scale's number of divisions, x = ...

$$Least\ count, L.C = \frac{Value\ of\ one\ main\ scale\ division}{Number\ of\ divisions\ on\ the\ vernier\ scale} = \frac{1\ m\ s\ d}{x} = \ldots\ldots\ldots\ldots cm$$

Trial No.	M S R (cm)	V S R	D = M S R + V S R × L C	Mean Diameter D (cm)

Axle's diameter, D = ... cm = ... m

Axle's radius, r = $\frac{D}{2}$ = ... m

Moment of inertia calculations

ss "m" kg hanging at one end of the thread	ght of the weight hanger measured from the ground up.	mber of string windings on the axle, "n"	Number of rotations made by the wheel following the peg's separation from the axle "N"			Time t, in seconds, between the peg coming loose and the wheel stopping			$\omega = \frac{4\pi N}{t}$	I $kg.m^2$
			1	2	Mean, N	1	2	Mean t sec		

Result:

Moment of inertia of the given flywheel, I = ... kg.m^2.

KNOW MORE

Leonhard Euler

Leonhard Euler (15 April 1707 – 18 September 1783) was a Swiss mathematician, physicist, astronomer, geographer, logician, and engineer who also made significant contributions to the fields of analytic number theory, complex calculations, and infinitesimal calculus. He founded the studies of graph theory and topology. Many modern mathematical terms and notations, such as the concept of a mathematical function, were introduced by him. He has made significant contributions to the fields of optics, astronomy, fluid dynamics, and music theory.

In addition to being the first to use the notation for a function's value, f(x), the imaginary unit $\sqrt{-1}$ expressed by letter i, summations expressed by the Greek letter **Σ** (capital sigma), finite differences denoted by the Greek letter **Δ** (uppercase delta), and lowercase letters to denote the sides of a triangle while using capital letters for the angles. Euler is also endorsed with popularising the Greek letter **π** (lowercase pi) that is used to indicate the circle's circumference to its diameter ratio. The constant e, currently known as Euler's number and the foundation of the natural logarithm, was defined by him in its current form.

The post-buckling load of an ideal strut is calculated using Euler's formula, which only considers the flexural stiffness and length of the strut. This formula is well-known in the field of structural engineering.

REFERENCES AND SUGGESTED READINGS

- Meriam, J.L. and Kraige, L.G., Engineering Mechanics: Dynamics, Wiley, 2011.
- Hibbeler, R.C., Engineering Mechanics: Dynamics, Pearson Prentice Hall, 2010.
- Dubey, N.H., Engineering Mechanics: Statics and Dynamics, McGraw Hill Education Pvt. Ltd., 2015.

4 Free Body Diagram

UNIT SPECIFIES

This unit discussed the following concepts:

1. Introduction of Free Body Diagram
2. Modelling of typical joints and supports
3. Discussion on Kinematics and Kinetics constraints
4. Analysis on the forces and moments of the Free Body Diagram

RATIONALE

A free body diagram (FBD) is a graphical representation used to visualize the forces and moments applied to a single body or a system of bodies. FBDs are foundational tools in solving statics and dynamics problems. Drawing a FBD helps in identifying all relevant forces and moments, including gravitational forces, normal forces, frictional forces, applied forces, and reaction forces. By accurately depicting all forces and moments, they provide a basis for applying Newton's laws of motion, enabling the systematic solution of equations for unknown forces and accelerations. By employing free body diagrams, engineers and scientists can systematically analyze mechanical systems, leading to more accurate and reliable solutions in both design and problem-solving contexts.

PRE-REQUISITES

1. Nil

UNIT OUTCOMES

U4-O1. Understanding the Free Body Diagram

U4-O2. Dividing the forces and analysis

U4-O3. Demonstrate the Kinematics and Kinetics constraints

U4-O4. Acting of moments and finding the forces

U4-O5. Resolving the forces and concluding resultant

Unit-4 Outcomes	EXPECTED MAPPING WITH COURSE OUTCOMES (*1- Weak Correlation; 2- Medium Correlation; 3- Strong Correlation*)				
	CO-1	CO-2	CO-3	CO-4	CO-5
U4-O1	3	2	3	3	1
U4-O2	3	2	3	3	1
U4-O3	3	3	3	3	1
U4-O4	3	2	3	3	1
U4-O5	3	2	3	3	1

4.1 INTRODUCTION

The rigid body or particle may be subject to known or unknowable forces and moments. These forces and moments can be identified by systematically representing them through a diagram for the initiation of equilibrium analysis. This diagram can be represented through a sketch (not a scaled drawing) of a particle or rigid body with a system of known or unknown forces and moments acting on that and free or isolated from surroundings is called a **Free Body Diagram (FBD)**. It shows a body or group of linked bodies along with all of the applied forces, times, and reactions that the body experiences. The body might be compact or have several interior pieces (like beams, truss members, etc.). A series of free bodies and other diagrams might be needed to provide solutions to challenging problems. For the solution of the mechanics' problem, the FBD is the most significant and first step.

4.1.1 Importance of FBD

1. The physical problem can be converted into the form of a mathematical problem for the analysis, FBD sketch is the important step.
2. The FBD for a particle or rigid body, a sketch of a body or other connected bodies entirely isolated or free from surrounding bodies, the force applied by all surrounding bodies on the body being considered.
3. FBD indicates all applied, reactive "forces and moments" acting on the body.
4. FBD provides to identify the known and unknown "forces, and moments" acting on a rigid body.
5. FBD will identify the type of force system for calculating the unknown forces and moments of proper condition of equilibrium.

4.1.2 Important steps for Drawing FBD

1. The outlined shape or sketch of the particle or rigid body can be drawn by assuming all supports removed.
2. One or more FBDs may be required to represent the force system for the complex problems.
3. On the FBD sketch, indicate all the required dimensions and angles for computing moments of forces.
4. Illustrate all forces (including forces caused by weights and cables) and moments acting on the body with point of application, direction, and magnitude, if known.
5. Identify all supports or constraints and corresponding reactive forces and moments.
6. Show all the reactive forces and moments (if applicable) for each support.
7. Assume, if the reaction force or moment direction is unknown. If the assumed direction is correct, a positive result will arrive. Whereas negative results arrive, the assumed direction should be opposite.
8. Wherever convenient use principle of transmissibility.

4.2 SUPPORT REACTIONS

A support can constraint the translation or rotation or both of a body in the given direction by exerting a reaction force or moment on the body in the opposite direction. The different types of supports and corresponding reactions are shown in Fig. 4.1 and described as follows:

1) **Roller Support**: This support prevents the body from translation (or provides restraint in one direction only) in the vertical direction.
2) **Hinged Joint or Support**: This joint or support also known as pin joint or support, prevents the body from translations in two directions i.e., the horizontal and vertical reactions. It does not offer any restraint from the rotation.
3) **Fixed Support**: This supports prevents the body from both translations (both horizontal and vertical) and rotation (moment) directions. Fixed support offers restraints in the horizontal, vertical and rotation directions.

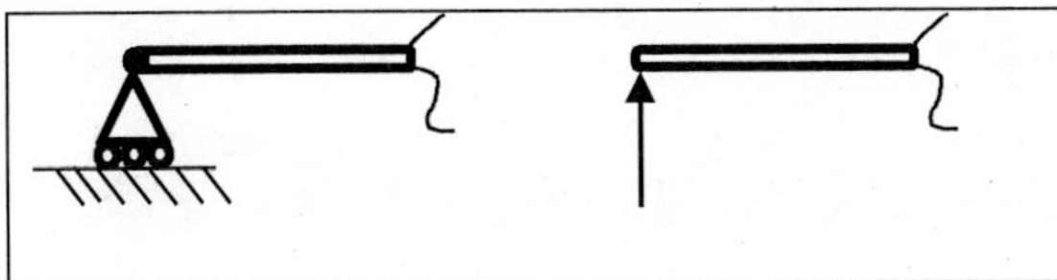	(a) Roller Support
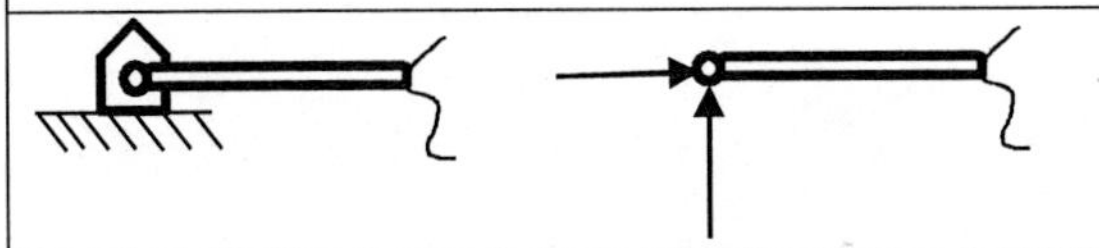	(b) Hinged or Pin Support
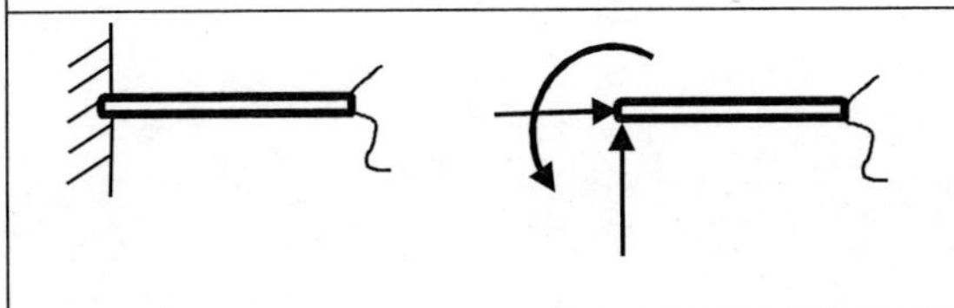	(c) Fixed Support
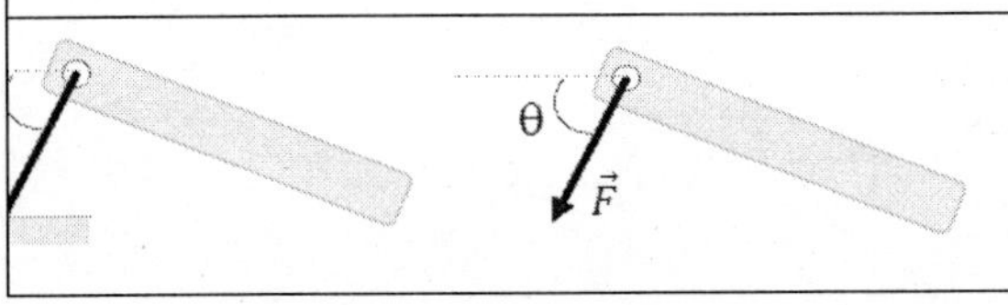	(d) Cable

Fig. 4. 1 Support reactions.

4.3 FORCE RESOLUTION

"A rigid body is considered to be in equilibrium if it is at rest or in motion with constant velocity." However, static equilibrium refers to an object at rest. The resultant of the given force-moment system must be zero in order for this equilibrium to exist. It is necessary to estimate the subsequent force and moment if the rigid body is subjected to an arbitrary force-moment system. The forces must resolve in the desired directions in order to calculate the "resultant force" or resultant moment.

Resolution of force is the division of a force into a number of components that are equivalent to the forces that are given.

4.3.1 Method of Projections

Method of projections for any force system:

- Resolve every force into horizontal and vertical components.
- Summation of all horizontal components $(\boldsymbol{\Sigma F_x})$
- Summation of all vertical components $(\boldsymbol{\Sigma F_y})$
- Find Resultant

$$R = \sqrt{(\sum F_x)^2 + \left(\sum F_y\right)^2} \tag{4.1}$$

- If Body/Point is in static equilibrium: $\sum F_x = 0$; $\sum F_y = 0$

1. $\overrightarrow{\sum Fx} = \overleftarrow{\sum Fx}$
2. $\overrightarrow{\sum Fy} = \overleftarrow{\sum Fy}$

Example 4.1: Solve the forces representation of the given Fig. E4.1(a) find the resultant and angle.

Solution: Resolving the forces as shown in Fig. E4.1(b)

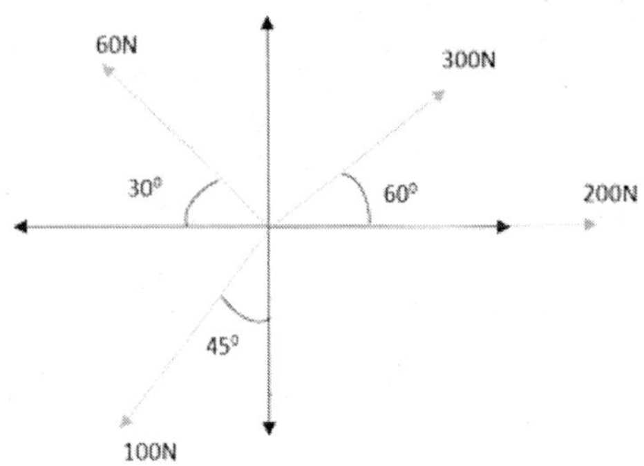

Fig. E4.1 (a)

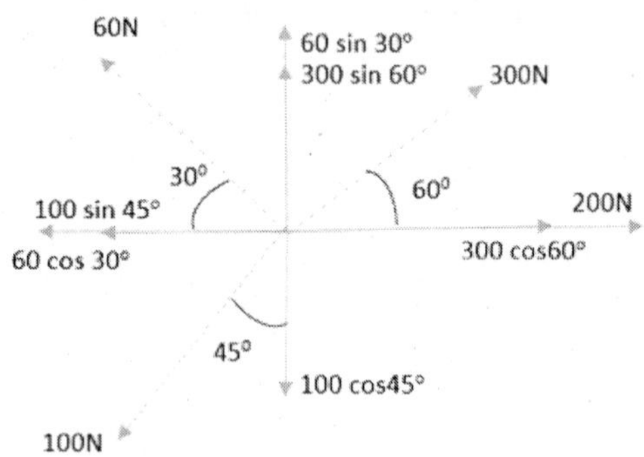

Fig. E4.1(b) Resolved forces

Pulley System

A wheel with a cord, chain, cable, belt, or flexible rope attached to its rim is referred to as a pulley. Both motion and energy are transmitted using it. In order to construct a simple machine, the pulley system is used to redirect the tension force that is applied to one side of the rope. These are some of the established and listed parameters:

- Assuming the pulley is a perfect simple machine, the direction of the applied force changes but not its magnitude. The rope experiences no resistance from the pulley.
- The energy of the rope and pulley is not a concern when using the "Law of Conservation of Energy".
- The mass of the pulley need not be considered when calculating the force's impact on the body linked to the pulley.
- A number of pulleys (Fig. 4.2 (a)) carries along the connected cord is to balance and distribute the weight by considering each pulley system in the free body diagram view.
- Should draw Free Body Diagram (Fig. 4.2 (b)) for where the effort is acting first.

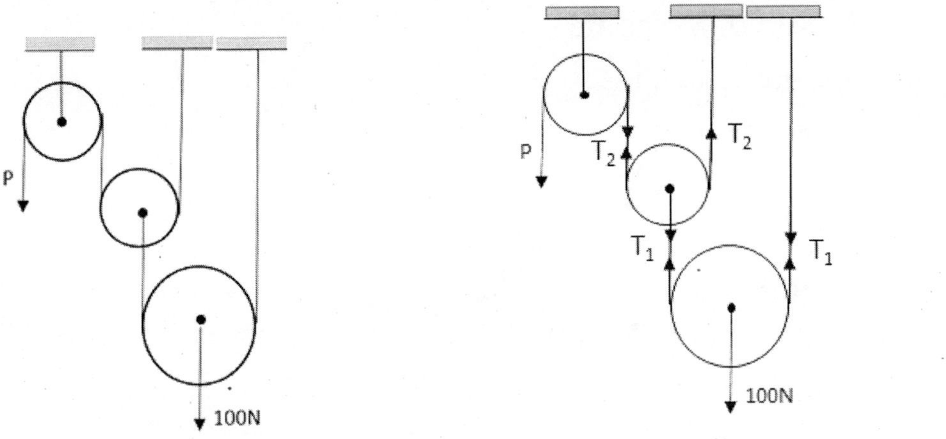

(a) (b)

Fig. 4. 2 Pulley system

Example 4.2: The spring balance and the strings as shown in Fig. E4.2 find the tension in string 1

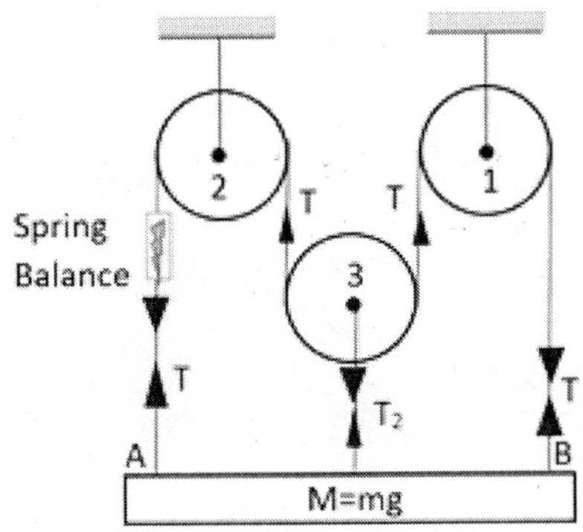

Fig. E4.2

Solution: Draw the FBD of plate and roller as depicted in Fig. E4.2 (a) and (b) respectively.

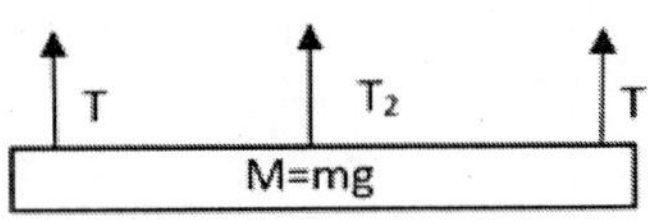

Fig. E4.2(a) FBD of plate

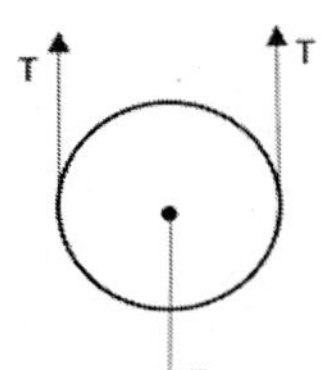

Fig. E4.2(b) FBD of roller 3

4.3.2 Method of Moments

Method of moments for any force system as follows:

- If a body or system is in static equilibrium the sum of moments of forces of any point is zero.
- Moment about any point causes rotation
- Moment of Force (M = F × shortest distance or perpendicular distance)
- If the system is in static equilibrium sum of moments, resultant moments is zero, about any point (Varignon's Principle)

Example 4.3: Find the force P, which is acting at the bottom of the plate Fig. E4.3 (a) shown below.

Solution:

At point O, horizontal (R_H) and vertical (R_v) force will occur due to hinge (Fig. E4.3 (b)). By imposing the method of moments, the all forces pass through that point O as illustrated in the Fig. E4.3 (c)

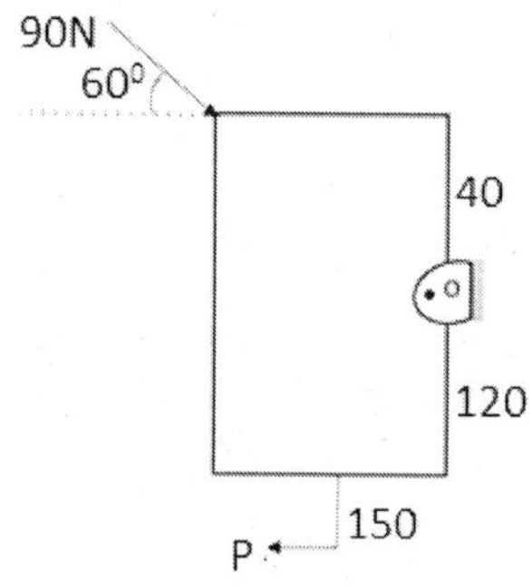

Fig. E4.3 (a)

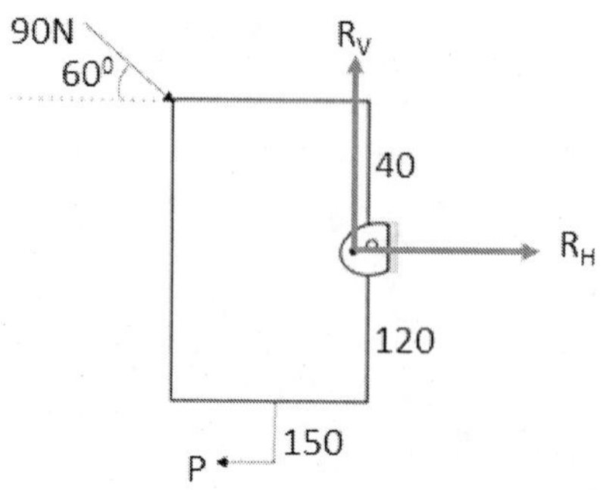

Fig. E4.3 (b)

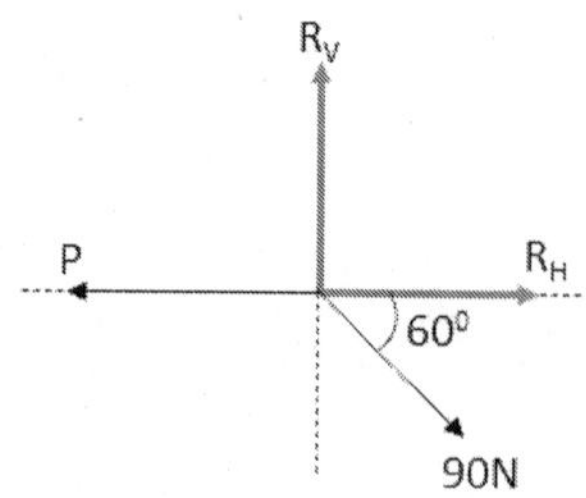

Fig. E4.3 (c)

4.3.3 Method of Joints

Method of joints are mainly used for the simple and plane trusses, if the total truss is in equilibrium, then all the joints are in equilibrium. The procedure of method joints as follows:

- Firstly, find the support reactions.
- At every point can have concurrent forces; hence two equations of equilibrium can be generated and two unknown forces can be joined.
- Consider Equilibrium of a suitable joint, which doesn't carry more than two unknowns.
- Consider all forces about that joint and find unknowns.
- Select suitable consequent joint which carries maximum two unknowns.

Example 4.4: Find the force of a tension rod of arm BC, from the given truss Fig. E4.4

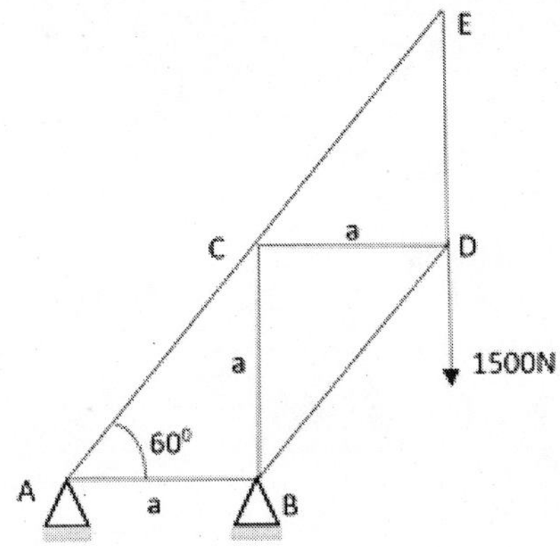

Fig. E4.4

Solution: The step-by-step procedure as follows:

- Find support reaction at point A and point B.
- Choose a point where less unknown will present.
- if any forces result in negative than arrow will be inward.
- In Fig. E4.4(a) forces are representing as tensile, if arrow shows in figure is inward then it is compressive in nature.

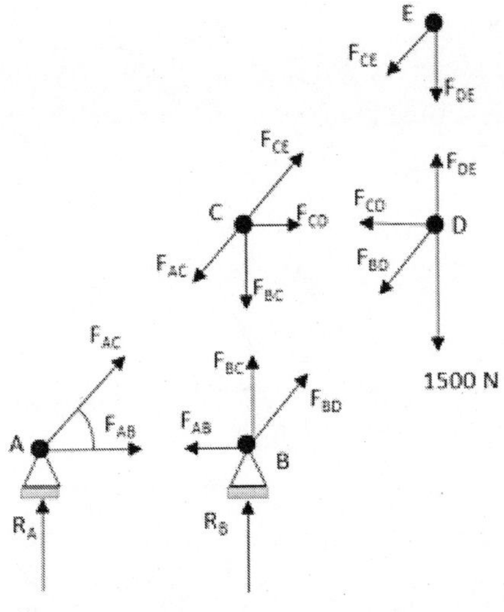

Fig. E4.4(a) FBD of truss at each point

Taking the point D (Fig. E4.4 (b)):

- Represent all forces present at point D.
- At point D, 4 forces are present F_{CD}, F_{ED}, F_{BD} and 1500N
- If a force having angle θ then it will resolve into two components.

Taking the point C (Fig. E4.4 (c)):

here we follow same steps similar as point D,

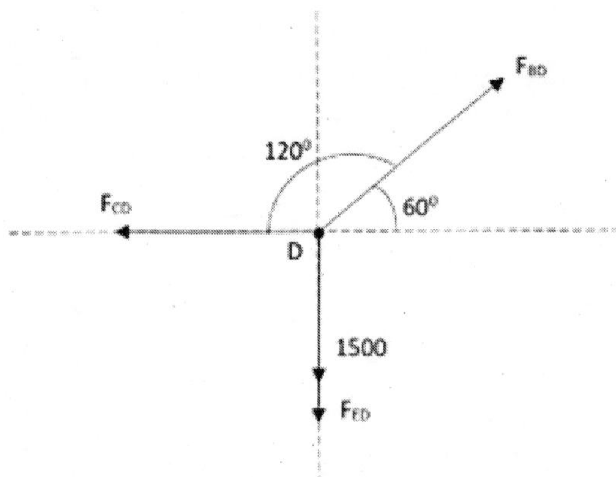

Fig. E4.4(b)

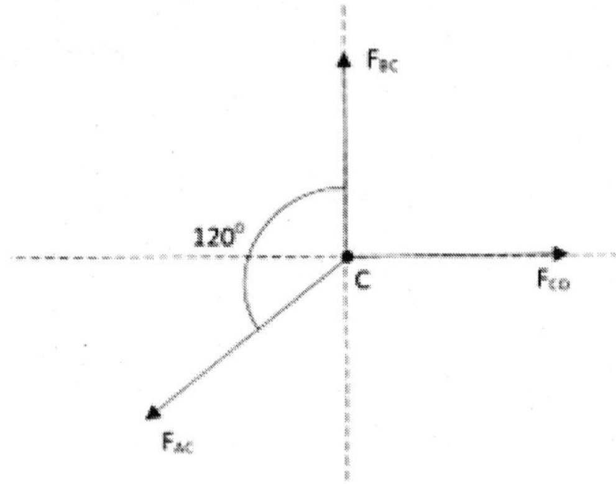

Fig. E4.4(c)

4.3.4 Method of Sections

If any truss member force is needed to be determined, it can be directly achieved by choosing the right section position, so it takes less time than the joint method. The procedure is as follows:

- Find support reactions
- Cut section into two parts
- Consider convenient as per at least one external point/Support.

While cutting Truss,

- The section plane should cut the members for which force are to be found.
- The section can cut maximum up to 3 members.
- The section plane should not pass-through joints.

Mandatory,

- Truss caries many members and it required to find forces in one or two members.
- All joints carry more than two unknown forces.

Example 4.5: Find the force of a member CD, from the given truss Fig. E4.5.

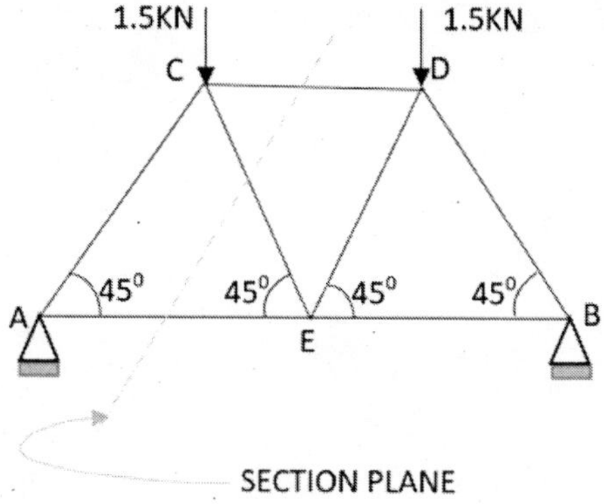

Fig. E4.5

Solution: The following steps to be followed to solve the problem.

- find all support reaction R_A and R_B.
- Cut into section as shown in Fig. E4.5 (a)

After sectioning the members with required conditions,

- Cut maximum three members at a time.
- Take a section, where a smaller number of unknowns are presents

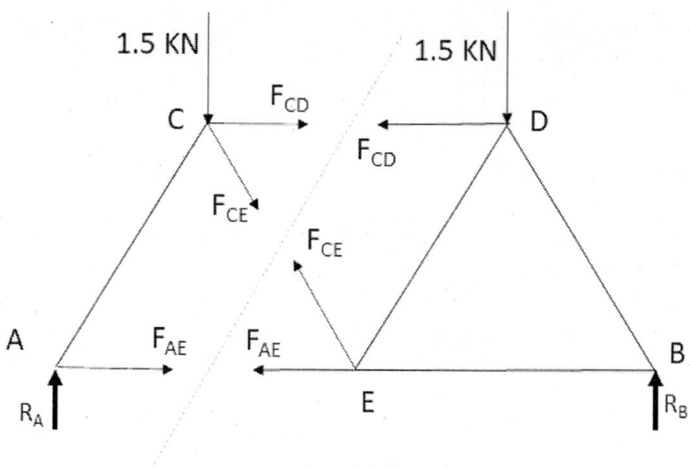

Fig. E4.5(a) Taking section of truss

Example 4.6: In the truss Fig. E4.6, identify the forces acting on each component.

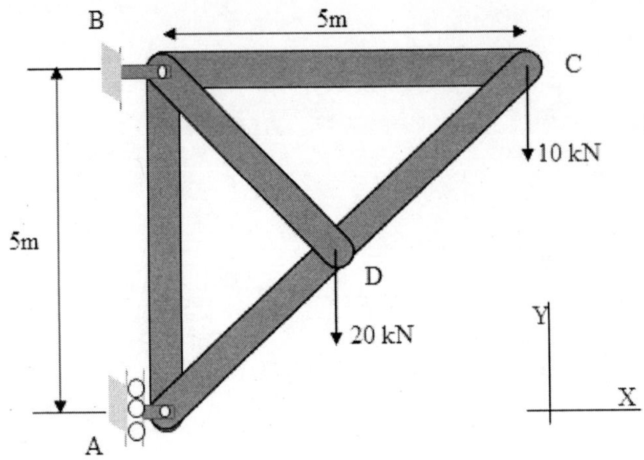

Fig. E4.6

Solution:

- Find all the support reaction R_{BX} and R_{AX}.
- Draw FBD at every point of the truss as shown in Fig. E4.6 (a).
- Member will induce either compression or tension due to load acting on it.
- If load appears in negative than load will be compressive or vice versa.

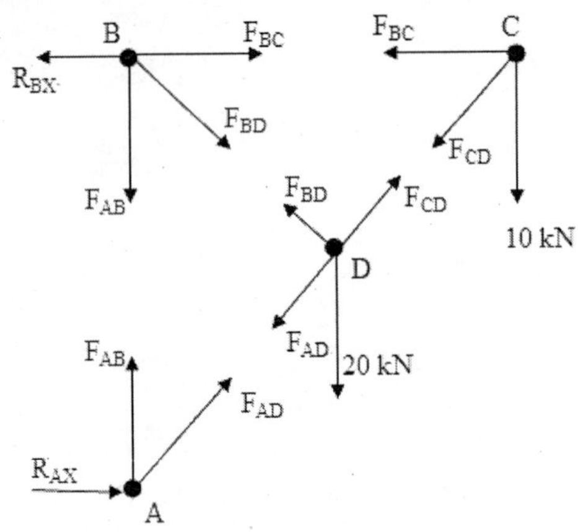

Fig. E4.6(a) FBD of truss

Example 4.7: A block of weight W is supported on a smooth inclined plane by applying the force P through a chord which passes over a smooth frictionless pulley as shown in Fig. E4.7. Draw the FBD

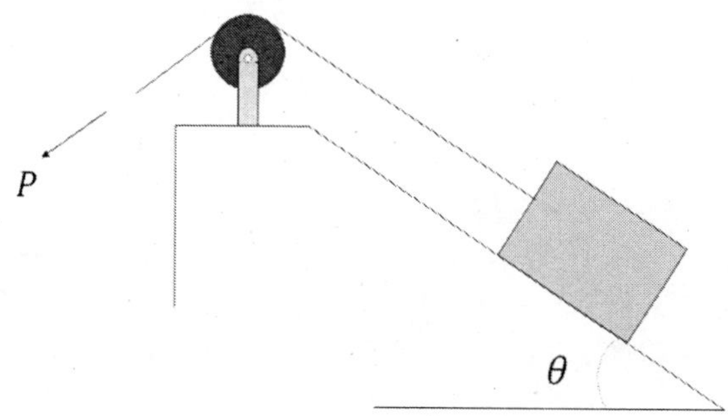

Fig. E4.7

FBD of Pulley (Fig. E4.7(a)):

- String attached to the block will induced tension in the string.
- Reaction force acting at centre of the spring.
- Normal force will act on the pulley due to string

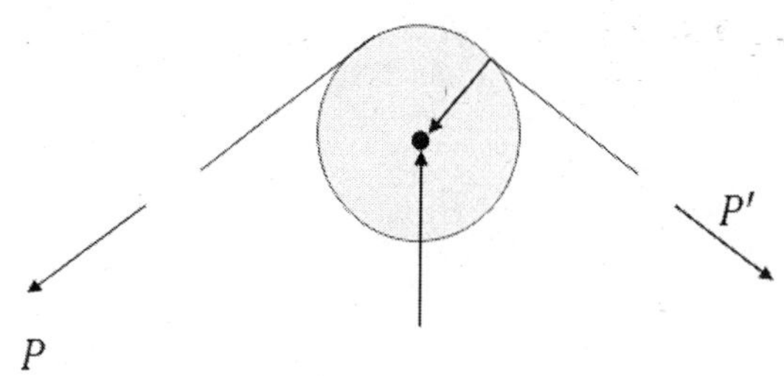

Fig. E4.7(a) FBD of Roller

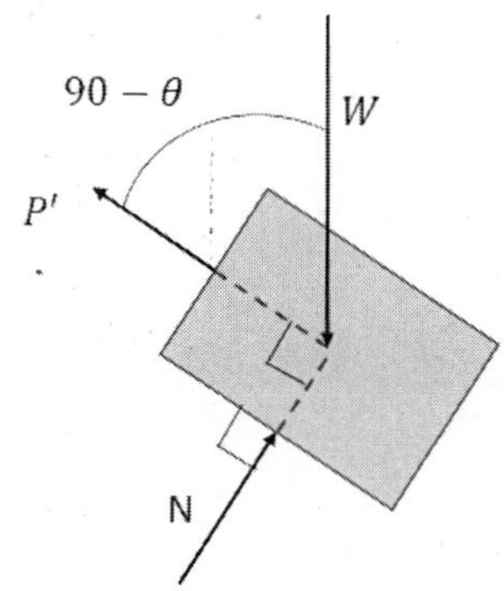

Fig. E4.7(b) FBD of Block

FBD of Block (Fig. E4.7(b)):

- Due to weight of the block downward force will act from the centre of the block.
- Normal force will act on the block
- Force P' will act due to string at angle θ.

As seen Fig. E4.7(b), the three coplanar forces (W, N and P′) acting on the same kind of body (pulling or pushing) are in equilibrium, each force acts in proportion to the sine of the angle formed by the other two forces is known as *Lami's theorem.*

By Lami's theorem, From Fig. E4.7 (b), the unknown forces can be calculated as shown below equation.

$$\frac{P'}{\sin\,(180^0+\theta)}=\frac{W}{\sin 90^0}=\frac{N}{\sin\,(90^0-\theta)}$$

Example 4.8: Draw the FBD of the roller and the beam as shown in Fig. E4.8. Assume all contact surfaces are smooth.

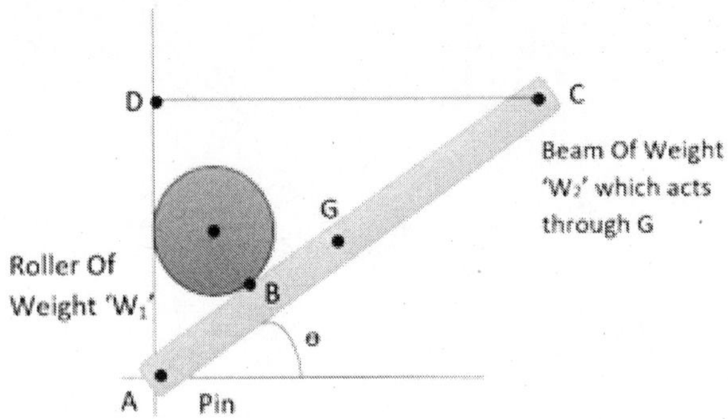

Fig. E4.8

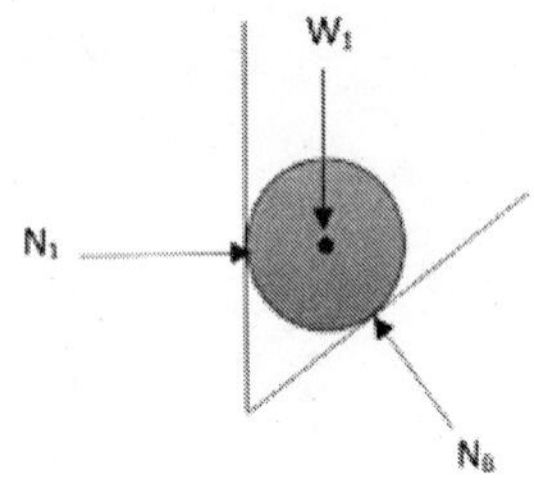

Fig. E4.8 (a) FBD of Roller

FBD of Roller (Fig. E4.8(a)):

- Since roller is resting on beam with help of support AD,
- Two Normal force will act toward centre of the roller.
- Weight of the roller W_1 will act downward from the centre.

FBD of Beam (Fig. E4.8(b)):

- at angle θ, weight of the beam W_2 will act downward of the beam.

The forces acting on the beam as shown in Fig. E4.8(c).

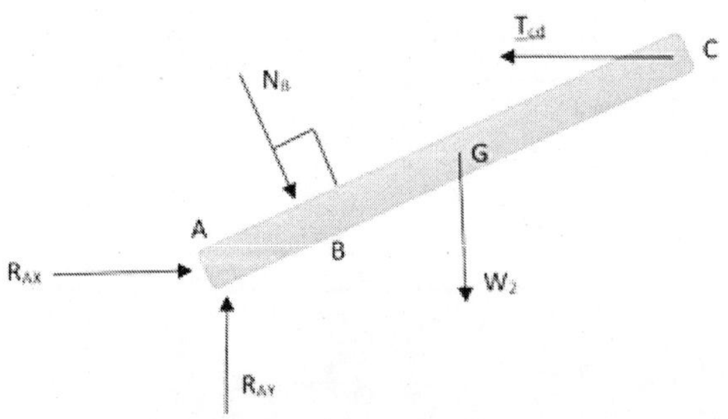

Fig. E4.8 (b) FBD of Beam

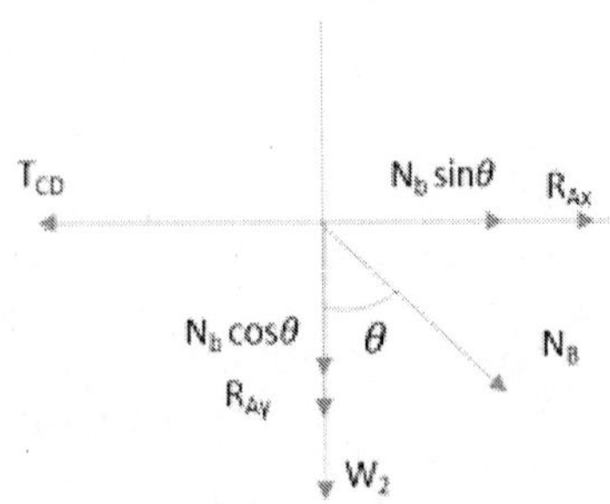

Fig. E4.8 (c)

Example 4.9: Draw the FBD of the water pipes each of weight 'W' and radius R and supported as shown in the following Fig. E4.9. Assume all contact surfaces to be smooth.

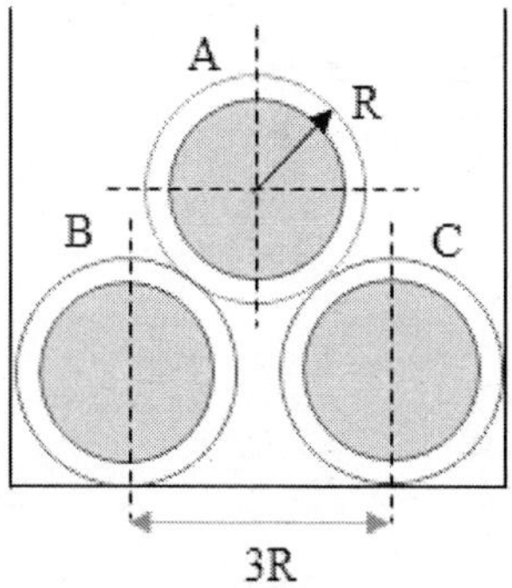

Fig. E4.9

Solution: The FBD of each water pipe as shown in Fig. E4.9 (a) to (c).

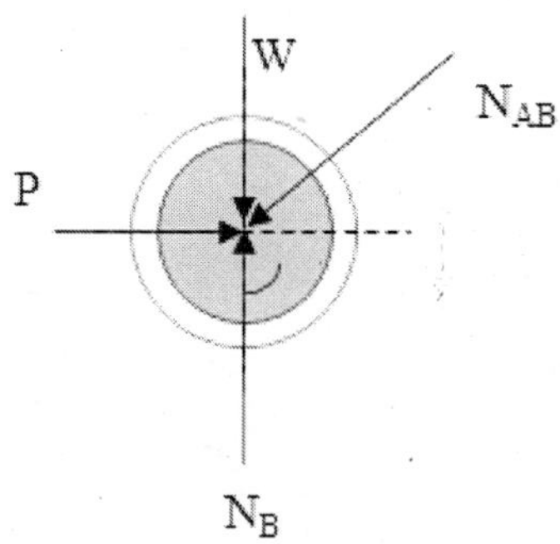

Fig. E4.9 (a)

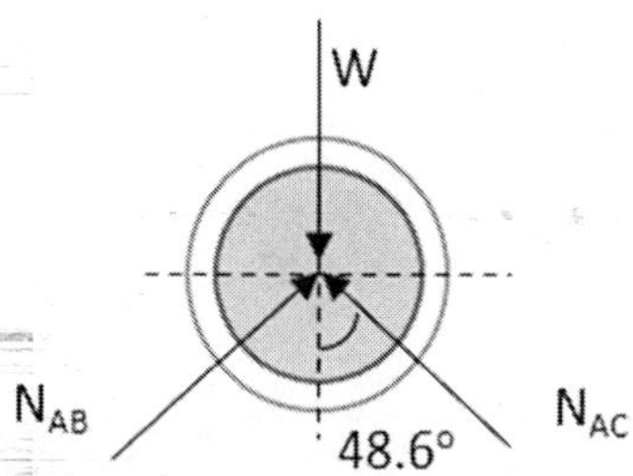

Fig. E4.9 (b)

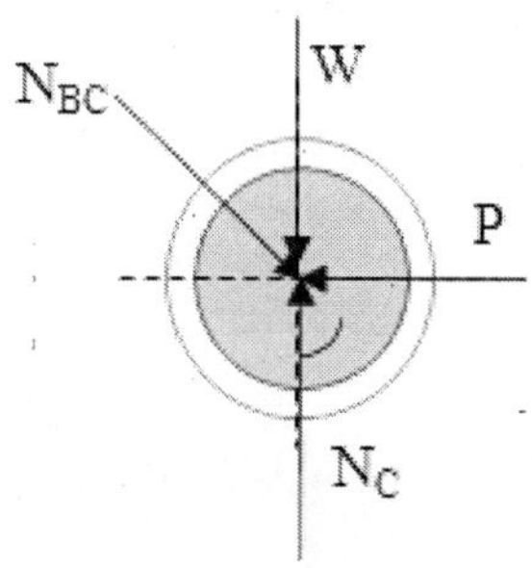

Fig. E4.9 (c)

4.4 KINETIC CONSTRAINTS

A force opposing the motion arises at the contact surfaces whenever one body moves over another body or has a tendency to do so. Tangential force is produced between contacting surfaces whenever there is a tendency for one contacting surface to slide along another one. Frictional force, or simply friction, is the term used to describe this force that opposes movement or the tendency to move. When the unlubricated surfaces of two materials come into contact while sliding or showing a tendency to slide, dry friction occurs.

As seen in Fig. 4.3, picture a weight block W that is resting on a flat surface. It is true that the contacting surface is quite rough. Let P denote the horizontal force that is being applied; it will continually range between zero and a force large enough to simply move the block and then maintain motion. The block's free body diagram displays both "active forces (such as applied force P and block weight W) and reactive forces (i.e., normal reaction N and tangential frictional force F)." The unique characteristic of frictional force F is that it can adapt itself to a magnitude that is equal to the applied force P until the "limiting equilibrium condition."

As the applied force P grows, the frictional force F grows in size and moves in the opposite direction. There is a point, though, beyond which the frictional force's size cannot be increased. One body will slide over the other if the applied force exceeds the maximum frictional force. As soon as the body starts moving, the frictional force F decreases from its greatest value under static conditions. Kinetic or dynamic friction F_K is the term used to describe the frictional force between the two surfaces when the body is moving.

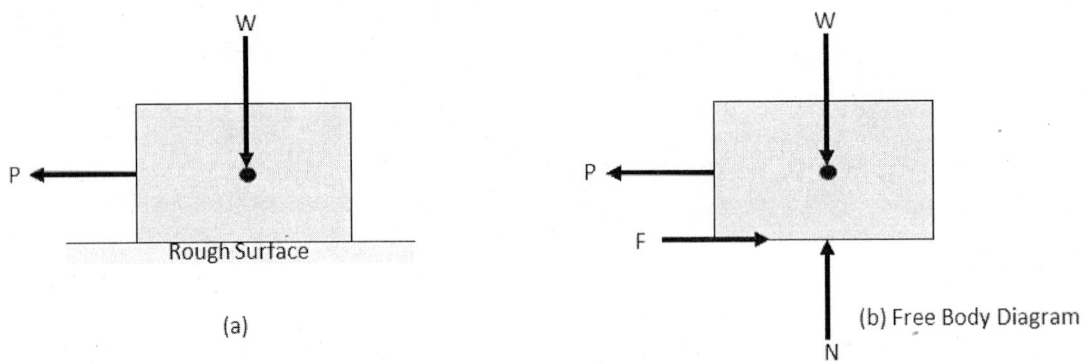

Fig. 4. 3 Friction mechanism

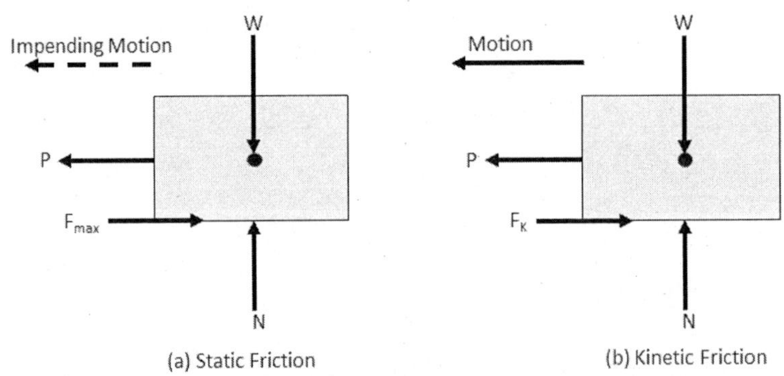

Fig. 4. 4 Static and kinetic friction

When a block is about to move (impending motion), the greatest frictional force that can be created (Fig. 4.4(a)) at the surface is known as the limiting frictional force (F_{max}). It has been demonstrated through experimentation that "limiting frictional force" is directly related to normal reaction.

$$F_{max} \propto N$$

Limiting frictional force to normal reaction ratio (F_{max}/N) is constant. The "static friction coefficient" is the name of this constant (μ_S).

$$\mu_S = \frac{F_{max}}{N}$$

if the body is moving (Fig. 4.4(b)),

$$F_K \propto N$$

The proportion of normal reaction to kinetic frictional force (F_K/N) is constant. The name of this constant is "kinetic friction coefficient" (μ_K). When compared to "limiting friction", "kinetic friction" is always less.

$$\mu_K = \frac{F_K}{N}$$

Types of kinetics problem

1. The kinematic conditions for the issue are known, and force and moment must be determined.

The magnitudes of force and moment are known. There are kinematic conditions to be discovered.

Example 4.10: Draw the FBD of the block shown Fig. E4.10

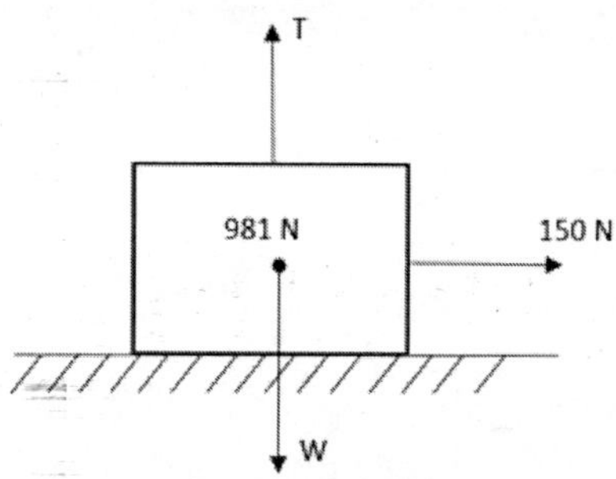

Fig. E4.10

Solution: FBD of block as shown in Fig. E4.10 (a)

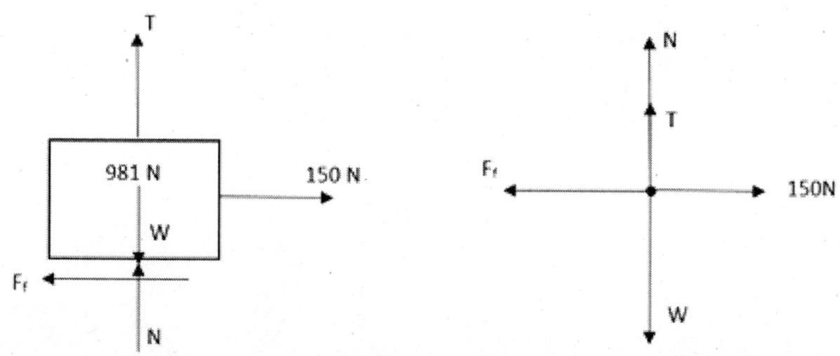

Fig. E4.10(a)

Example 4.11: Draw the FBD of the Block shown below Fig. E4.11.

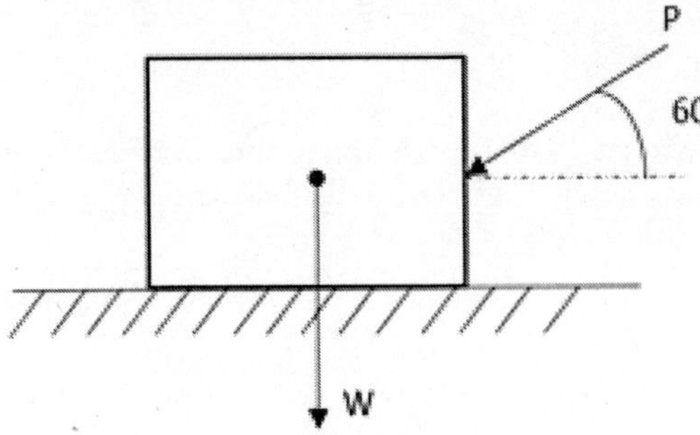

Fig. E4.11

Solution: FBD of block as shown in Fig. E4.11(a) and (b).

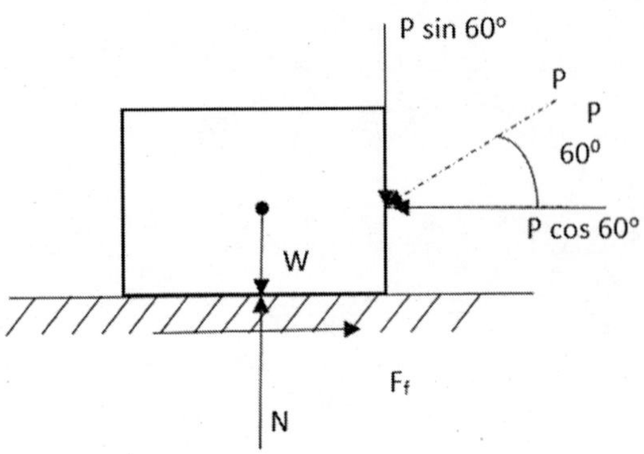

Fig. E4.11(a)

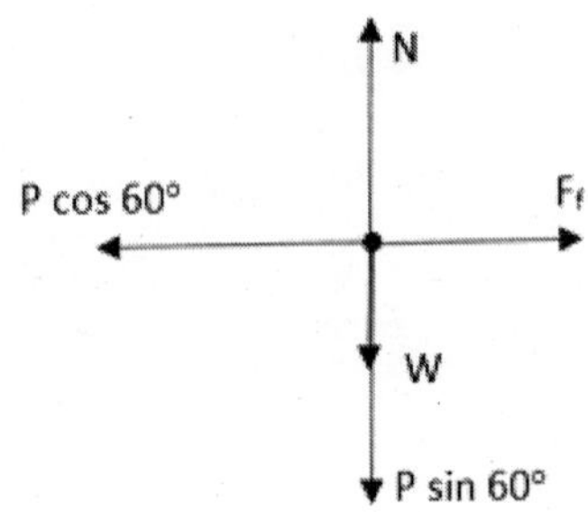

Fig. E4.11(b)

Example 4.12: Draw the FBD of the blocks in the given kinetics motion of the bodies as shown in Fig. E4.12.

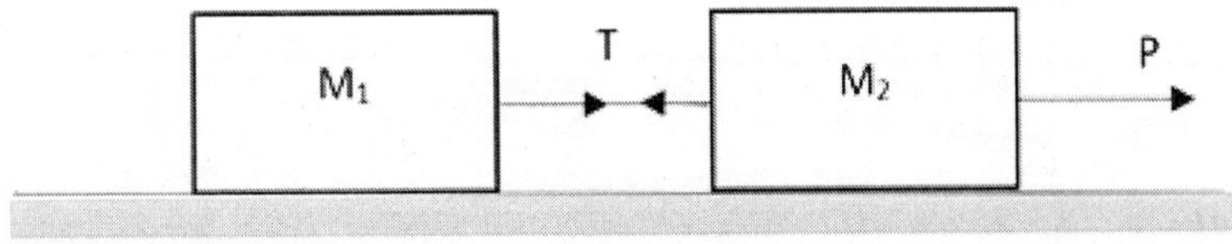

Fig. E4.12

Solution: FBD of block 1 (Fig. E4.12(a)):

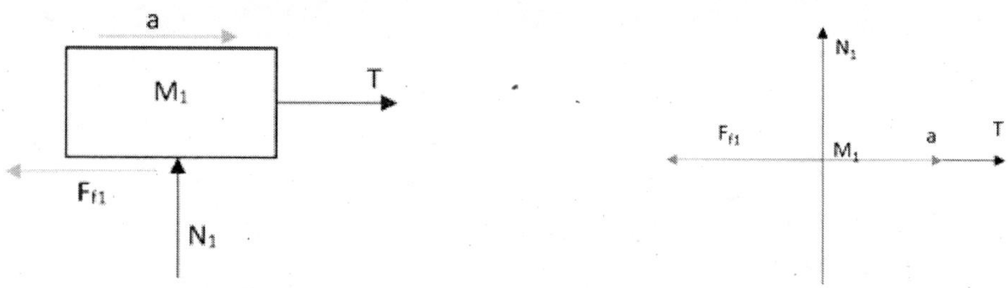

Fig. E4.12(a)

FBD of block 2 (Fig. E4.12(b)):

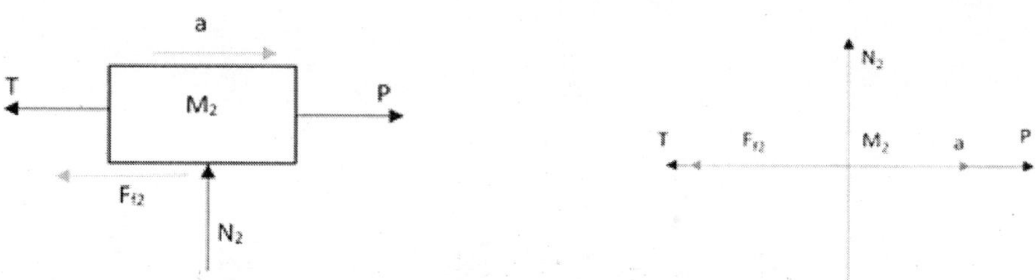

Fig. E4.12(b)

Example 4.13: Draw the FBD for finding the reactions at the point A for the given Fig. E4.13 shown below.

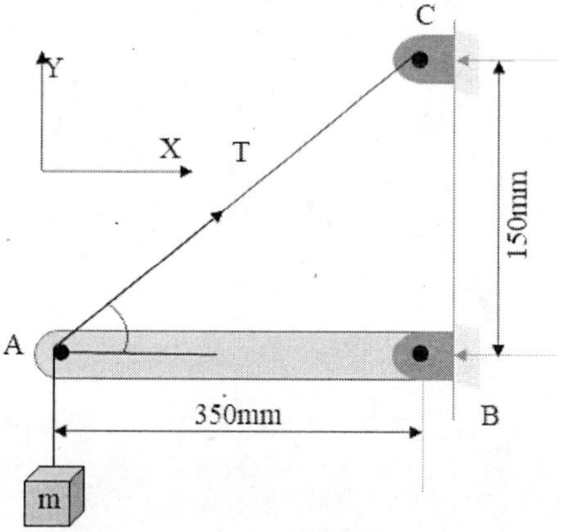

Fig. E4.13

Solution: FBD at A is given in the Fig. E4.13 (a)

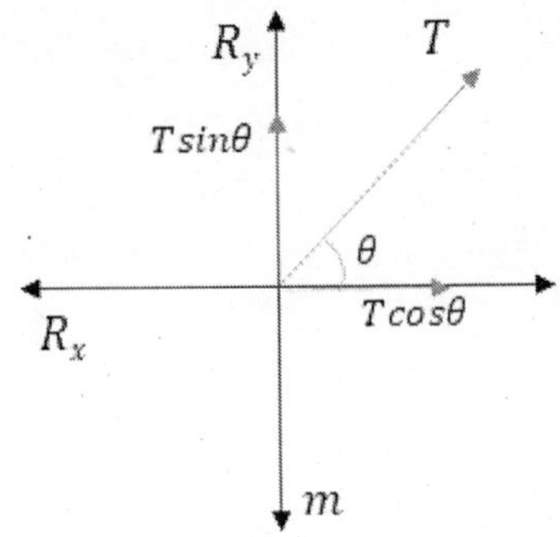

Fig. E4.13 (a)

Example 4.14: Draw the FBD of the blocks which are in kinetic motion as shown Fig. E4.14.

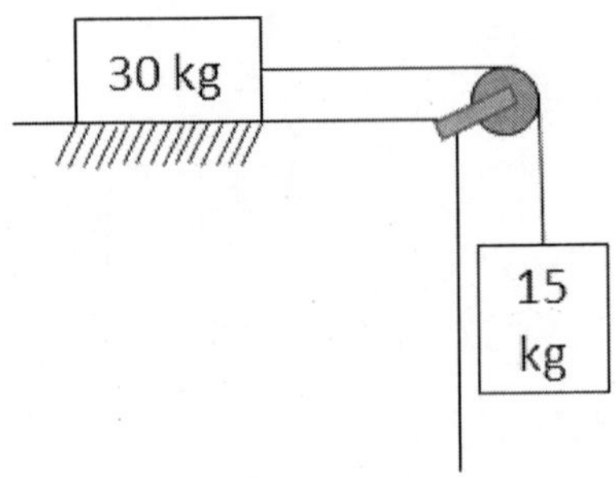

Fig. E4.14

Solution: FBD of Fig. E4.14 (a) and (b) as shown in Fig. E4.14 (a) & (b)

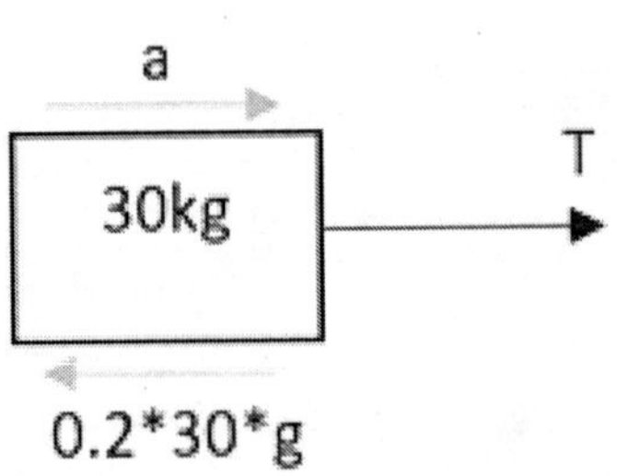

Fig. E4.14(a) FBD of 30kg Block:

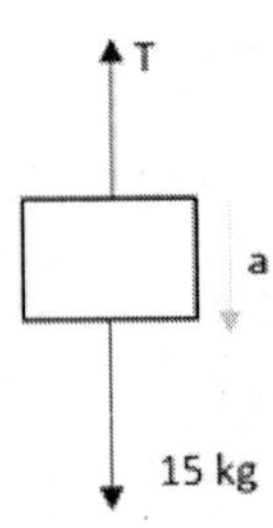

Fig. E4.14 (b) FBD of 15 kg Block

Example 4.15: A horizontal surface with a kinetic friction coefficient of $\mu_k = 0.3$ supports the 60 kg crate in Fig. E4.15 Calculate the velocity of the container in 3 seconds if it is being towed by a 500 N force as illustrated.

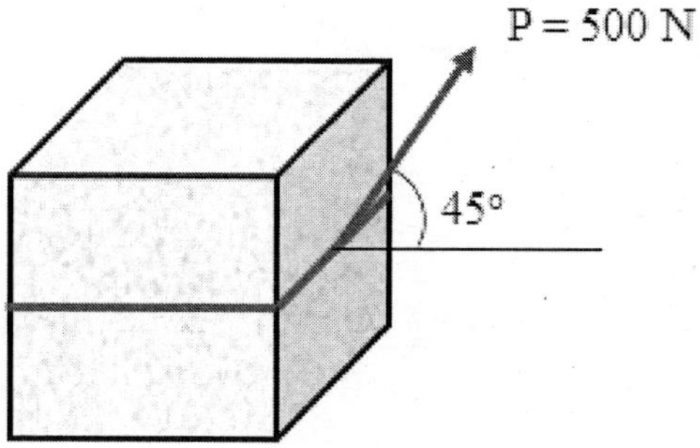

Fig. E4.15

Solution: FBD of the block as shown in Fig. E4.15(a), $\overset{+}{\rightarrow} \Sigma F_x = ma_x$;

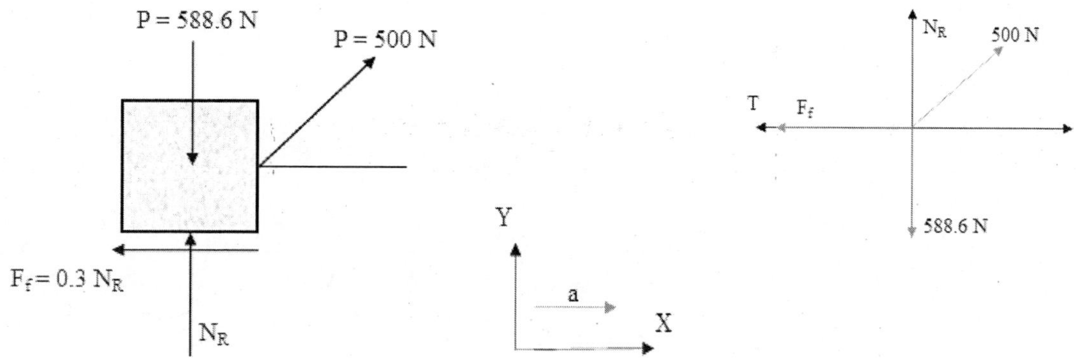

Fig. E4.15(a) FBD of block

$+\uparrow \Sigma F_y = ma_y$;

Draw FBD of the block as shown in Fig. E4.15(b).

As a result of the constant applied force P, observe that the acceleration is constant. The starting velocity of crate is zero, hence its speed after three seconds equals.

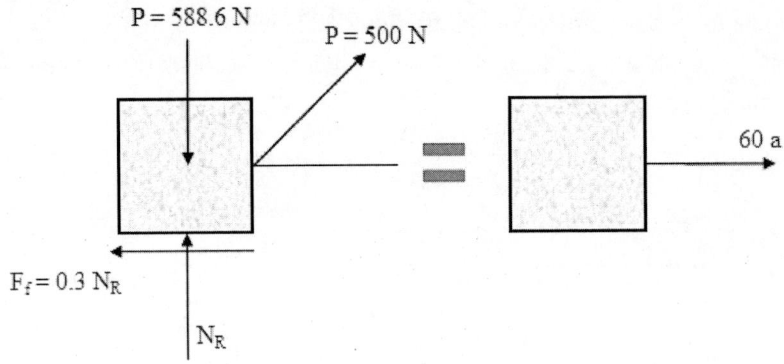

Fig. E4.15(b)

Example 4.16. If the driver of a car travelling down a straight, level road suddenly applies the brakes, causing the car to slide for 2 sec and travel 10m before coming to a halt, what would be the coefficient of friction between the wheels and the road?

Solution: Draw the FBD of the automobile as shown in Fig. E4.16

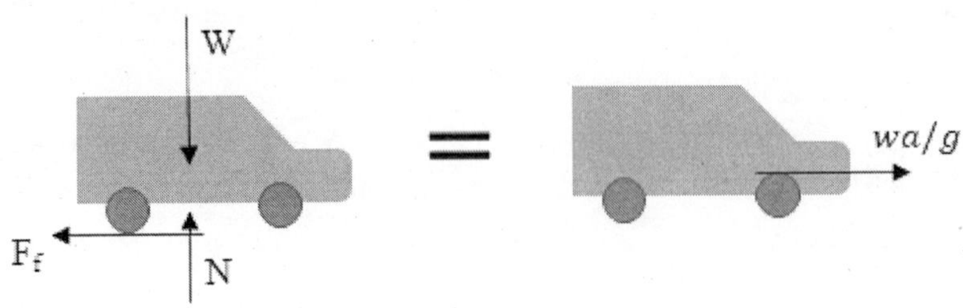

Fig. E4.16

Example 4.17: An initial velocity of $v_a = 5\ ms^{-1}$, a block P weighing 250N moves 8 m down the ramp (Fig. E4.17). What point will it hit the ground at C and how long would it take to go from A to C if μ_k for the block and the ramp are both 0.25?

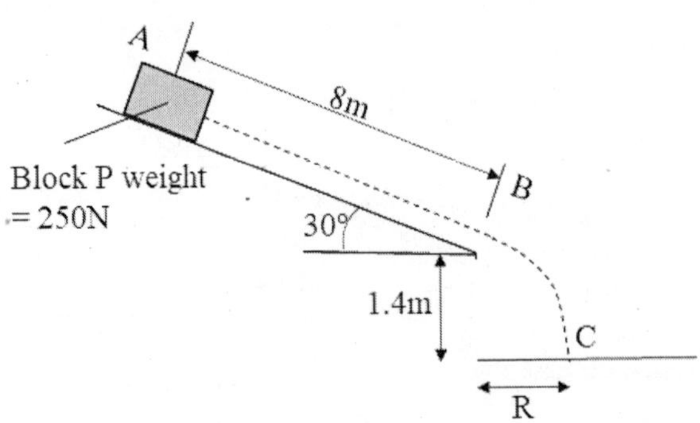

Fig. E4.17

Solution: Draw the FBD (Fig. E4.17(a))

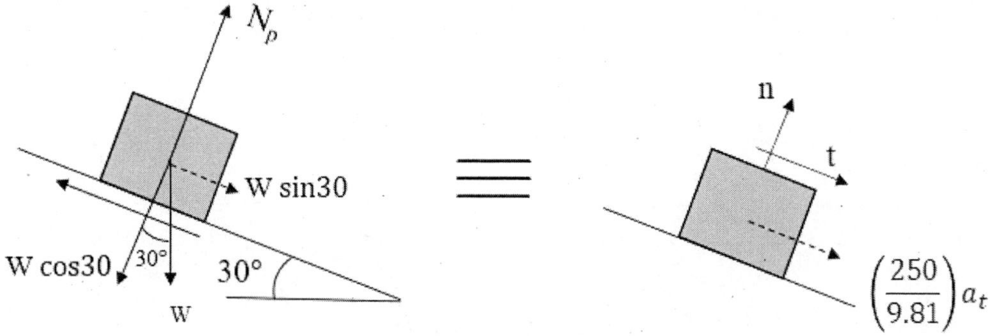

Fig. E4.17 (a)

$$\Sigma F_n = 0; \Rightarrow N_P = 250 \cos 30° = 216.5 \text{ N}$$

$$\Rightarrow F = 0.25 \text{ } N_P = 54.125 \text{ N}$$

$$\Rightarrow \Sigma F_t = ma_t; \Rightarrow 250 \sin 30° - 54.125 = \left(\frac{250}{9.81}\right) a_t$$

$$\Rightarrow a_t = 2.78 \text{ ms}^{-2}$$

$$\Rightarrow v_B^2 = v_A^2 + 2a_t s; \; v_B = 8.33\text{ms}_{-1}$$

Time taken by P to reach B from A = t_{ab}

$$v_B = v_A + a_t t_{AB} \Rightarrow t_{AB} = \frac{v_B - v_A}{a_t}$$

$$t_{AB} = 1.197s$$

Solve for the downward motion from B to C

Example 4.18: Find the acceleration of block B (Fig. E4.18), if block A is subjected to force P = 80 N. Avoid friction. The corresponding weight of block A and block B are 40 N and 50 N, respectively.

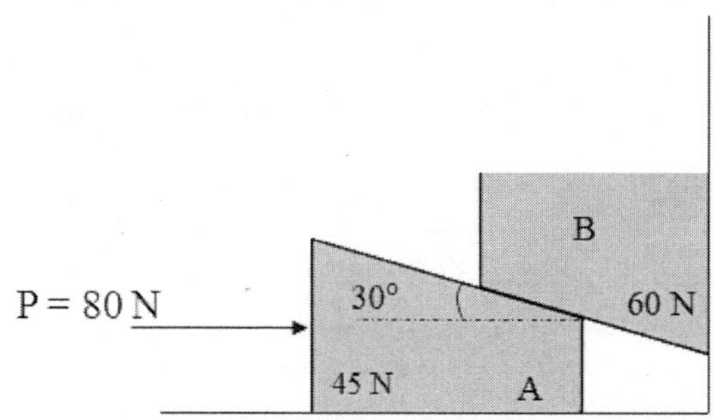

Fig. E4.18

Solution: Draw the FBD of block A (Fig. E4.18(a)) and block B (Fig, E4.18(b))

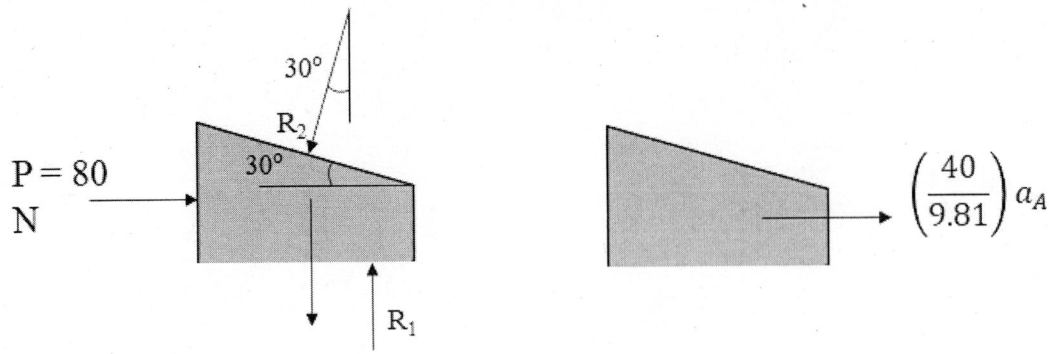

Fig. E4.18(a)

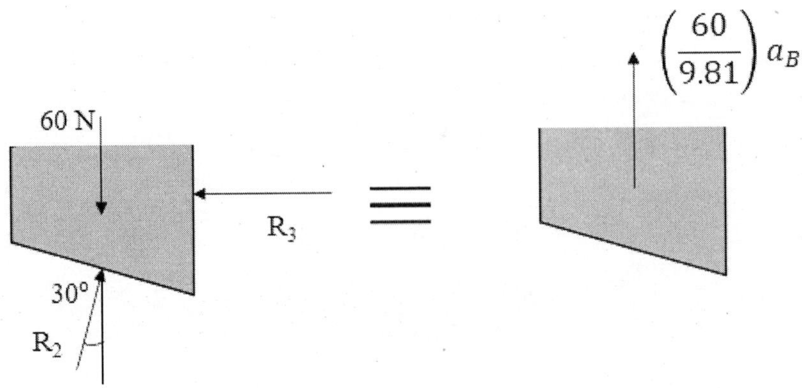

Fig. E4.18(b)

Example 4.19: The rounded guide rod and arm AB allow the smooth, 0.5-kg double-collar in Fig. 4.19 to slide smoothly. Calculate the force the arm applies to the collar at the instant that $\theta = 45°$ if the arm rotates at a constant angular velocity of 3 rad/s. Horizontal plane motion is present.

Solution: Draw the FBD (Fig. E4.19(a))

- The circular guide rod's normal reaction N_c and the arm's force F act on the collar in the plane of motion.
- Keep in mind that F acts perpendicular to arm AB's axis, or in the direction of the θ axis,
- Nc behaves perpendicular to the circular path's tangent at $\theta = 45°$ angle $\theta = 45°$. The four unknowns are N_C, F, a_r , a_θ.

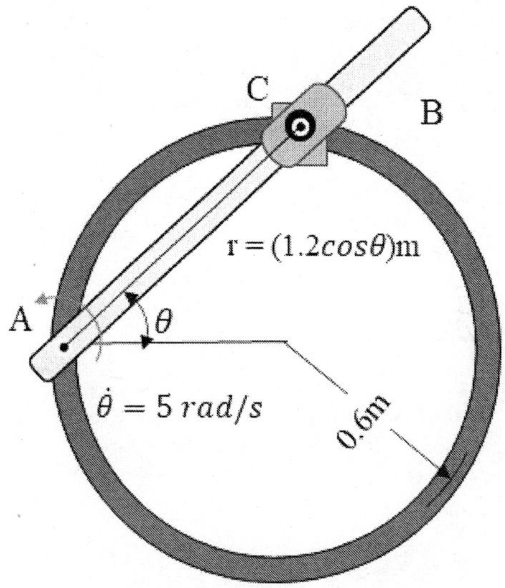

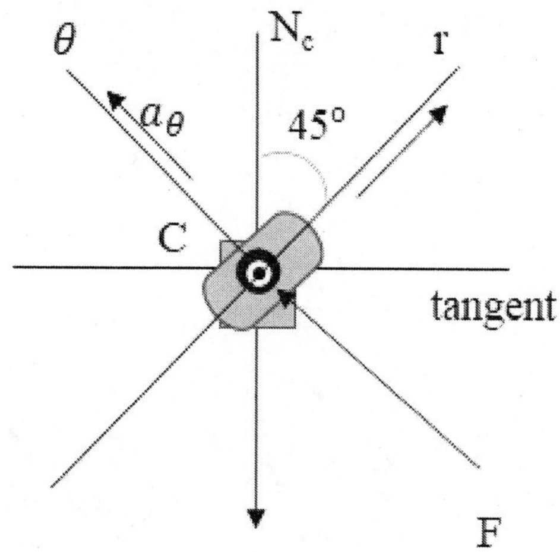

Fig. E4.19 **Fig. E4.19(a)**

Example 4.20: The block (Fig. E4.20) on a rotating platform has a mass of 75 kg. Due to rotation, his speed is increased at the rate of $\dot{v} = 0.5\ m\ s^{-2}$. If $\mu_s = 0.3$, find the time required to cause block to slip assume that platform starts at rest.

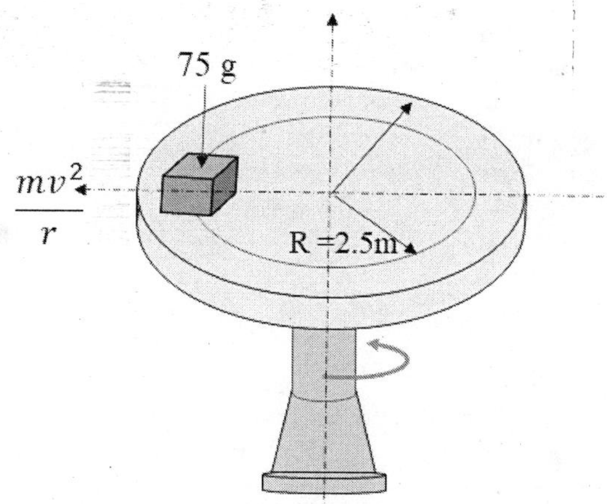

Fig. E4.20

Solution: Draw the FBD as Fig. E4.20 (a)

Normal reaction offered by the platform acting in the block = R = 75g N

Friction force = F = $\mu_s R$ = $(0.3)(75)(9.81)N = 220.725$ N

By dynamic equilibrium =

$$(ma_t)^2 + \left(\frac{mv^2}{r}\right)^2 - (\mu_s R)^2 = 0$$

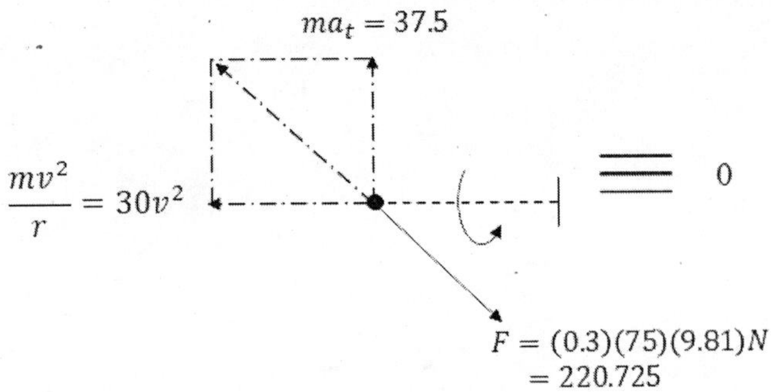

Fig. E4.20(a)

UNIT SUMMARY

In this unit, acquire the knowledge regarding the free body diagrams using various methods like method of Projections, Joints, Sections and moments. Focusing on the certain system and point of the body and analysing the forces acting on that point in both kinematic and kinetic parameters and to find out the resultant and force of the members and to have the general bodies to will have the max mode of the efficient rate to find out the forces.

EXERCISES

Multiple Choice Questions:

1. The force exerted by the ________on the ________is the cart propels the horse forward when it pulls a horse-drawn carriage.

a) cart, horse b) ground, horse

c) ground, cart d) horse, ground

2. The resultant of two forces, one of which is twice as strong as the other and weighs 260 N. The outcome is decreased to 180N if the direction of the stronger force is reversed while the other remains unchanged. The forces' strength and the angle at which they interact are ________

a) 150N, 300N, 70^0 b) 50N, 100N, 40^0

c) 100N, 200N, 64^0 d) 70N, 140N, 55^0

3. A tripod system carrying a camera is an example of force system representing,
   a) Collinear Concurrent forces
   b) Coplanar Concurrent forces
   c) Non-Coplanar Concurrent forces
   d) Non-Coplanar Non-Concurrent forces

4. R is the resultant of the two forces P and Q. The new result is perpendicular to P if Q is doubled, therefore,

a) P = R b) Q = R c) P = Q d) Q = 2R

5. The body is subjected to only two forces, then the requirement for it to be in static Equilibrium is ______.
   a) the force must be equal in magnitude
   b) the force must be equal in magnitude and opposite in direction
   c) the force must be collinear, equal in magnitude and opposite in direction
   d) the force must be equal in magnitude and perpendicular to each other
6. The tensions T_{ab} and T_{ac} have the following values as shown in Fig. P4.1 if point A is in equilibrium with the applied forces.
   a) 250 N and 433 N
   b) 433 N and 250 N
   c) 500 N and 250 N
   d) 250 N and 500 N

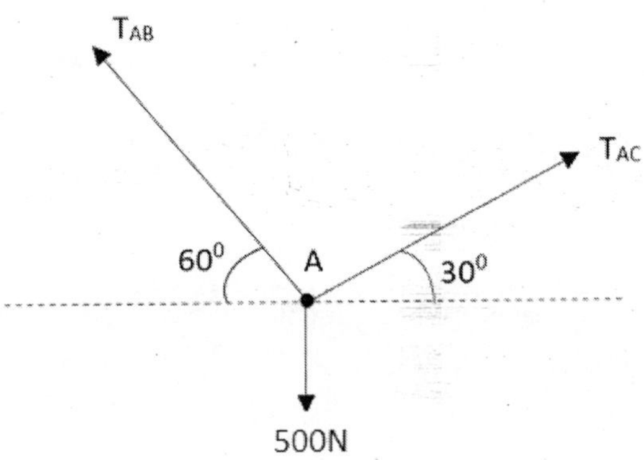

Fig. P4.1

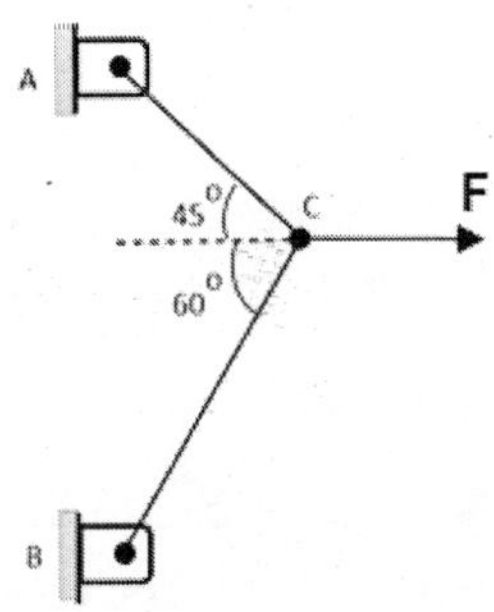

Fig. P4.2

7. A horizontal force F is applied to two steel truss members as shown in Fig. P4.2. Hinges are used in every joint. The vertical reaction force generated at point B in kN is of the magnitude F = 2kN.

a) 1.268 b) 0.328

c) 1.208 d)1.468

8. The greatest force F in kN as shown in Fig. P4.2 that can be used at C such that any truss member's axial stress does not exceed 150MPa. AC and BC, each of which has a 150 mm^2 cross sectional area. Hinges are used in every joint.

a) 2 b)12.5
c)11.15 d) 25

9. At the hinge support, there are ___________ response components.

a) Three b) Two c) One d) Zero

10. A cable is always represented by a ___________ force in FBD.

a) tensile b) compressive c) normal d) spring

Answers of Multiple-Choice Questions

1(d) 2(c) 3(c) 4(b) 5(c) 6(b) 7(a) 8(d) 9(b) 10(a)

Short and Long Answer Type Questions

1. With examples, describe the free body diagram.
2. A free-body diagram FBD is used for what purpose?
3. What explains the free-body diagram's basic theory?
4. Which forces are acceptable for a free-body diagram?
5. A free-body diagram has how many drawn vectors?
6. Explain FBD and the role it plays in problem analysis.
7. Describe the various support types and list the unknown effects they have.
8. What are the many kinds of loads?

Numerical Questions

1. To hold the container in the position illustrated in the Fig. P4.3, identify the force the person applied to the rope. The container weighs 350 kg.
2. A 50 kg mass is suspended from a bar that is not weighted and is held up by a cable CB and a pin at A, as depicted in Fig. P4.4. On the bar AB, the pin reactions are at A.

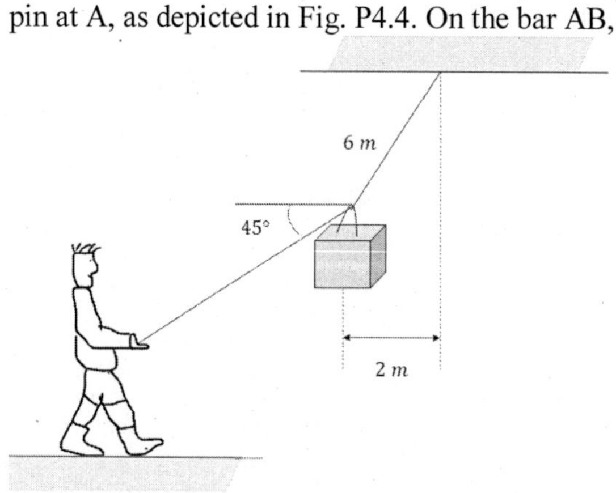

Fig. P4.3

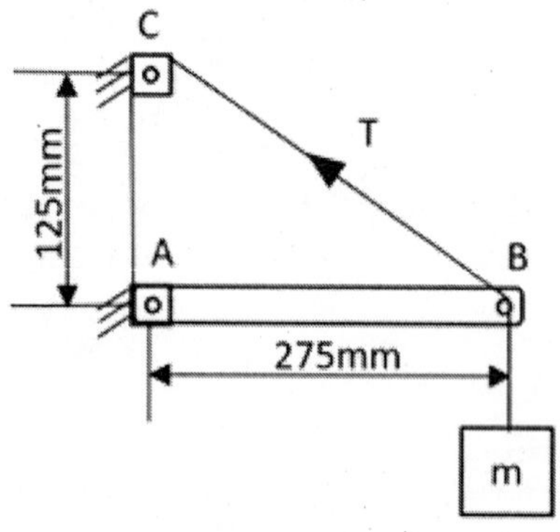

Fig. P4.4

3. The illustration depicts in Fig. P4.5 a pin-jointed plane truss that has a 100kg mass suspended at point M to load the structure. Determine the force is applied to the truss member LN.
4. Think about the truss PQR in the Fig. P4.6, which is loaded at P with a force F. Ascertain the member QR's tension.

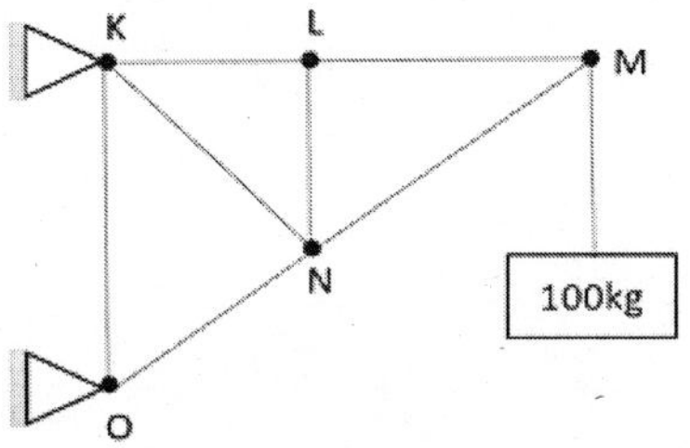

Fig. P4.5

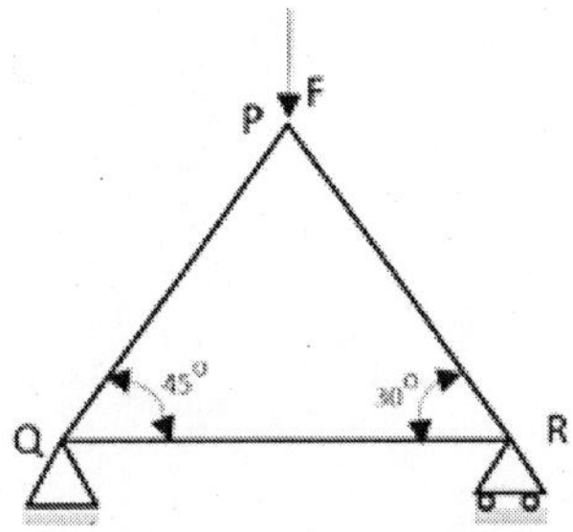

Fig. P4.6

5. As seen in the Fig. P4.7, a sphere with a weight of Q = 445N is at rest in a trough. if all surfaces are absolutely smooth, find the forces acting at A and B on the trough's sides.
6. As seen in the illustration in Fig. P4.8, two cables support a weight "W." Find the value of Θ, if there will be the least amount of tension in the cable when it forms the angle Θ.
7. A string is used to suspend a hard ball that weighs 100 N. A horizontal force F pulls on the ball in such a way that the string produces a 30° angle with the vertical (Fig. P4.9). Find the strength of force F.
8. A vertical wall and an inclined plane support two identical rollers, each weighing Q = 445N, as illustrated in the Fig. P4.10. Assuming smooth surfaces, it was established what response was present at support point A.

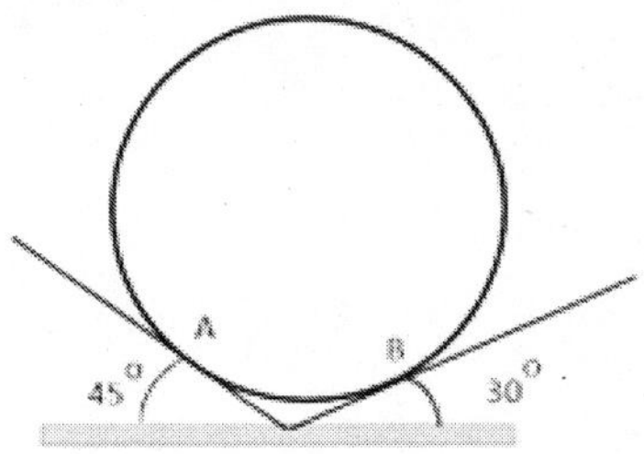

Fig. P4.7

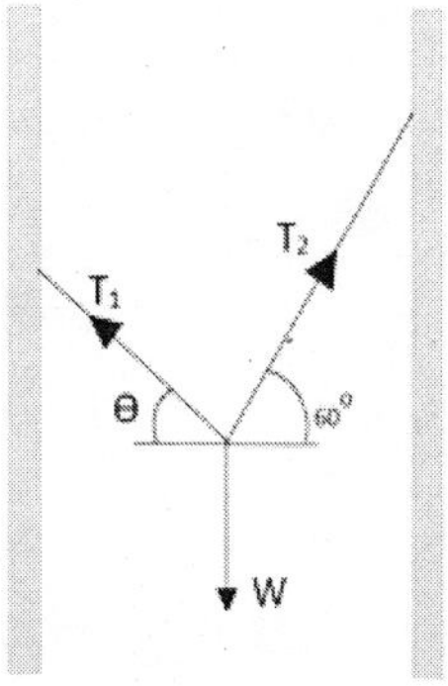

Fig. P4.8

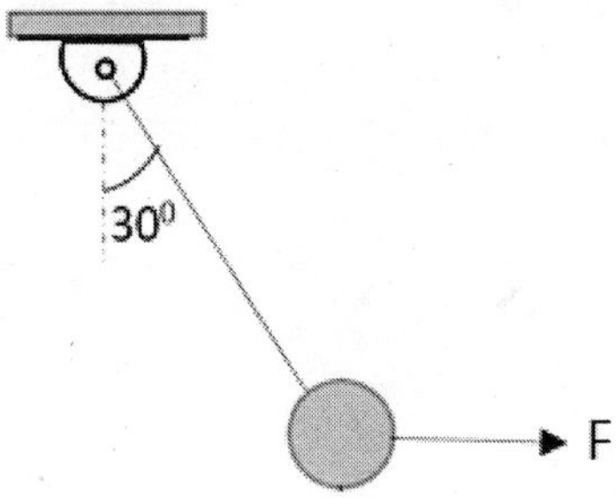

Fig. P4.9

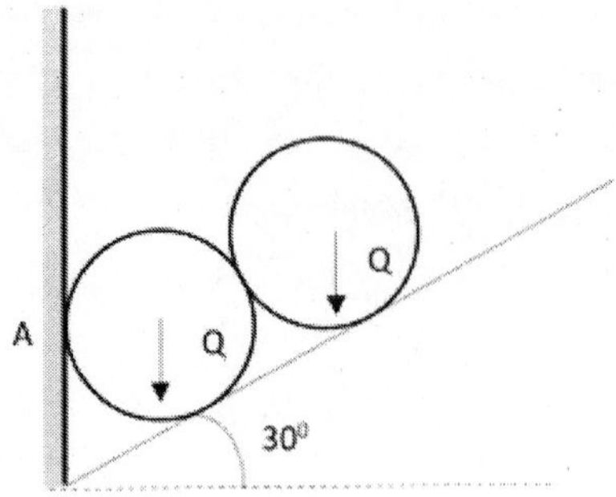

Fig. P4.10

9. If the friction coefficients for the 40 kg block A and the 120 kg cart B are equal to 0.50 (Fig. P4.11), calculate the acceleration of each component If P = 80 N and P = 65 N.

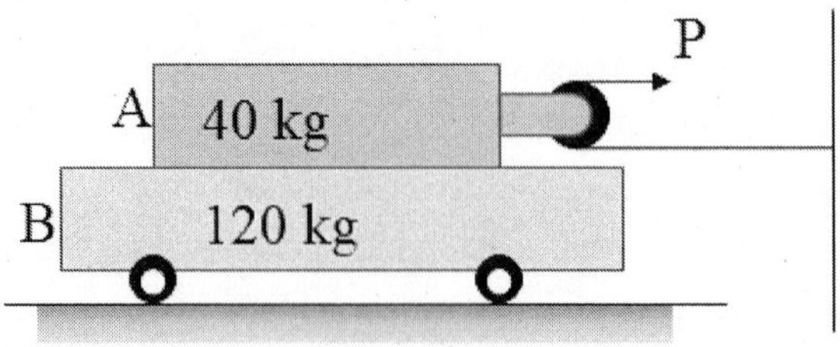

Fig. P4.11

10. The beam AB is pivoted at A and carried by a cable that crosses a frictionless pulley in the illustration P4.12. Determine the distance x from point A at which a load Q must be placed to maintain the balance of the beam. Disregard the weight of the beam.

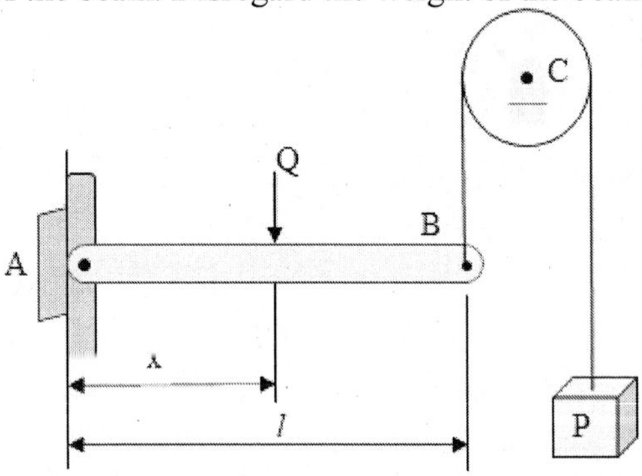

Fig. P4.12

11. The 150 kg load B is raised by the tractor using a 25-m-long cable, Boom, and pulley system (Fig. P4.13). Find the "tension" in the cable at the moment $S_A = 6$ m if the tractor is moving to the right at a speed of 3m/s and undergoing a 5 m/s^2 acceleration. Whenever S_A=0, S_B=0.

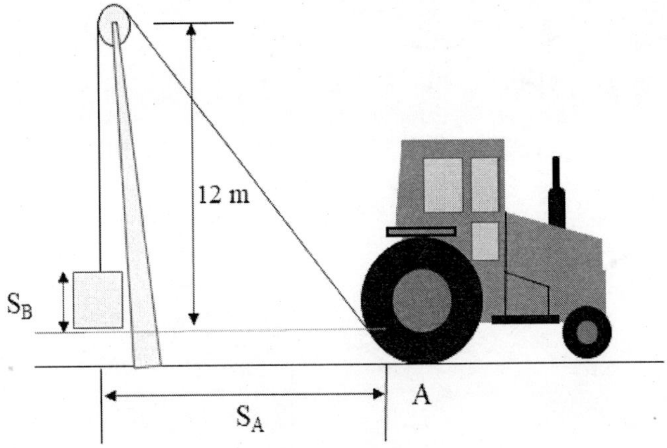

Fig. P4.13

12. A system's acceleration and each cable's tension should be determined. The horizontal surface and block C have a smooth inclined plane as shown in Fig. P4.14 and a coefficient of kinetic friction of $(\mu_k)_c = 0.2$.

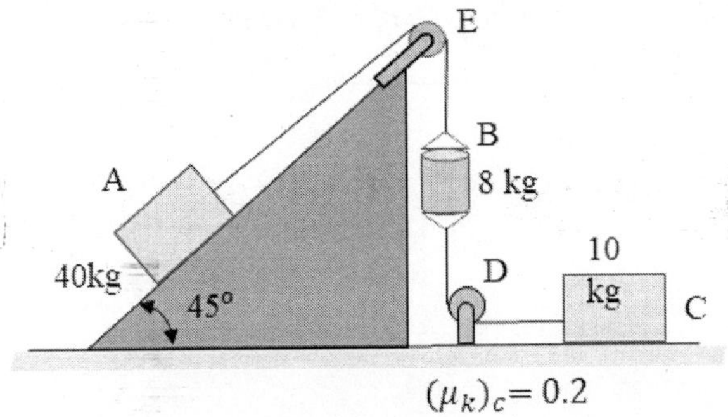

Fig. P4.14

13. Neglect the mass of the box and cable determine the angular velocity θ of the vertical shaft if $\phi = 30°$ (Fig. P4.15).

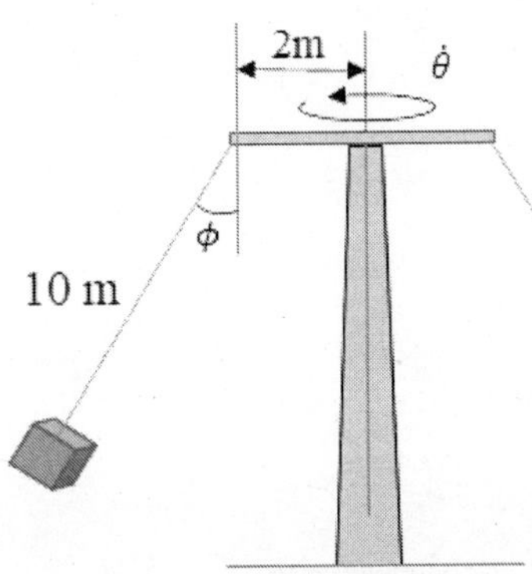

Fig. P4.15

14. Find the highest continuous rotation velocity of the platform that will prevent the block (Fig. P4.16) from slipping if the "static friction coefficient" between the mass m block and the turntable is μ_s.

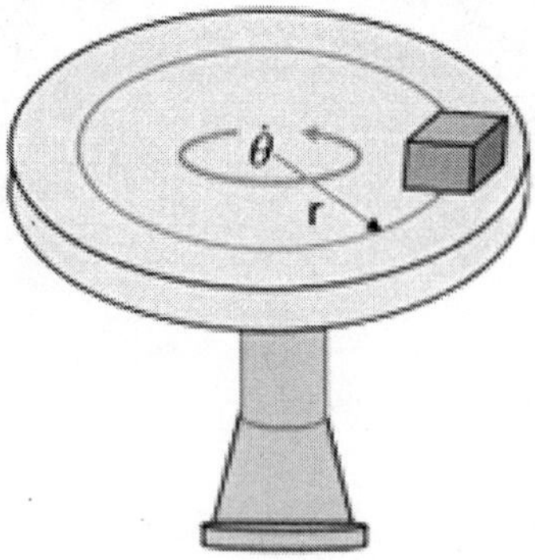

Fig. P4.16

15. The hinge at O and the roller at A hold the uniform 30-kg bar OB in place vertically to the accelerating frame (Fig. P4.17). Calculate the force F_A on the roller and the horizontal component of the force maintained by the pin at O if the frame's horizontal acceleration is a = 20 m/s^2.

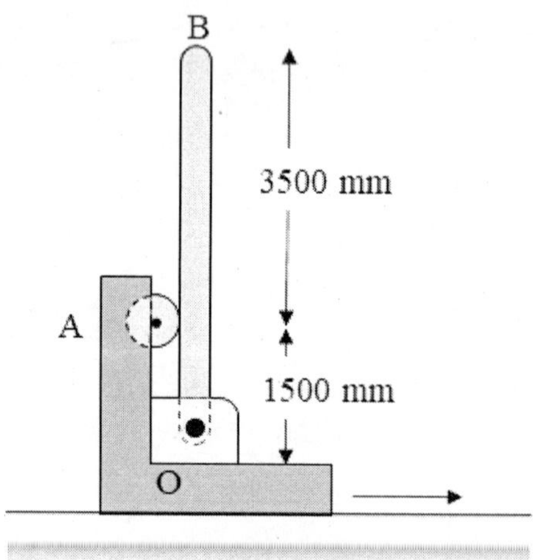

Fig. P4.17

16. Find the maximum constant angular velocity $\dot{\theta}$ without causing the block to slip upward (Fig. P4.18) take the coefficient of "static friction "between the block and the surfaces $\mu_s = 0.3$.

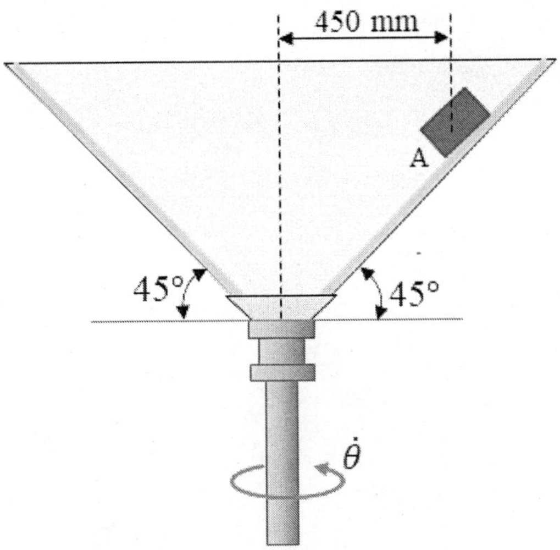

Fig. P4.18

17. The 10-kg block is positioned against a spring with a rigidity of k = 450 N/m (Fig. P4.19). Find the block's velocity at s = 0.6 m if the force applied is F = 600 N. The block and spring are both at rest when s = 0. It has a smooth touch surface.

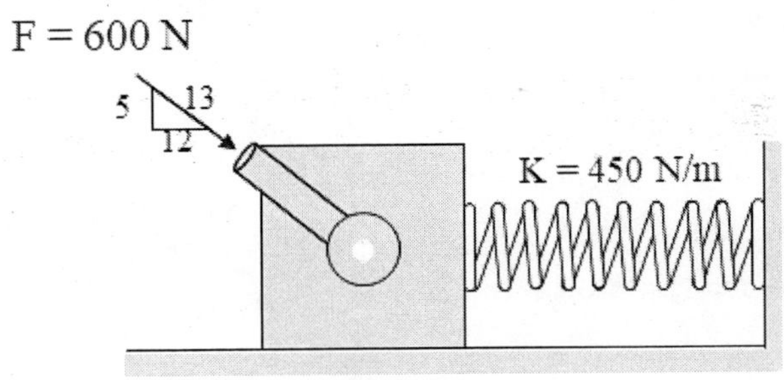

Fig. P4.19

PRACTICAL

AIM: To confirm the polygon of forces law.

Theory: The sides of a polygon taken in order can be used to represent a set of coplanar forces operating at a location in equilibrium.

Apparatus: The "Force Table" is the name of the experiment's tool as shown in Fig. P4.20, which consists of a circle of aluminium with a perimeter that is graded in degrees to indicate the direction of forces. Three levelling screws can be used to level the table as illustrated in the illustration. On the graded circular plate, there are five moveable jaws that may be set in any location. Over the knob is retained a circular plate. The jaws are covered by five strings, each of which ends is linked to a load hanger. The knob is located in the centre of the ring when five forces are in balance. i.e., the ring and knob are equally spaced apart.

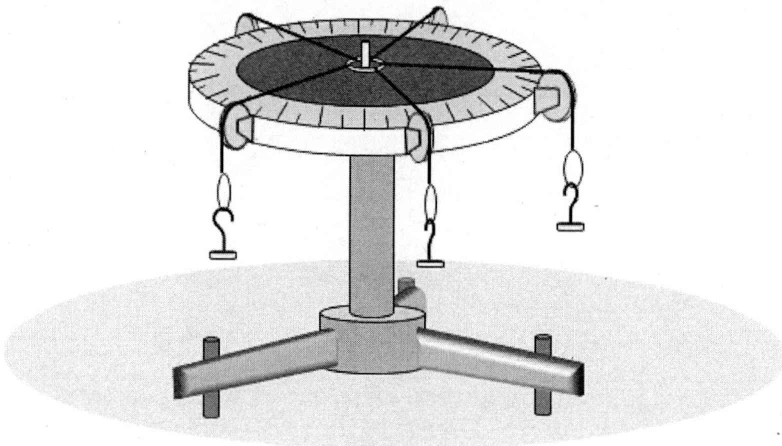

Fig. P4.20

Procedures:

1. Set one of the jaws to a 0-degree angle.
2. Assign various weights to each hanger.
3. Change the four remaining jaws' positions such that the ring seems to be in balance.
4. Write down the weights' angular location and size.
5. Adjust the hanger weights and jaw angles, excluding the jaw that is set at zero degrees.
6. Perform the test multiples time.
7. Note the readings at least five times.
8. Fill out the following table with the observation.

Observation table:

Reading	Set				
	1	2	3	4	5
Force F_2 magnitude =................ Angular position = 0 Angle = α..................					

Reading	Set				
	1	2	3	4	5
Force F_3 magnitude =................ Angular position = O Angle = β..................					
Force F_4 magnitude =................ Angular position = 0 Angle = y..................					
Force F_5 magnitude =................ Angular position = ... Angle = Θ..................					
Closing error in grams					

Results: Measure the closing error in each case

Discussion:

1. Why polygons are not closed when the force system are in equilibrium?
2. What is the law of parallelogram of forces?
3. What is the law of triangle of forces?
4. Are the forces are coplanar of the concurrent?
5. What does the closing error indicate?

Precautions:

1. Carefully put the weights on the weight hangers.
2. Apparatus should be placed at plane surface.
3. Note the readings carefully.

KNOW MORE

To visualise the loads and moments occurring on a body and to calculate responses, free body diagrams (FBD) are employed in mechanics problems. These diagrams are frequently used to calculate internal forces within a structure and to establish the loading of certain structural elements.

Features: A free body diagram is a diagram, not a scaled drawing. Depending on how a body is represented, a free body diagram may utilise different symbols.

In free body diagrams, there are:

1. A simplified form of the body (often a dot or a box)
2. The direction of the forces acting on the body is shown by straight arrows.
3. Moments are represented as curves with arrowheads or by vectors with two arrowheads indicating in the direction in which they operate on the body.
4. At least one reference coordinate system
5. Conventionally, responses to applied load are shown by hash marks through the vector's stem.

Establishing body models:

Three models can be used to depict a body:

Particle: The graphic becomes a collection of concurrent arrows when the body is replaced with a tiny symbolic blob. A bound vector is a force acting on a particle.

Stiff Extended: Rotational effects are more interesting than stresses and strains. The direction in which the force arrow is located along the line of force is unimportant. A "force acting on a rigid body" in extension is a sliding vector.

Non-rigid extended: The diagram needs to depict the force's point of application since it becomes crucial. A bound vector is a force acting on a flexible body. Some people use the arrow's tail to denote the application point. Others utilise the tip.

The forces that are not aligned to axis: If all of the forces and moments are in line with the coordinate axes, calculating the total of them is simple; but, if any of them are not, it becomes more difficult. It is more practical to employ the forces' constituents, in which case ΣF_x and ΣF_y are used in place of ΣF. (the variable M is used for moments).

How to create a force diagram:

1. Determine the item for which you will create a diagram. (You will need to create many diagrams if there are several things of interest.)
2. List all the forces that are being applied directly to the item and the source of those forces. The two objects will be in direct touch, with the exception of gravity and a few other forces that aren't
3. To symbolise the subject of interest, draw a dot.
4. Develop a vector for each force. Draw the object in the direction that the force is being applied after labelling the type of force, the object applying the force, and the item receiving the force (which will be you object of interest).
5. If the object is stationary or moving at a constant pace, the vector sum should graphically equal 0. The vector formed by adding the individual ones should move in the same direction if the object is travelling more quickly.

Steps to draw force diagram:

Determine the direction of all forces and acceleration.

1. To reduce the number of objects (forces and acceleration) that need to be divided into components, particularly unknown quantities, use a coordinate system.
2. When a force or acceleration does not lie along the X or Y axis, draw the components for it and note the angle (or being looked for).
3. List each force or component of a force flowing in a certain direction, designating it with a positive or negative sign to indicate the direction it is flowing in.
4. Assign the mass multiplied by the angular acceleration in that direction as the constant for the total forces acting in that direction. (Acceleration will be 0 if not moving or going in that direction at a steady speed.)
5. Repeat for different direction.

REFERENCES AND SUGGESTED READINGS

- Meriam, J.L. and Kraige, L.G., Engineering Mechanics: Dynamics, Wiley, 2011.
- Hibbeler, R.C., Engineering Mechanics: Dynamics, Pearson Prentice Hall, 2010.
- Chanda, A. and Nag, D., Engineering Mechanics, Wiley, 2018.

5 General Motion

UNIT SPECIFIES

This unit discusses the following concepts:

1. *General motion*
2. *General 3-D motion*
3. *Brief view on the free precession*
4. *Definition of gyroscope and gyroscopic effects*
5. *Rolling coin*

RATIONALE

The study of general motion, which encompasses both translational and rotational movements of bodies, is a fundamental aspect of classical mechanics. Understanding general motion is essential for a comprehensive analysis of how objects move in real-world scenarios. Engineers and designers need to understand general motion to create and optimize mechanical systems, such as vehicles, machinery, and robotics. Precession, gyroscopic effects and 3D motion are essential to understand for designing systems that can move efficiently and effectively, meeting performance requirements and safety standards. General motion involves both translational movement (change in position) and rotational movement (change in orientation) of a body. This combination is common in most real-world scenarios, such as a rolling coin or ball or a rotating satellite.

PRE-REQUISITES

Vector algebra

UNIT OUTCOMES

List of outcomes of this unit is as follows:

U5-O1: Understanding the general planar motion

U5-O2: Describe the general 3D-motion

U5-O3: Explain precession angular motion

U5-O4: Apply gyroscopic couple to solve complex problems

U5-O5: Effects and cause of the rolling coin

Unit - 5 Outcomes	EXPECTED MAPPING WITH COURSE OUTCOMES (*1- Weak Correlation; 2- Medium Correlation; 3- Strong Correlation*)				
	CO-1	**CO-2**	**CO-3**	**CO-4**	**CO-5**
U5-O1	1	2	3	1	-
U5-O2	3	3	3	3	1
U5-O3	3	3	2	3	1
U5-O4	3	3	3	3	2
U5-O5	3	3	3	3	2

5.1 INTRODUCTION

If a body tends to move from its rest position with a velocity and an acceleration, it is basically known as the motion of a body. The simultaneous translation and rotation of a certain body at a given moment is called general motion. The forces and moments must therefore be related to the acceleration of the body's centre of mass (i.e., the system's average location, weighted based on each component's mass.) and to the angular accelerations, respectively. The free body diagram as depicted in Fig. 5.1 of the moving body in order to evaluate a body that is experiencing general planar motion. Locate the body's centre of mass (G) and all forces, both known and unknown, as well as all moments that the body is experiencing. According to Newton's second law of motion, the following equations of motion can be used to evaluate a body:

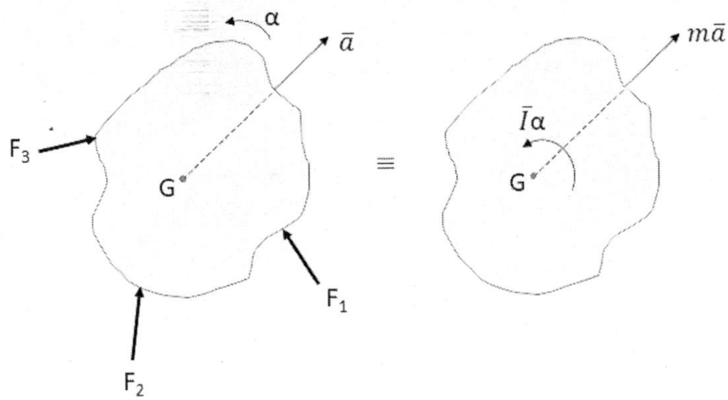

Fig. 5. 1 General motion

$$\sum \vec{F} = m\vec{a}_G$$

$$\sum \overrightarrow{M_G} = I_G\vec{\alpha}$$

The above general equation of motion written in any reference coordinate system is as follows:

$$\sum F_x = m\, a_{Gx}$$

$$\sum F_y = m\, a_{Gy}$$

$$\sum M_G = I_G \alpha$$

If the body moves in both two-axis directions either or both x-axis and y-axis it is generally called as the planar motion (Fig. 5.2), and if it causes only one rotation, the degree of freedom (DOF) for the general planar motion is three.

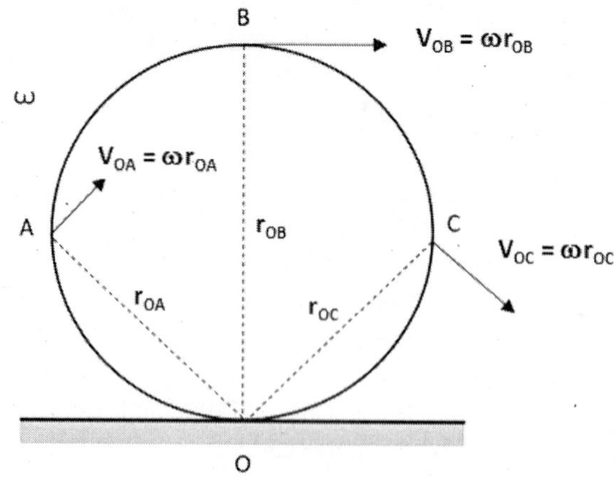

Fig. 5. 2 General Planar Motion

Example 5.1 A disc is moving with an angular acceleration α, and an angular velocity ω with specific points as shown in the Fig. E5.1, find total accelerations at every point on the disc?

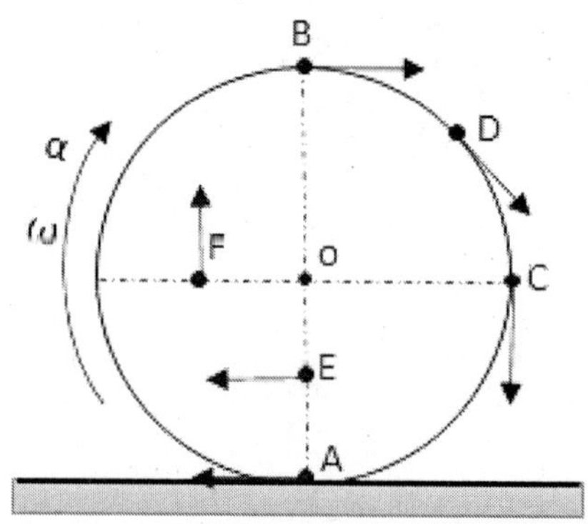

Fig. E5.1

Solution:

Point O:

Translational acceleration: $r\alpha\rightarrow$

Tangential acceleration: 0 = radial dist $\times\ \alpha$

Centripetal acceleration: $r\omega^2 = 0*\omega^2$

Total acceleration: $r\alpha\rightarrow$

Point A:

Translational acceleration: $r\alpha\rightarrow$

Tangential acceleration: $r\alpha\leftarrow$

Centripetal acceleration: $r\omega^2\uparrow$

Total acceleration: $r\omega^2\uparrow$

Point B:

Translational acceleration: $r\alpha\rightarrow$

Tangential acceleration: $r\alpha\rightarrow$

Centripetal acceleration: $r\omega^2\downarrow$

Total acceleration: $\sqrt{(r\omega^2)^2 + (2r\alpha)^2}$

Point C:

Translational acceleration: $r\alpha\rightarrow$

Tangential acceleration: $r\alpha\downarrow$

Centripetal acceleration: $r\omega^2\leftarrow$

Total acceleration: $\sqrt{(r\alpha - r\omega^2)^2 + (r\alpha)^2}$

Point D:

Translational acceleration: $r\alpha\rightarrow$

Tangential acceleration: $r\alpha$

Centripetal acceleration: $r\omega^2$

Total acceleration: Resolving forces

Point E:

Translational acceleration: $r\alpha\rightarrow$

Tangential acceleration: $\frac{r\alpha}{2}\leftarrow$

Centripetal acceleration: $\frac{r\omega^2}{2}\uparrow$

Total acceleration: $\sqrt{\left(\frac{r\alpha}{2}\right)^2 + \left(\frac{r\omega^2}{2}\right)^2}$

Point F:

Translational acceleration: $r\alpha\rightarrow$

Tangential acceleration: $\frac{r\alpha}{2}\uparrow$

Centripetal acceleration: $\frac{r\omega^2}{2}\rightarrow$

Total acceleration: $\sqrt{\left(\frac{r\alpha}{2}\right)^2 + \left(\frac{r\omega^2}{2} + r\alpha\right)^2}$

Example 5.2 A 15 kg cylinder with a radius of 0.2 m is positioned at an angle of 20^0 on a ramp as shown in Fig. E5.2(a). What is the cylinder's initial angular acceleration if it is released from rest and rolls without slipping? How long will it take the cylinder to roll 4.5 m?

Solution: Given data: Cylinder mass, $m = 15$ kg, radius, $r = 0.2$ m, $\theta = 20^0$, initial velocity, $Vi = 0$;

The free body diagram as shown in Fig E5.2(b).

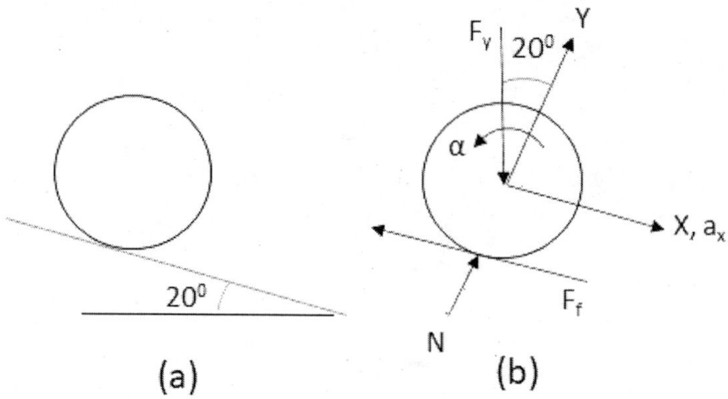

Fig. E5.2

$\sum F_x = m\, a_{Gx}$ m g sin 20^0 - F_f = m a_x => 15 x 9.81 x sin 20^0 - F_f = 15 x a_x ----- (5.1a)

$\sum F_y = m\, a_{Gy}$ N - mg cos 20^0 = 0

$\sum M_G = I_G \alpha$ -F_f x 0.2 =($\frac{1}{2}$ x 15 x 0.2^2) α ------ (5.1b)

Since rolling without slipping, a_x = - r α = -0.2 α and substitute the same in equation (5.1a):

F_f = 50.33 + 3 α and further F_f is substituted in equation (5.1b) then:

-(50.33 + 3 α) x 0.2 = 0.3 α => α = - 11.18 rad/sec^2

a_x = -0.2 (-11.18) = 2.24 m/sec^2

The time required for the cylinder center to move 4.5 m is

$x = \frac{1}{2}\, a_x t^2$ 4.5 = 0.5 x 2.24 x t^2 then t = 2 sec

5.2 GENERAL 3-D MOTION

If a body moves in all three directions (X, Y and Z-axis), it is generally called as the 3-D motion of a body, if it causes three rotations and three translations with respect to the three axes, the degree of freedom for the general 3-D motion is six.

Let us say a particle is navigating a three-dimensional space between two points. A position vector is needed to characterise this particle's position. These vectors are always in relation to the origin of the reference frame. To accurately explain the behaviour of a particle travelling in a plane, the following variables must be present:

a) Position
b) Velocity
c) Acceleration

5.2.1 Position Vector

A particle can be located anywhere in a three-dimensional space; it cannot be defined by a single coordinate. In this instance, it serves as a reference to the starting place and the path one should take to locate that point. Because of this, the position must be described using a vector. The position vector is the one that represents the direction and location of the particle with respect to the origin (Fig. 5.3). A particle's location vector is represented as,

$$\vec{r} = x\hat{\imath} + y\hat{\jmath} + z\hat{k} \tag{5.2}$$

Where x, y and z are their components along the x, y and z axis.

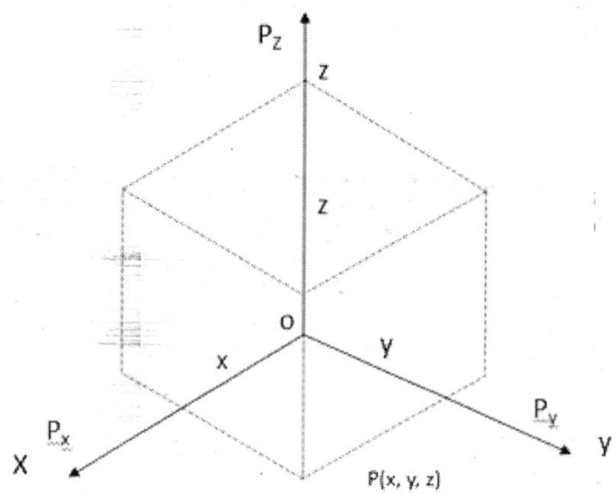

Fig. 5. 3 Position vector

5.2.2 Velocity

There are two ways to define a particle's velocity while it is moving across three dimensions: average velocity and instantaneous velocity. Every second the particle changes its velocity when it is being accelerated. Therefore, a velocity cannot be given a single value. In these situations, the instantaneous velocity is recommended since it captures the velocity and direction at a certain moment. It comes from,

$$\vec{v} = lim_{\Delta t \to 0} \frac{\Delta \vec{r}}{\Delta t}$$

$$\vec{v} = \frac{dr}{dt} \tag{5.3a}$$

Also, the velocity can be expressed as,

$$v = \frac{dx}{dt}\hat{\imath} + \frac{dy}{dt}\hat{\jmath} + \frac{dz}{dt}\hat{k} \tag{5.3b}$$

The ratio of overall displacement to overall time is the average velocity. Suppose a particle goes from $\vec{r}$ to $\vec{r'}$ in a total time of Δt

The velocity is given by

$$\vec{v} = \frac{\vec{r} - \vec{r'}}{\Delta t} \tag{5.3c}$$

Example 5.3 The equation r = 8.0ti + 5.0t²j + 6.0k describes the location of an electron (where t is in seconds and the coefficients have the proper units for r to be in meters). What is the electron's v (t)? (b) What does v at t = 3.0 s look like in unit-vector notation? (c) What were v's current magnitude and direction?

Solution: The velocity vector v is the time–derivative of the position vector r

$$v = \frac{dr}{dt} = \frac{d}{dt}(8.0ti - 5.0t^2 j + 6.0k)$$

$$v = 8i - 10tj$$

Where t is in seconds, v is given in m/s.

(b) at t = 3.0 s, the value of v is

$$v(t = 3.0s) = 8i - 10(3)j$$

That is, Velocity is $= 8i - 30j$

(c) magnitude of v at t = 3.0s from (b)

$$v = \sqrt{v_x^2 + v_y^2 + v_z^2} = \sqrt{(8.0m/s)^2 + (-30.0m/s)^2 + (0)^2} = 30.265\ m/s$$

Even though this is a three-dimensional problem, we observe that the velocity vector is on the xy plane, allowing us to represent its direction using a single angle, the conventional angle, which is measured counter clockwise in the xy plane from the x axis. For this perspective, we get:

$$tan\theta = \frac{v_y}{v_x} = -56.25 \Rightarrow \theta = tan^{-1}(-56.25) = -88.98^0$$

Always make sure we have chosen the correct quadrant for before taking the inverse tangent. Since vy is negative and vx is positive in this instance, the answer -88.98° is correct.

5.2.3 Acceleration

The rate at which a body's velocity changes determines how quickly it accelerates when travelling in a plane. In this situation, average acceleration and instantaneous acceleration are the two possible cases, just like with velocity. The ratio of the object's net change in velocity to the total time taken provides the average acceleration. Let the starting (v_i) and end (v_f) velocities be represented by

$$\vec{a} = \frac{\vec{v_f}-\vec{v_i}}{\Delta t} \quad (5.4a)$$

When the body's acceleration varies with the passage of time, instantaneous acceleration is utilised.

$$\vec{a} = \ lim_{\Delta t \to 0} \frac{\Delta \vec{v}}{\Delta t}$$

$$\vec{a} = \frac{dv}{dt} \quad (5.4b)$$

Taking this apart and analysing,

$$a = \frac{dv_x}{dt}\hat{\imath} + \frac{dv_y}{dt}\hat{\jmath} + \frac{dv_z}{dt}\hat{k}$$

$$a = \frac{d^2x}{dt^2}\hat{\imath} + \frac{d^2y}{dt^2}\hat{\jmath} + \frac{d^2z}{dt^2}\hat{k}$$

Utilizing the velocity's component parts, it is also possible to determine the acceleration's magnitude.,

$$a = \sqrt{a_x^2 + a_y^2 + a_z^2} \quad (5.4c)$$

Relative Motion in Three Dimensions: When viewed from one frame of reference, relative motion depicts a body's velocity. These ideas have existed in first- and second-dimensional spaces, respectively. But three-dimensional spaces can also benefit from the application of similar ideas. the illustration is below and depicted in Fig. 5.4.

Consider a particle P along with reference frames S and S'. The positions of the particle P with respect to the frame S', the frame of reference S, and the position of the frame S' as measured in S are provided by the variables $r_{ps'}, r_{ps}, and\ r_{s's}$ respectively.

The Fig. 5.4 demonstrates that

$$r_{ps} = r_{ps'} + r_{s's} \quad (5.5a)$$

Using the aforementioned equation to differentiate, these vectors also provide us with the formula for relative velocities.

$$\frac{d}{dt}(r_{ps}) = \frac{d}{dt}(r_{ps'} + r_{s's})$$

$$\vec{v}_{PS} = \vec{v}_{PS'} + \vec{v}_{S'S} \quad (5.5b)$$

According to common sense, a particle's velocity with regard to S is determined by adding the velocities of S' and the particle. After further differentiating this equation, the acceleration equation is provided by,

$$\frac{d}{dt}(\vec{v}_{PS}) = \frac{d}{dt}(\vec{v}_{PS'} + \vec{v}_{S'S})$$

$$\vec{a}_{PS} = \vec{a}_{PS'} + \vec{a}_{S'S} \tag{5.5c}$$

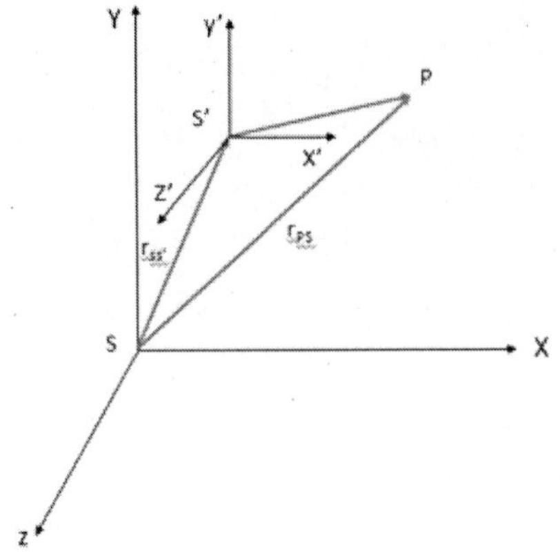

Fig. 5. 4 Three-dimensional space

A particle's acceleration with respect to S is determined by adding the acceleration of S' with respect to S and the acceleration of the particle with respect to S.

Example 5.4 If a particle is travelling in a plane and its location is known as v(t) = 2ti + 4t^{3}j + k, what is its average acceleration between t = 0 and t = 3?

Solution: Given:

v(t) = 2ti + 4t^{3}j + k

The velocity vector changes with time. The average acceleration is given by the formula,

$$\vec{a} = \frac{\vec{v_f} - \vec{v_i}}{\Delta t}$$

At t = 0; v(0) = v_i = 0i + 0j + k

At t = 3; v(3) = v_f = 2x3 i + 4x3^3 j + k = 6i + 108j + k

The v_i and v_f values substitute in the above equation, then $\vec{a} = \frac{(6i+108j+k)-(0i+0j+k)}{(3-0)} = 2i + 36j$

5.3 GYROSCOPE

A rigid body that has the ability to rotate rapidly in three dimensions about any axis that passes through a fixed point known as the centre, which may or may not be its centre of mass, is referred to as a gyroscope. The fixed point of a child's toy top, which is the point at which it contacts the table or floor on which it spins, satisfies this criterion and is a type of gyroscope. A toy top's weight acts downward from its centre of mass, while the ground's reaction force pushes up on it where it comes into touch with the support. The top precesses ("Precession is a variation in the direction of a rotating body's rotational axis") as a result of the torque created by these two opposing forces.

The typical gyroscope has a mechanical design with a heavy rim on a rotor that spins rapidly around a fixed point on the rotor axis. When a vehicle changes direction of motion, gyroscope or gyroscopic forces are acting in engines with rotating components with a large moment of inertia. Gyroscopic forces act on rotating components including the crankshaft, flywheel, clutch, transmission gears, propeller shaft, and wheels when automobiles rotate at high speeds. In turns and pullouts, gyroscopic effects have an impact on an airplane's engine components, propeller, and gear reduction system. Ships and locomotives are both impacted in the same way.

5.3.1 PRECESSION ANGULAR MOTION

Consider a disc that is revolving about OZ-axis with an angular velocity ω as shown in Fig. 5.5 (a). The revolving axis is also known as the spin axis and the disc rotation is in clockwise direction with respect to the view direction as depicted in Fig. 5.5 (a).

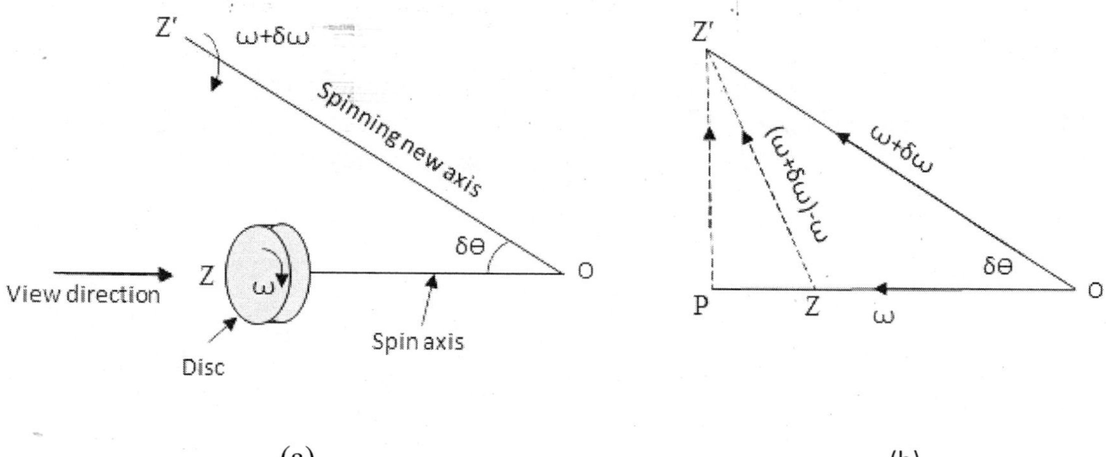

Fig. 5. 5 Free Precession

After sometime, the disc tends to spin about the new axis OZ' with an angle δΘ and with the new angular velocity (ω+δω). According to Fig. 5.7 (b), the vector OZ represents the disc's initial angular velocity ω and the vector OZ'represents the disc final angular velocity (ω+δω) after taking time into account. The vector ZZ' represents the angular acceleration of the disc, it gauges the rate at which angular velocity changes with relation to time. δt. This can be divided into two pieces, a parallel component and a perpendicular part. with reference to OZ.

The parallel component which is angular acceleration along the OZ direction is,

$$\alpha_t = \frac{ZP}{\delta t} = \frac{OP - OZ}{\delta t} = \frac{OZ' \cos\delta\theta - OZ}{\delta t}$$

$$= \frac{(\omega + \delta\omega)\cos\delta\theta - \omega}{\delta t} = \frac{\omega\cos\delta\theta + \delta\omega\cos\delta\theta - \omega}{\delta t}$$

Since δΘ is quite little, we get the following result when we substitute cos δΘ = 1,

$$\alpha_t = \frac{\omega + \delta\omega - \omega}{\delta t} = \frac{\delta\omega}{\delta t}$$

In the limit, when $\delta t \to 0$,

$$\alpha_t = \underset{\delta t \to 0}{\text{Lt}} \left(\frac{\delta\omega}{\delta t}\right) = \frac{d\omega}{dt} \tag{5.6a}$$

The perpendicular to OZ component of angular acceleration is, also known as gyroscopic acceleration.

$$\alpha_c = \frac{PZ'}{\delta t} = \frac{OZ' \sin\delta\theta}{\delta t} = \frac{(\omega + \delta\omega)\sin\delta\theta}{\delta t} = \frac{\omega\sin\delta\theta + \delta\omega.\sin\delta\theta}{\delta t}$$

Since δΘ is quite little, we get the following result when we substitute sin δΘ = δΘ,

$$\alpha_c = \frac{\omega.\delta\theta + \delta\omega.\delta\theta}{\delta t} = \frac{\omega.\delta\theta}{\delta t}$$

In the limit when $\delta t \to 0$, gyroscopic acceleration, $\alpha_c = \underset{\delta t \to 0}{\text{Lt}} \left(\frac{\omega.\delta\theta}{\delta t}\right) = \omega \times \frac{d\theta}{dt} = \omega.\Omega$

(5.6b)

Where $\frac{d\theta}{dt} = \Omega$ is the angular velocity of precession, which is perpendicular to the plane is about to rotate the spin's axis. The rotational axis is identified as axis of precession. Precessional angular motion refers to the angular rotation of the spin axis about the precession axis. If the disc's angular velocity is constant throughout all orientations of the spin axis, then $\frac{d\theta}{dt} = 0$; and $\alpha_c = 0$.

5.3.2 GYROSCOPIC COUPLE

Assume a disc revolving at an angular velocity of ω rad/s around the spin axis OZ, which is rotating clockwise with reference to the direction shown in Fig. 5.6. The YOX plan is spinning plan because it is parallel to the disc rotation. The spin axis revolves around an axis OY (precession axis) in a precession plane XOZ with an angular velocity Ω rad/s.

The vector $(\overrightarrow{OZ})$ can be used to represent an angular momentum, which is equal to product of mass moment of inertia with respect to OZ and angular velocity of the disc. When viewed from the top, the axis of spin OZ also rotates clockwise around the axis OY. Allow the axis OZ to be rotated a small angle δθ radians in time δt seconds of new axis is OZ′, in the plane XOZ. The vector ox′ will now be used to represent the angular momentum, if the angular velocity ω is constant.

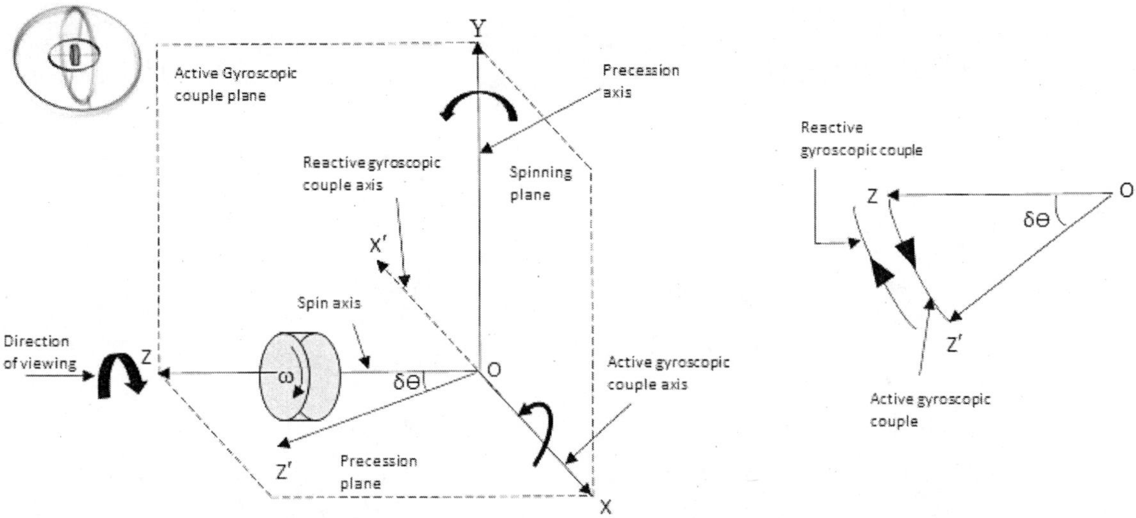

Fig. 5. 6 Gyroscopic Couple

The rate of change in angular momentum

$$= \frac{\overrightarrow{OZ'} - \overrightarrow{OZ}}{dt} = \frac{\overrightarrow{ZZ'}}{dt} = \overrightarrow{OZ}.\frac{\delta\theta}{dt} = I.\omega.\frac{\delta\theta}{dt}$$

Precession is caused by the couple (C, unit is N-m) applied to the disc because the couple application will result in a change in the angular momentum with respect to time,

$$C = \underset{\delta t \to 0}{\text{Lt}}\ I.\omega.\frac{\delta\theta}{\delta t} = I.\omega.\frac{d\theta}{dt} = I.\omega.\Omega \tag{5.7}$$

Where I is mass moment of inertia (kg-m^2).

The active "gyroscopic couple" (ZZ′) which is equal to $I.\omega.\Omega$ must be used over the disc whenever the spin axis is made to revolve with angular velocity Ω about the precession axis. The horizontal plane, or XOZ plane, is where the vector ZZ′ is located. If there is only a small displacement δθ, the vector ZZ′ will be perpendicular to the ZOY plane. As a result, the couple responsible for this shift in angular momentum will be located in the plane ZOY. A clockwise couple in the plane ZOY is represented by the vector ZZ′, as shown in Fig. 5.6. As a result, the axis OX perpendicular to the ZOY plane, which is referred as the axis of active "gyroscopic couple" and the plane ZOY is known as active "gyroscopic couple" plane.

When the spin axis rotates with an angular velocity of Ω, the disc is subject to a reactive couple ($I.\omega.\Omega$) which is the same as the active couple's but moves in the opposite direction. Reactive "gyroscopic couple" is the name given to the reactive couple that the disc experiences as the spin axis revolves around the precession axis. In Fig. 5.6, OX′ stands in for the reactive gyroscopic couple's axis.

The bearings that hold the shaft are typically used to apply the gyroscopic coupling. The bearings will withstand an opposite and equal couple. The gyroscope is an instrument or toy that employs the gyroscopic concept. Ships are equipped with gyroscopes to lessen the effects of pitching and rolling caused by waves. The gyroscope is also used in gyrocompasses, monorail cars, aeroplanes, and other devices.

Example 5.5 An arm of length 1000 mm with a uniform disc placed on one end and a free-rotating universal bearing at the other end. What will be the precession speed about the vertical axis if the disc, with a mass of 10 kg and a diameter of 600 mm, rotates with a speed of 600 rpm about the arm as viewed from the front?

Solution: Given data: m = 10kg; d = 600 mm = 0.6 m; L = 1000 mm = 1m; N = 600 rpm;

$$\omega = \frac{2\pi\,(600)}{60} = 62.83\;rad/s$$

mass moment of inertia of the disc, $I = \frac{1}{2}mr^2 = \frac{10(0.3)^2}{2} = 0.45\;kg.m^2$

couple, $C = I\,\omega\,\Omega$ = 0.45 x 62.68 x Ω (i)

But couple due to disc's mass, $C = m.g.L = 10 \times 9.81 \times 1 = 98.1\;N.m$ (ii)

Now equating both the equations of (i) and (ii)

$$98.1 = 0.45 \times 62.68 \times \Omega$$

$$\therefore \quad \Omega = 3.45\;rad/s$$

Example 5.6 A homogeneous disc with a diameter of 300 mm and mass 10 kg. Its axle is kept in a horizontal plane by bearings positioned in the centre of the vehicle. The disc spins equally at 600 rpm around its axle, whereas the axle rotates uniformly at 120 rpm about the vertical axis. The rotational axes are as depicted in Fig. E5.3. Find the impact of mass and gyroscopic effects at each bearing if the space between the bearings is 400 mm.

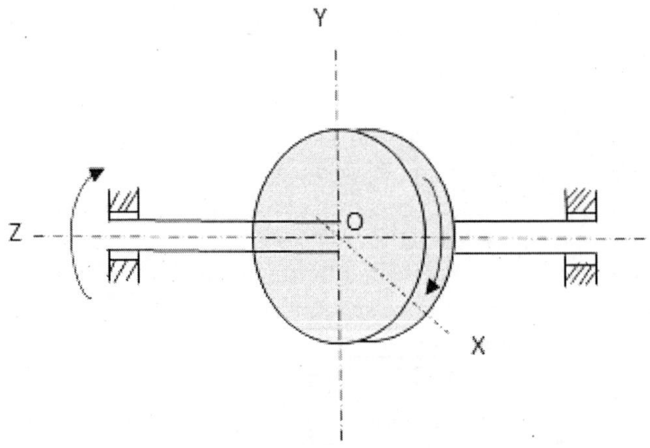

Fig. E5.3

Solution: Given data: m = 10kg; d = 300 mm = 0.3 m; L = 400 mm = 0.4 m; N = 600 rpm; Na = 120 rpm;

Angular velocity of the disc, $\omega = \frac{2\pi\,(600)}{60} = 62.83\ rad/s$ (clockwise)

Angular velocity of axle or precession $\Omega = \frac{2\pi\,(120)}{60} = 12.57\ rad/s$ (clockwise)

mass moment of inertia of the disc, $I = \frac{1}{2}mr^2 = \frac{10(0.15)^2}{2} = 0.1125\ kg.m^2$

Gyroscopic couple, $C = I.\omega.\Omega = 0.1125\ \times 62.83\ \times 12.57 = 88.85\ N.m$

Fig. E5.4 depicts the reactive gyroscopic coupe's direction. Let F be the gyroscopic coupling force at each bearing (P and Q) and it will act opposite directions at each.

$$F = \frac{C}{L} = \frac{88.85}{0.4} = 222.12N$$

The bearings at P and Q will act as simply support shaft with centrally loaded disc, Therefore,

$$R_P = R_Q = \frac{10\ X\ 9.81}{2} = 49.05N$$

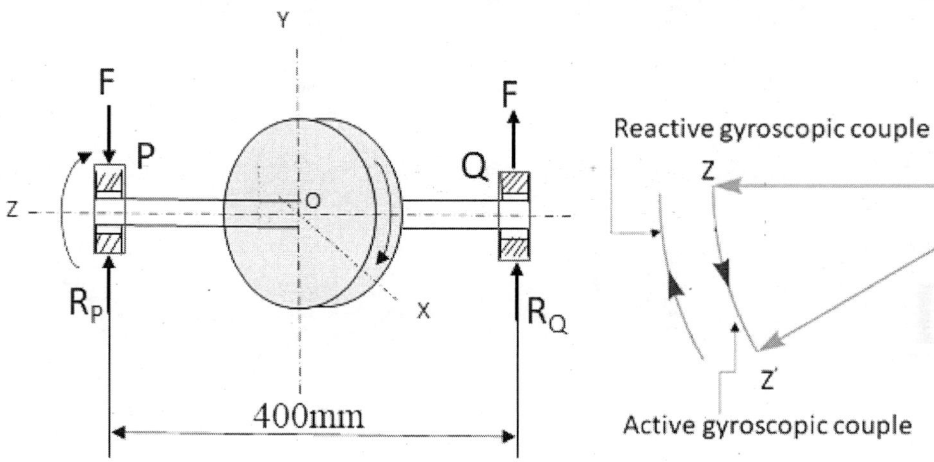

Fig.E5.4

Net reactions (R_{Pn} and R_{Qn}) at P and Q are as follows:

$R_{Pn} = R_P - F = 49.05 - 222.12 = 173.07$ N (downwards)

$R_{Qn} = R_Q + F = 49.05 + 222.12 = 271.17$ N (upwards)

Example 5.7 A plane travelling at a velocity of 600 km/h does a full half circle with a radius of 100 m to the left. The aircraft's rotary engine and propeller have a combined mass of 600 kg and a 0.6 m gyroscopic radius. When viewed from tail, the engine turns at a clockwise speed of 3000 rpm. Find the aircraft's gyroscopic pair and describe its impact.

Solution: Given data: m = 600 kg; k = 0.6 m; N = 3000 rpm; v = 600 km/h = 166.667 m/s; R = 100 m;

The engine angular velocity, $\omega = \frac{2\pi\,(3000)}{60} = 314.16\ rad/s$

Angular velocity of precession, $\Omega = \frac{v}{R} = \frac{166.667}{100} = 1.667\ rad/s$

Aircraft's gyroscopic couple, $C = I\ \omega\ \Omega = (m\ k^2)\ \omega\ \Omega = 600$ X $(0.6)^2\ X\ 314.16\ X\ 1.667$

C = 113120.22 = N.m = 113.12 kN.m

The "gyroscopic couple" causes the nose to rise and the tail to fall when the plane turns to the left with reference to pilot.

Example 5.8 A ship's turbine rotor masses 3000 kg and turns anticlockwise at a speed of 1800 rpm when viewed from the stern. The rotor's radius of gyration is 0.6 m. Calculate the "gyroscopic couple" and its impact when

The ship steers to the left while travelling at a speed of 7.75 m/s in a curve with a radius of 90 m.

The bow is lowering at its greatest speed while the ship pitches 60 above and 60 below its normal position. The pitching motion has a 50 s periodic time and is simple harmonic.

The ship is rolling clockwise at an angular velocity of 0.5 rad/s when observed from the stern. Find the pitching's maximum angular acceleration as well.

Solution: m = 3000 kg; N = 1800 rpm; k = 0.6 m;

$\omega = \frac{2\pi\ (1800)}{60} = 188.5\ rad/s\ ;\ I = m\ k^2 = 3000\ X\ (0.6)^2 = 1080\ kg.m^2$

v = 7.75 m/s; R = 90 m; $\Omega = \frac{v}{R} = \frac{7.75}{90} = 0.086\ rad/s;$

$$C = I\ \omega\ \Omega = 1080\ X\ 188.5\ X\ 0.086 = 17508\ N.m = 17.5\ kN.m$$

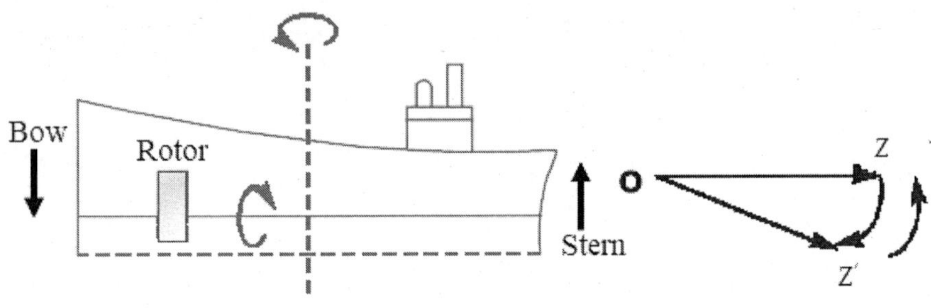

Fig. E5.5

The result is a stern is raised and a bow is lowered as shown in Fig. E5.5.

$\theta = 6^0 = 6\ X\ \frac{\pi}{180} = 0.105\ rad;$ T = 50 s

$\Omega_0 = \frac{2\pi}{50} = 0.126\ rad/s;\ \Omega = \theta\ \Omega_0 = 0.105\ X\ 0.126 = 0.013\ rad/s$

$$C = I\ \omega\ \Omega = 1080\ X\ 188.5\ X\ 0.013 = 2646.54\ N.m = 2.65\ kN.m$$

During pitching, the ship would turn to the right or starboard as the bow descended.

$\Omega = 0.5\frac{rad}{s};$

$$C = I\ \omega\ \Omega = 1080\ X\ 188.5\ X\ 0.5 = 101790\ N.m = 101.79\ kN.m$$

There is no gyroscopic effect when rolling.

When pitching, the angular acceleration is maximum.

$$\alpha_{max} = \theta\, \Omega_0^2 = 0.105\, X\, (0.126)^2 = 0.00167\, rad/s^2$$

5.4 ROLLING COIN

On a table surface, a coin with a radius of *r*, mass of m, and is positioned to roll in a circle with radius R (R >> *r*). As the coin rolls with a given spin angular velocity, it is observed to tilt at a slight angle to the vertical, as seen in Fig. 5.7 (a). Let's assume the coin won't slip, gravity always acting downward direction.

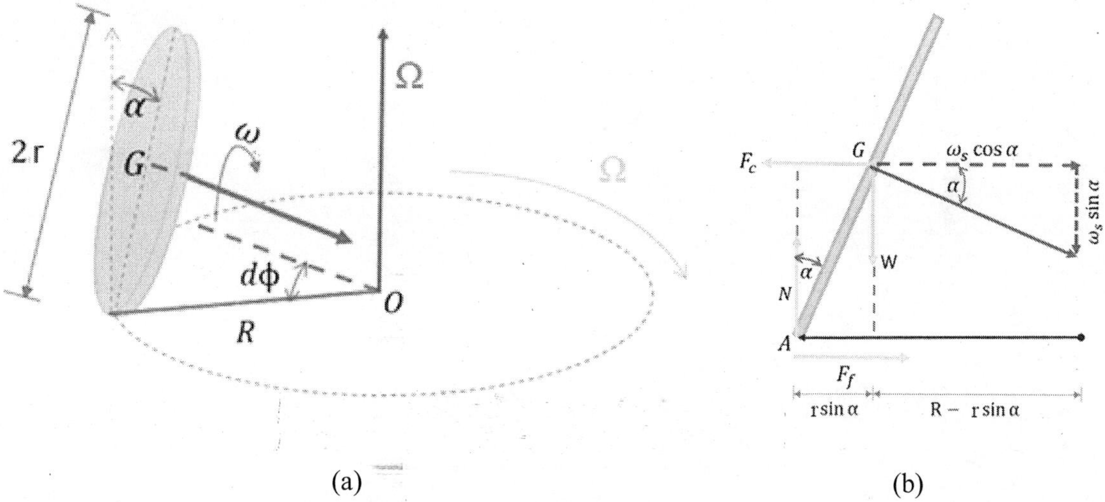

Fig. 5. 7 Rolling Coin

If the disc does not slip, implies that the arc length travelled by the coin along its path is the arc length traversed along the edge of the coin per unit time, else, one of the surfaces would be slipping in relation to the other. Using this condition, the precession angular velocity, Ω of the rolling motion can be calculated.

The magnitude of tangential velocity of coin at any point is R Ω = r ω

$$\Omega = \omega \frac{r}{R} \tag{5.8a}$$

The coin is revolving about O, a centripetal force, F_c acting at centre, G of the coin, weight of coin, W act downward direction at G, as the coin is contacting to the ground at A, the normal force, N or reaction force will act perpendicular to the surface at A. The coin is rotating without slipping means a friction force F_f is resisting to slip the coin and it is acting towards O (Fig.5.7 (b)).

There is no linear displacement of the coin in the horizontal direction, so F_f and F_c balance each other. Radius of rotation of the coin = R - r sin α

$$F_f = F_c = m\Omega^2(R - r\sin\alpha) \tag{5.8b}$$

In case of spinning the centrifugal force due to spin is balance each other and no net forces are working. Hence in this equation centrifugal force due to precession is only under consideration.

The total torque can be expressed at centre of mass, G is

$T = N\,r\sin\alpha - F_f\,r\cos\alpha = mg\,r\,\sin\alpha - m\,\Omega^2(R - r\sin\alpha)\,r\cos\alpha$

$$T = mg\,r\,\sin\alpha - m\Omega^2(R - r\sin\alpha)\,r\cos\alpha \qquad (5.8c)$$

The coin's angular moment about centre of mass, G is calculated by taking components of its angular velocity along its principal axes. The principal axes of the coin are passing through G, and perpendicular to each other. Fig. 5.8 (a) illustrates the axes (2) and (3) are plane of the coin and coming out of coin is axis (1).

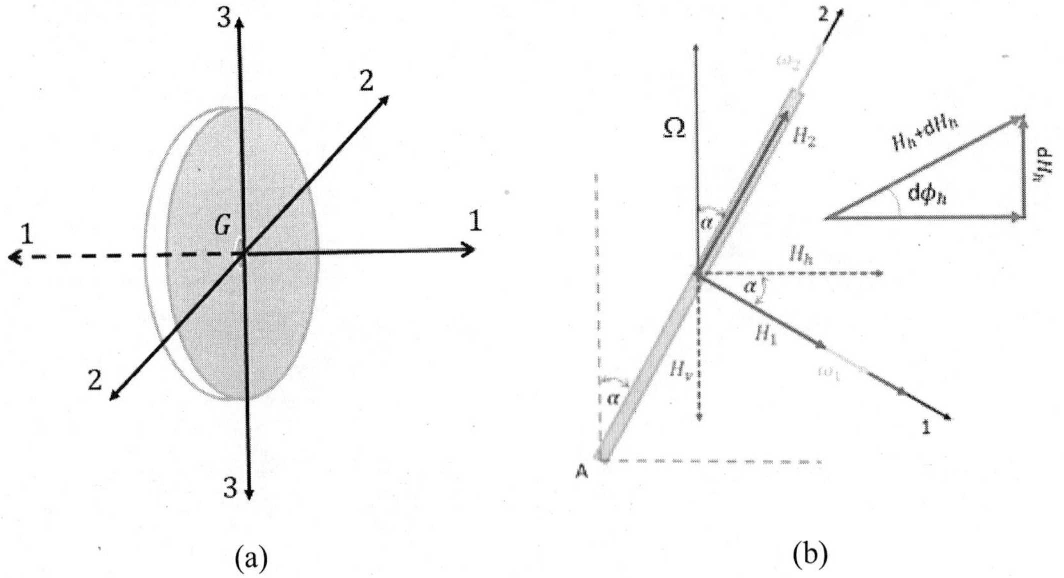

Fig. 5. 8 Principal Coordinates

The component of angular velocity in the direction of principal axes (Fig. 5.8 (b))

$\omega_1 = \omega - \Omega\sin\alpha$

$\omega_2 = \Omega\cos\alpha$

$\omega_3 = 0$

(5.8d)

The components mass moment of inertia in the direction of principal axes

$$I_1 = \frac{1}{2}mr^2\,; \quad \text{Using perpendicular axis theorem}, I_1 = I_2 + I_3$$

Again, due to symmetricity of the axes (2) and (3), $I_2 = I_3$ hence, $I_2 = I_3 = \frac{1}{2}I_1 = \frac{1}{4}mr^2$

The angular momentum (H) in the principal axis directions (H_1, H_2, and H_3) are as follows:

$H_1 = I_1\,\omega_1 = \frac{1}{2}\,mr^2\,(\omega - \Omega\sin\alpha)$

$H_2 = I_2\,\omega_2 = \frac{1}{4}\,mr^2\,\Omega\cos\alpha$

$$H_3 = I_3\,\omega_3 = 0$$
(5.8e)

The angular momentum components of horizontal (H_h) and the vertical (H_v) at G in terms of H1 and H2 are given below:

$$H_h = H_1 \cos\alpha + \; H_2 \sin\alpha$$

$$= \frac{1}{2}\mathrm{mr}^2\,(\omega\; - \Omega\sin\alpha)\cos\alpha + \frac{1}{4}mr^2\Omega\cos\alpha\sin\alpha$$

$$H_h = \frac{1}{2}\mathrm{mr}^2\omega\cos\alpha - \frac{1}{4}mr^2\;\Omega\cos\alpha\sin\alpha$$
(5.8f)

$$H_v = H_1 \sin\alpha - \; H_2\cos\alpha$$

$$H_v = \frac{1}{2}\mathrm{mr}^2\;(\omega\; - \Omega\sin\alpha)\sin\alpha - \frac{1}{4}mr^2\Omega\cos\alpha\cos\alpha$$

Replacing ω by using equation (1) in equation (6),

$$H_h = \frac{1}{2}\mathrm{m\,r}\,R\,\Omega\cos\alpha - \frac{1}{4}m\,r^2\;\Omega\cos\alpha\sin\alpha$$
(5.8g)

We see that the coin goes around; its horizontal component changes the direction by rotating in a circle with angular velocity while its vertical component remains unchanged. Thus, the change in angular momentum is then given by

$$\left|\frac{dH}{dt}\right| = T$$

$$\frac{1}{2}\mathrm{m\,r}\,R\,\Omega^2\,\cos\alpha - \frac{1}{4}m\,r^2\,\Omega^2\cos\alpha\sin\alpha \; = mg\,r\sin\alpha \; - m\Omega^2(\mathrm{R} - \mathrm{r}\sin\alpha)\,r\cos\alpha$$

$$\Omega^2\,\left(\frac{1}{2}\mathrm{m\,r}\,R\cos\alpha - \frac{1}{4}m\,r^2\cos\alpha\sin\alpha + m(\mathrm{R} - \mathrm{r}\sin\alpha)\,r\cos\alpha\right) = mg\,r\sin\alpha$$

$$\Omega^2\,\left(\frac{1}{2}R\cos\alpha - \frac{1}{4}\;r\cos\alpha\sin\alpha + (\mathrm{R} - \mathrm{r}\sin\alpha)\cos\alpha\right) = g\sin\alpha$$

$$\Omega^2\,\left(\frac{3}{2}R\cos\alpha - \frac{5}{4}\;r\;\cos\alpha\sin\alpha\right) = g\sin\alpha$$

$$\Omega^2\,(6R\cos\alpha - 5r\;\cos\alpha\sin\alpha) = 4g\sin\alpha$$

$$\Omega^2 \; = \frac{4g\sin\alpha}{(6R\cos\alpha - 5\,r\cos\alpha\sin\alpha)} = \frac{4g\;\tan\alpha}{(6R - 5r\;\sin\alpha)}$$

$$if,\, r \ll R\; then\; (6R - 5r\sin\alpha) = 6R$$

$$\Omega^2 \; = \frac{4g\tan\alpha}{6R} = \frac{2g\tan\alpha}{3R}$$

From equation (5.8a), $\Omega^2 \; = \omega^2\frac{b^2}{R^2}$

$\therefore \omega^2 \; = \frac{2\,gR\tan\alpha}{3\,r^2}$; Therefore, $\tan\alpha = \frac{3r^2\omega^2}{2gR}$

$$\boldsymbol{\alpha = \tan^{-1}\left(\frac{3r^2\omega^2}{2gR}\right)}$$
(5.8h)

Example 5.9 Grain is grounded in a grain mill by a large wheel that revolves without slipping on a flat, horizontal millstone and is driven by a vertical shaft. The rotating wheel has mass m, radius b, and is confined to rotating at an angular speed Ω in a horizontal circle with radius R (Fig. E5.6). The lower millstone is pressed downward by the wheel with a force that is equal to twice its weight. Mass of the wheel axle is ignored. Find the precessional angular frequency Ω?

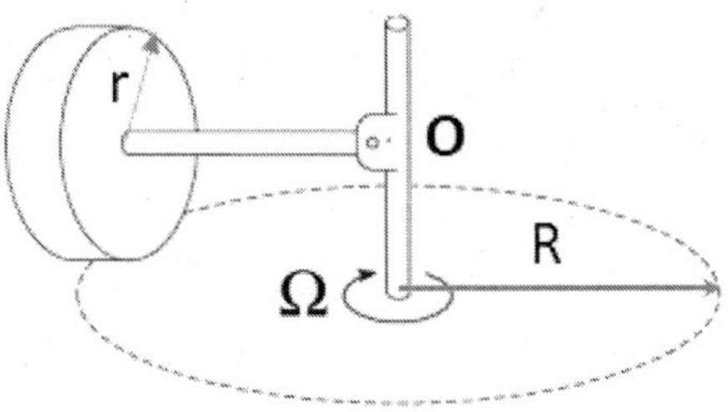

Fig. E5.6

Solution: The magnitude of tangential velocity of wheel at any point is R Ω = r ω

$$\Omega = \omega\frac{b}{R}$$

The wheel is rotating without slipping means a friction force F_f is resisting to slip the wheel and it is acting towards O. There is no linear displacement of the wheel in the horizontal direction, so F_f and F_c balance each other as shown in Fig. E5.7.

$$F_f = F_c = m\Omega^2 R$$

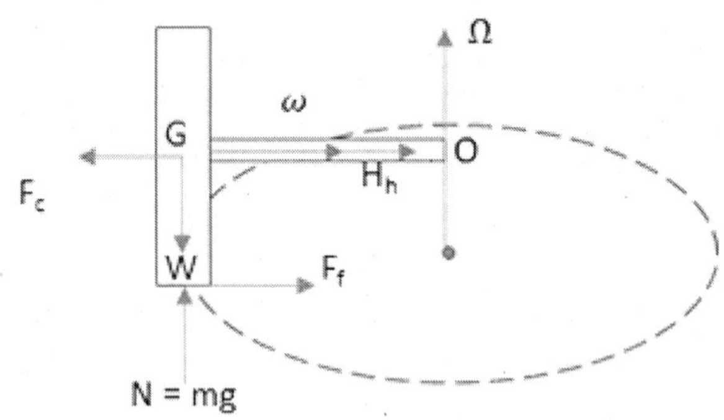

Fig. E5.7

The total torque can be expressed at the centre of mass, O is

$$T = F_f\,\mathrm{r} - \mathrm{N\,R} = m\Omega^2 R\,r - mgR$$

The components mass moment of inertia in the horizontal direction

$$I = \frac{1}{2}mr^2$$

We see that the wheel goes around; its horizontal component changes the direction by rotating in a circle with angular velocity while its vertical component remains unchanged. the magnitude of the spin angular momentum about the center of mass, O is

$$H = I\,\omega = \frac{1}{2}mr^2\,\omega = \frac{1}{2}m\,\mathrm{r}\,\mathrm{R}\,\Omega$$

Thus, the change in angular momentum is then given by

$$\left|\frac{dH}{dt}\right| = T$$

$$\frac{1}{2}m\,\mathrm{r}\,\mathrm{R}\,\Omega^2 = m\Omega^2 R\,r - mgR$$

$$\frac{1}{2}\mathrm{r}\,\Omega^2 = \mathrm{g}$$

$$\Omega = \sqrt{\frac{2g}{r}}$$

UNIT SUMMARY

By this unit, one can acquire the knowledge regarding the general motion using various conditions like planar and 3-D motion. Focusing on the certain system and point of the body and analysing the forces acting on that system in both kinematic and kinetic parameters and to find out the velocities and angular motions, brief knowledge regarding the precession, "gyroscopic couple" and understanding through the various example and problems. Determine the rolling coin inclination with respect the vertical axis and its concept with examples.

EXERCISES

Multiple Choice Questions

1) A disc is rotating about its axis of spin with an angular velocity of ω rad/sec. The couple will be the disc that is producing precession------------------

a) $\frac{1}{2}I.\omega.\Omega$ b) $I.\omega^2$ c) $\frac{1}{2}I.\omega^2$ d) $I.\omega.\Omega$

2) When a torque of 200 N-m is applied about an axis normal to a disc that is rotating on its axis at 20 rad/s and the disc's mass moment of inertia is 2 kg-m^2, the disc will precession -------rad/s

a) 20 b) 10 c) 5 d) 2

3) The planes of spin, gyroscopic coupling, and precession are -----------------in the case of the gyroscopic effect

a) in same plane b) in different planes

c) any two in perpendicular planes and third in different plane d) mutually perpendicular

4) A flywheel's radius of gyration is 0.1 m, and it mass 100/π kg. It is spin at 120 rpm around its horizontal axis. At a speed of 8 rad/sec, the entire assembly rotates around a vertical axis. The couple's gyroscopic experience will be---------- (N.m)

a) 32 b) 18 c) 20 d) 19

5) When viewed from the tail, an airplane's engine turns clockwise and the plane makes a left turn. The "gyroscopic couple" will have the following effects on the aircraft:

a) a tail dip and a nose raise. b) to raise the tail and dipping the nose.

c) to raise the tail and nose. d) to droop the tail and nose

6) A motor vehicle turns left on a curved path while travelling at a specific speed. Due to centrifugal forces --, if the engine and wheels rotate in the same direction.

a) the outside wheel's reaction decrease, whereas the inner wheel's reaction increases.

b) the inside wheel's reaction decrease, whereas the outside wheel's reaction increases.

c) the front wheel's reaction increases, whereas the back wheel's reaction decreases.

d) the back wheel's reaction increases, whereas the front wheel's reaction decreases.

7) From the stern, a ship's rotor seems to rotate in a clockwise direction as it makes a left turn. The "gyroscopic couple" will have the following effects on it:

a) the stern and bow to be raised b) to bring the stern and bow down

c) the stern to be lowered and the bow raised d) to raise the stern and lower the bow

8) Which of the following impacts is more hazardous for ships?

a) Rolling b) Pitching c) Steering d) Waving

9) The gyroscopic torque generated by a truck turning left on a highway

a) abrupt in their behaviour b) the forces on the outer wheel are increased

c) the forces on the outer wheel are decreases d) has no impact on the forces acting on the outer wheel.

10) A bicycle is kept steady when moving through a bend because

a) curvature action b) gyroscopic action c) centrifugal action d) reaction action

Answers of Multiple-Choice Questions

1. (d) 2. (c) 3. (d) 4. (a) 5. (a) 6. (b) 7. (c) 8. (a) 9. (b) 10. (c)

Short and Long Answer Type Questions

1. In rigid body planar kinematics, what is a general motion?
2. What does "gyroscopic couple" mean to you? Derive the equation to determine the "gyroscopic couple" magnitude.
3. Describe the impact of the "gyroscopic couple" on a vehicle's four wheels as they navigate a curve.
4. What impact will the "gyroscopic coupling" have on a disc that is fixed at a specific angle to a rotating shaft?
5. Describe how gyroscopic concepts are applied to aeroplanes, ships and automobiles.
6. Describe the differences between applied and reactive torques.

7. What exactly are gyroscopic, precession, and spin planes?
8. Give an explanation of the gyroscopic couple's impact on a disc that is attached at a specific angle to a rotating shaft.
9. Describe how a "gyroscopic couple" affects ships when they are rolling and pitching.
10. How does the rider of a two-wheeler tilt to one side due to the effects of the "gyroscopic couple" and centrifugal force? Establish a relationship to determine the vehicle's limiting speed.
11. What does it indicate when the coin spins?
12. Why does a coin that is rolling continue to roll until it is almost completely stopped?

Numerical Questions

1) The cable spool displayed ass shown in Fig. P5.1 weighs 20 N and has an inertia moment of 0.38 kg.m^2. Assume the spool rolls without slipping. How much resistance does the ground present to the spool's motion and how quickly is the spool's mass centre moving?

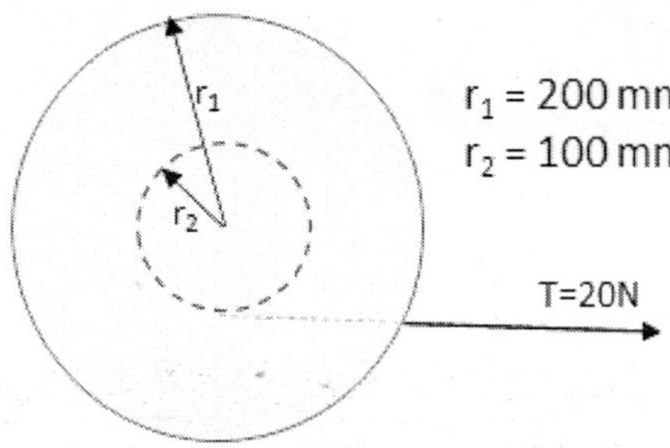

Fig. P5.1

2) The block shown in Fig. P5.2 has been set up so that A is attached to a rope with tension T. Block A weighs 12 kg and measures $h_2 = 0.8$ m in height and 0.4 m in width. The static and kinetic friction coefficients are provided as $\mu_s = 0.35$ and $\mu_k = 0.3$, and the rope links at a height of $h_1 = 0.4$ m. Analyse whether Block A slips or tips first and the force T necessary for this to happen.

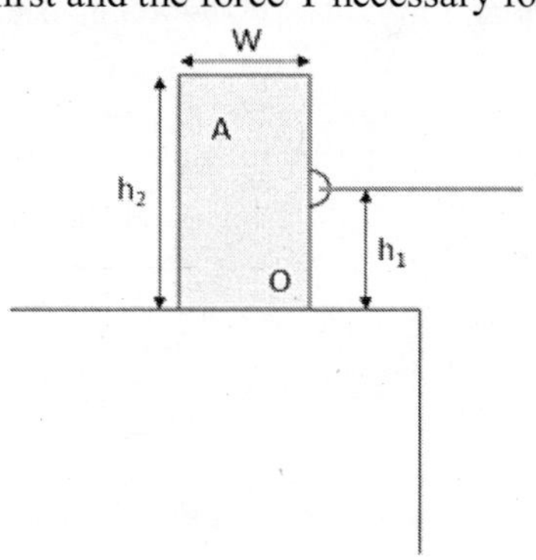

Fig. P5.2

3) When a particle is travelling in a plane and its position is known, $r = t\,i + 6t^3\,j + t^2\,k$, calculate the particle's velocity and average acceleration between t = 0 and t 4sec, also find the instantaneous acceleration at t=1.

4) The mid-point C is the pivot point for a 1 m long horizontal axle AB. It supports a 40 N weight at A and a 100 N wheel at B. When viewed from the front, the wheel is designed to rotate at a speed of 800 rpm (clockwise). Calculate the system's angular velocity of precession along the vertical axis through C, assuming that the flywheel's weight is distributed uniformly around the rim with a mean diameter of 0.6 m.

5) An aircraft is cruising at a speed of 120 m/s takes a right turn on an arc of 50 m radius. The turbines and propellers have a total mass of 600 kg with radius of gyration of 300 mm. Engine rotates at 2400 rpm. Calculate the magnitude of the "gyroscopic couple" generated.

6) Fig. P5.3 depicts a 2 m diameter of disc rolling on the ground without slipping. The disc's centre is travelling at a specific instant with an acceleration of 20 m/s^2 and a velocity of 20 m/s. Identify the amount of acceleration at which point P on the disc instantly touches the ground.

7) At point P, as shown in Fig. P5.4, a particle enters a circular loop with a smooth surface and no friction. If g is the acceleration caused by gravity, what speed the particle must enter the circular loop such that it can complete one full rotation of the circle.

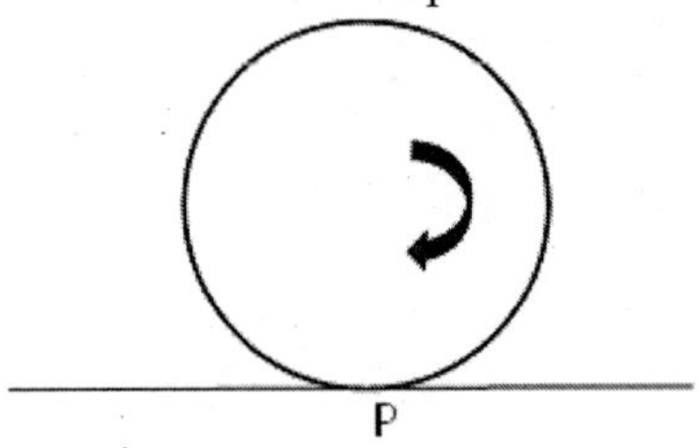

Fig. P5.3

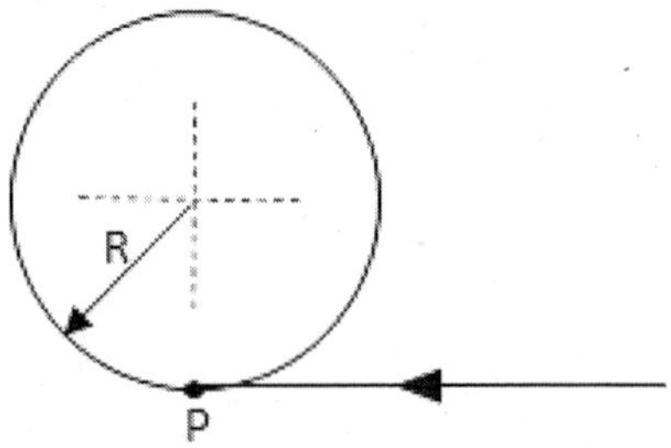

Fig. P5.4

8) The compound disc as shown in the Fig. P5.5 rolls on the horizontal floor. The linear velocity of the cord leaving A is 12 m/s rightward and the disk rotates counter clockwise at 6rad/s. Find the total velocity at the point B.

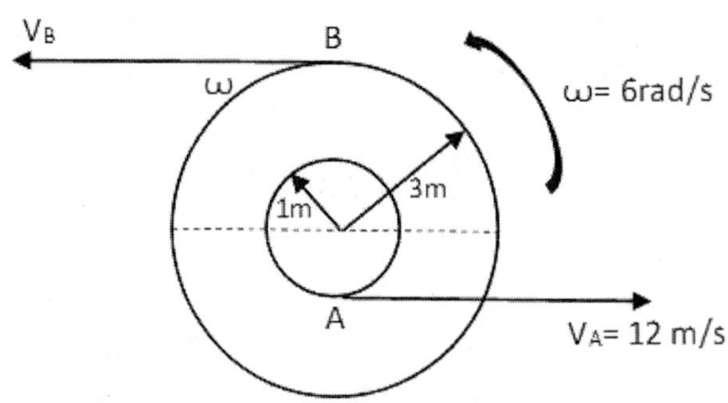

Fig. P5.5

9) A car is travelling at a speed of 30 m/s along a horizontally curving road with a radius of 150 m. When viewed from the front of the car, the engine's spinning masses rotate at a 200 rad/s angle in a

clockwise orientation. The rotating masses of total moment of inertia is 20 kg.m^2. Calculate the gyroscopic moment's size (in N-m).

10)The combined weight of a motorcycle and its rider is 2500 N, and while the motorcycle is upright, its centre of gravity is 650 mm from the ground. Each road wheel has a rolling diameter of 650 mm and a moment of inertia of 1.030 kg.m^2. The engine rotates 5 times as fast and in the same direction as the road wheels. The rotating engine components' moment of inertia is 0.165 kg.m^2. If the vehicle is travelling at 60 km/h around a 30 m curve, calculate the necessary angle of heel.

11)A vehicle with a rear engine is moving along a track with a mean radius of 120 m. The effective diameter of each of the four road wheels is 0.75 m, and each has a moment of inertia of 2.5 kg.m^2. The engine's rotational components have a 1.2 kg.m^2 moment of inertia. The crank shaft rotates in the same direction as the road wheels, and the engine axis is parallel to the rear axle. The back axle speed to engine speed ratio is 3:1. With a 1800 kg weight, the vehicle's centre of gravity is 600 mm above the ground. The vehicle's track is 1.5 m wide. Calculate the vehicle's maximum speed around the curve at which all four wheels must remain in touch with the pavement. Assume that there are no cambers in the road and that the vehicle's centre of gravity is centred in relation to its four wheels.

12)Derive an equation to determine the maximum angle of heel (θ) that a two-wheeled vehicle can have without skidding.

13)An aeroplane air screw, while viewed from the front, it rotates at a clockwise speed of 1200 rpm and a moment of inertia of 25 kg.m^2. If the flight travels at a 300 km/h, calculate the air screw's gyroscopic reaction when the plane's left-handed turn along a 200 m radius track.

14)A marine turbine's rotor rotates at 3600 rpm clockwise when viewed from the rear and has an 80 kg.m^2 moment of inertia. If the ship pitches with an angular simple harmonic motion with a periodicity of 15 seconds and an amplitude of 0.1 radians, calculate the rotor axis' maximum angular velocity, the maximum "gyroscopic couple", and the gyroscopic effect when the bow dips.

15)A locomotive's driving wheels have an axle moment of inertia of 250 kg.m^2. The wheel treads have a diameter of 1.8 metres and there are 1.6 metres between the wheel centres. A locomotive travelling at 120 km/h on a level track experiences a poor ballast that causes one wheel to drop by 6 mm and raise again in a total of 0.1 seconds. Find the "gyroscopic couple" created and the resultant reaction between the wheel and rail if the displacement of the wheel occurs with simple harmonic motion.

16)A motor cycle's road wheels have a 680 mm diameter and a moment of inertia of 1.2 kg.m^2. When the motorcycle is upright, the combined centre of mass of the rider and motorcycle, which weighs 250 kg, is 650 mm above the ground. The rotating engine components have a 0.2 kg.m^2 moment of inertia. In the same sense, the engine rotates five times as quickly as the wheels on the road. Calculate the required heel angle for a 90 km/h turn on a motorbike with a 40 m radius.

17)When viewed from the stern, the turbine rotor turns at 1800 rpm of a sea vessel with a mass of 1000 kg, With an angular velocity of 1.8 rad/s, the vessel pitches. What "gyroscopic couple" will the rising bow communicate to the hull if the rotor's gyration radius is 360 mm.

PRACTICAL

Aim: To determine "gyroscopic couple" on Motorized Gyroscope.

Apparatus required: Motorized Gyroscope as shown in Fig. P5.6.

Theory: A force in the direction of centripetal acceleration, also known as centripetal force, must be delivered externally over a body in order for it to move along the necessary curved path when it moves with a uniform linear speed over a curved path. Active force is the term used to describe this applied external force. A body experiences centrifugal force that is directed radially outwards while it is moving along a circular track at a constant linear speed. The term "reactive force" refers to this centrifugal force. Active "gyroscopic couple" is the name for the change in angular momentum (I, ω, Ω). When the spin axis rotates with an angular velocity of Ω, the disc is subject to a reactive couple whose magnitude (I ω Ω) is the same as the active couple's but which moves in the opposite direction.

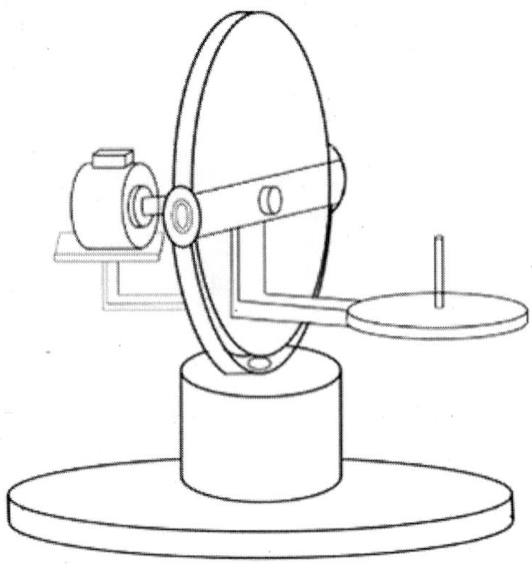

Fig P5.6: Motorized Gyroscope

Procedure:

1. Switch on the AC mains.
2. Set the variable transformer to its highest resistance level before turning on the AC mains.
3. Reduce the resistance of the variable transformer to gradually raise the speed of the steel disc to a tolerable level. Wait a while for the disc speed to stabilise. Utilize a tachometer to measure the RPM.
4. Place a known weight on the setup arm. The setup begins spinning about the vertical (zz) axis as soon as the weight is added. As the arm travels downward due to the weight, the rotation speed gradually slows down. This is due to the decrease in the torque component which causes precession due to the arm's increased inclination.
5. Take note of the circular scale's reading as soon as the setup begins to rotate across a range of 30^0 to 50^0. Take a stopwatch to record the time in addition to the movement.

Observation:

1. Weight of rotor:
2. Rotor Diameter:
3. Rotor thickness:
4. Distance of bolt centre of weight pan from disc centre:

S. No.	Disc speed (N), RPM	Angular Speed, (ω), rad/s	Circular scale reading			Angular Speed of precession, (Ω), rad/s	Calculated torque, (T=I ω Ω)	Applied Torque (T_a)	% error, $(T_a-T)/T_a$
			Rotation in		Time (s)				
			degrees	radians					

$I = mk^2 =$ _________ kg-m^2

Power = _________ watt

Applied Torque = $\frac{Power}{\omega}$

Result:

Gyroscopic couple is = -------------- N-m.

Precautions:

1. Use the apparatus carefully while experimenting.
2. Increase disc speed slowly.
3. Place weight carefully while disc is rotating.

KNOW MORE

Léon Foucault, sometimes known as Jean Foucault, or in full Jean-Bernard-Léon Foucault, was a French physicist who was born in Paris on September 18, 1819, and died there on February 11, 1868. His "Foucault pendulum" gave experimental evidence that the Earth revolves on its axis. Additionally, he invented and contributed to the development of a method for precisely determining the absolute speed of light.

Although Foucault received a medical education, his interests eventually shifted to experimental physics. He started a series of experiments into light and heat with Armand Fizeau. He proved in 1850 that light moves more slowly in water than it does in the air. He measured the speed of light in the same year and came up with a result that is within 1% of the actual number.

He established that Earth revolves about its axis in 1851 by analysing the motion of a large iron ball swinging from a cable that was 67 m long. A "Foucault pendulum" of this kind always oscillates in the same vertical direction. The pace and direction of this vertical plane's movement on a spinning Earth, however, depends on the pendulum's latitude. Foucault was appointed physical assistant at the Imperial Observatory, Paris, and awarded the Copley Medal of the Royal Society of London in 1855 for this demonstration and another similar one with a gyroscope.

He built a better reflecting telescope mirror, established the existence of eddy currents, or "Foucault currents," in a copper disc rotating in a strong magnetic field, and in 1859 created a straightforward but incredibly precise technique for examining telescope mirrors for surface flaws.

REFERENCES AND SUGGESTED READINGS

- Meriam, J.L. and Kraige, L.G., Engineering Mechanics: Dynamics, Wiley, 2011.
- Hibbeler, R.C., Engineering Mechanics: Dynamics, Pearson Prentice Hall, 2010.
- Rattan, S.S., Theory of Machines, Tata McGraw Hill Education pvt. Ltd, .2009

6 Beam Reactions, Shear Force and Bending Moment

UNIT SPECIFICS

Through this unit we have discussed the following aspects:

- *Classification of structures;*
- *Types of beams and loadings;*
- *Internal loads in beam structures;*
- *Shear force and bending moment sign conventions and calculations;*
- *Relationship between shear force and bending moment;*
- *Problems involving calculation of shear force and bending moments;*

RATIONALE

This unit on internal forces of beam structures helps students to understand about the types of structures, beam loadings, and shear force and bending moment calculations. These are relevant to the students since the knowledge gained shall be applied to derive the bending equations of the beams. The strength calculations, stresses and deflections induced in the beams and combined loading of mechanical structures wherein bending is one of the components of deformation shall require adept skill and knowledge of deriving the shear force and bending moment equations and calculating respective values.

PRE-REQUISITES

Calculus

Mechanics

UNIT OUTCOMES

List of outcomes of this unit is as follows:

U6-O1: Distinguish and classify structural elements

U6-O2: Classify types of beams and their loads

U6-O3: Describe internal loads generated in beams

U6-O4: Derive the beam shear forces and bending moments for different load combinations

U6-O5: Realize the application of shear force and bending moment calculations practically

U6-O6: Apply the knowledge gained to solve for stresses and deflections of beams.

Unit - 6 Outcomes	EXPECTED MAPPING WITH COURSE OUTCOMES (*1- Weak Correlation; 2- Medium Correlation; 3- Strong Correlation*)					
	CO-1	CO-2	CO-3	CO-4	CO-5	CO-6
U6-O1	3	3	3	-	3	1
U6-O2	1	1	2	2	1	-
U6-O3	2	1	3	1	2	1
U6-O4	-	-	3	1	2	2
U6-O5	3	3	3	-	3	1

6.1 INTRODUCTION

In structural mechanics, members are identified and distinguished based on the loads that these are carrying. A member, of slender dimensions, if loaded axially, is usually called *bar* or *rod* and may be under tension or compression, depending upon the direction of the load applied along the axis. Similarly, if a structural member of slender dimensions is loaded by relative twisting at its two ends, then it is called a *shaft* under torsion, whereas a similar structure when loaded transversely or laterally, is called a *beam* and is subjected to bending. Hence, a beam resists laterally applied loads and primarily deflects by bending. The applied lateral loads give rise to reaction forces at the beam's end supports and results in shear forces and bending moments, thus causing deflections, internal stresses and strains in the beam.

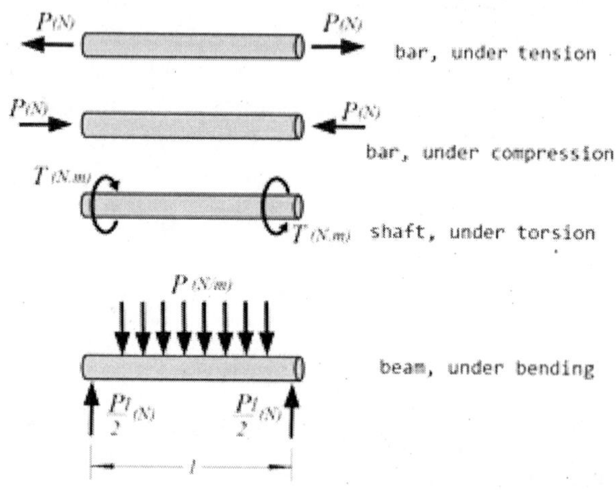

Fig. 6. 1 Classification of structural members

Beams are classified by end support types, cross-sectional area, length, and material. One must remember that in all the cases, as discussed above, the geometry of the slender structural member may or may not be same, as shown in Fig.6. 1. However, how much slender the structure must be to be called a slender structure depends on the slenderness ratio. Slenderness ratio for a structure is defined as length to thickness ratio (***l/h***). For slenderness ratio (***l/h***) > 20, the beams are treated as Euler beams and such beams are known to obey the simplified assumptions of beam kinematics as the deformations of such beams is commonly known to be caused by the bending stresses only. In contrast to Euler beams, beams with slenderness ratio (*l/h*) < 10 are much shorter and are commonly known as Timoshenko beams wherein the beam kinematics are treated with different set of assumptions due to the considerable presence of shear stresses along with bending stresses. The intermediate range of 10<*l/h*<20 is an uncertain range wherein the simplified assumptions of elementary beam theory does not remain valid.

6.2 TRANSVERSE LOADING ON BEAMS

As noted above, any structural member of slender dimensions (i.e., of sufficient axial length in comparison to the lateral dimensions), when supported and loaded transverse to its longitudinal axis is usually referred to as a beam and the transverse loads acting on the beam are oriented to exist in a plane defined by a cross-sectional axis of symmetry. Generally, the beam is assumed to be oriented horizontally and consequently, the applied loads are considered in a vertical (orthogonal to the beam axis) plane passing through the axis of symmetry. The extent of transverse load that the beam can carry depends upon (a) the intermediate distance between the end supports, (b) length of overhang from support, (c) nature of end supports, and (d) beam material properties.

Due to transverse loading, for the beam to be in equilibrium, the ends of the beam need to be supported. Hence beams are classified according to their end support conditions and are named as cantilever beams, simply supported beams, overhanging beams or fixed beams. These beams may carry external loads or moments and the different combinations of load acting on different types of beams are shown in Fig.6. 2.

Cantilever is a beam projecting freely from a fixed end as shown in Fig.6. 2(b). The requirement of such beams arises when the slope at one end of the beam must be maintained to zero and hence it becomes necessary to fix that end. For fixing the end of the beam, it is generally built in the wall and mathematically, in the free body diagram, the fixing of end of the beam is represented by showing a 'fixing moment' along with other support reactions. A simply supported beam (SSB) is supported at two ends using roller/knife or/and hinge (pin) supports as shown in Fig.6. 2(a). In case a beam is supported along its span at a certain distance from the end points on one or both sides, then such beams are called overhanging beams (Fig.6. 2(d)). Apart from these, other beam configurations are also possible depending upon the application and load supported, such as fixed beam or *encastrè* as shown in Fig 2(c) or a propped cantilever, wherein the free end of a cantilever is supported or 'propped' by means of a simple support.

The reader may observe carefully and assess the various mathematical possibilities of combinations of loads on beams and are also required to observe their surroundings to identify these combinations in physical world. The beam, under the transverse load bends as shown in Fig.6. 3.

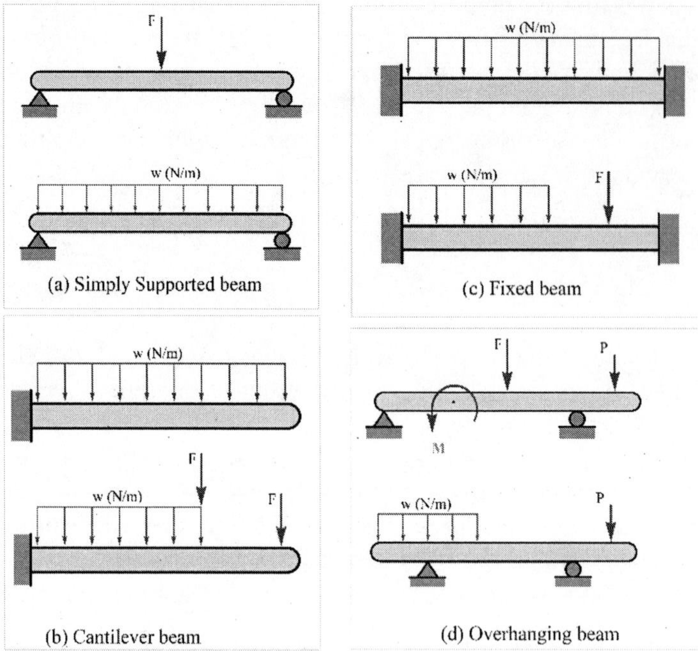

F (N) and ***P*** (N) are the point or concentrated loads,
M (N.m) is the external moment and ***w*** (N/m) is the distributed load.

Fig. 6. 2 Types of beams and loadings

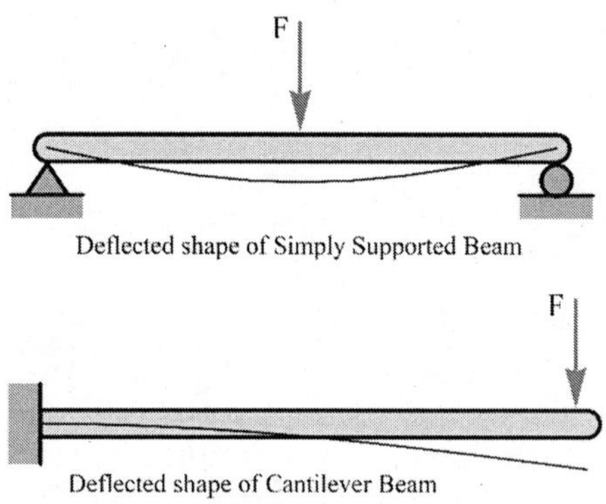

Fig. 6. 3 Deflected centrelines of beams

6.3 EQUILIBRIUM OF STRUCTURES

Structures, under external loads, must exist in equilibrium. The conditions of equilibrium for structures in two dimensions (assuming in X-Y plane) and three dimensions are mathematically expressed in form of Eq. 1 and Eq. 2 respectively as:

$$\sum F_x = 0;\ \sum F_y = 0;\ \sum M_z = 0 \tag{6.1}$$

$$\sum F_x = 0;\ \sum F_y = 0;\ \sum F_z = 0;\sum M_x = 0;\ \sum M_y = 0;\ \sum M_z = 0 \tag{6.2}$$

In these equations, $\sum F$ refers to the algebraic summation of the external forces acting along a certain axis (X, Y or Z), represented by the subscript and $\sum M$ refers to the algebraic summation of all the external moments acting along a given plane, the orthogonal axis to which is indicated by the subscript letter. For a structure loaded externally, if the above conditions are met satisfactorily, then it is understood that the structure is being restrained under some supports. Expressed in a roundabout way, structures which are supported sufficiently to prevent any kind of displacement or rotation, under the action of external forces and moments, shall satisfy the condition of equilibrium. Consequently, the reader must now realize the support types and conditions since the information of support type provided is very important to ensure its stability. Supports usually join the member to some fixed surface such as ground or other non-moving part of the structure. The characteristics of some of the supports shown below in Fig.6. 4.

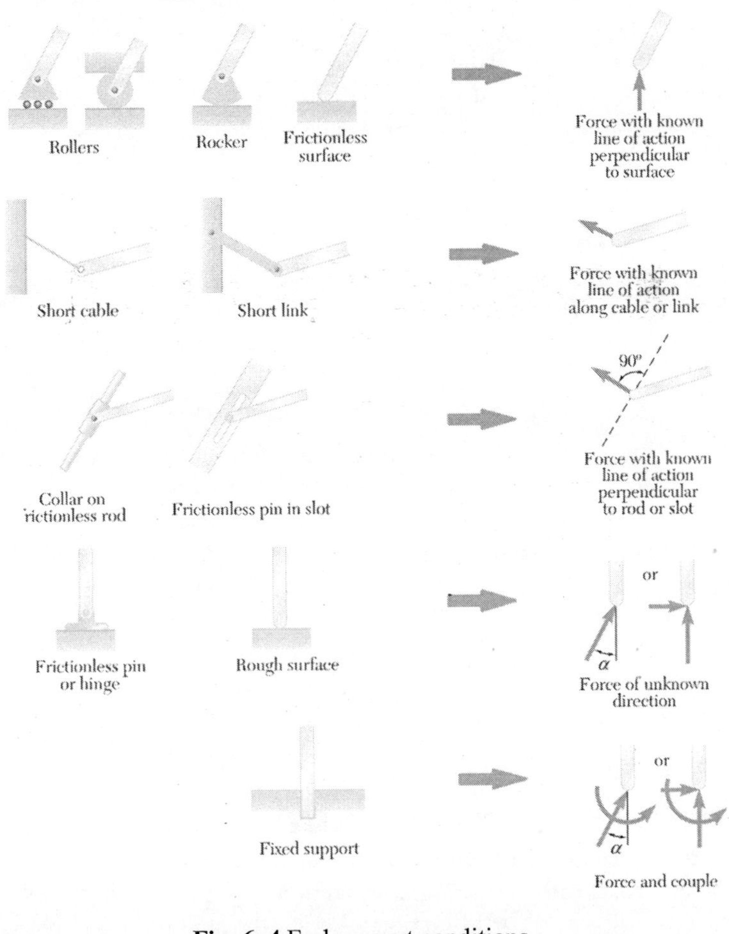

Fig. 6. 4 End support conditions

6.4 INTERNAL LOADS: SHEAR FORCE AND BENDING MOMENT IN BEAMS

Internal loads are always present at every point inside the rigid body but are not visible as they always occur in equal-and-opposite pairs and cancel each other. A bar, under axial pull or push has axial internal loadings, either tension (or compression) as shown in Fig.6. 5(a). However, a transversely loaded beam, under support conditions, undergoes two additional but different kinds of internal loadings, namely, internal shear forces and internal bending moments (as shown in Fig.6. 5(b)).

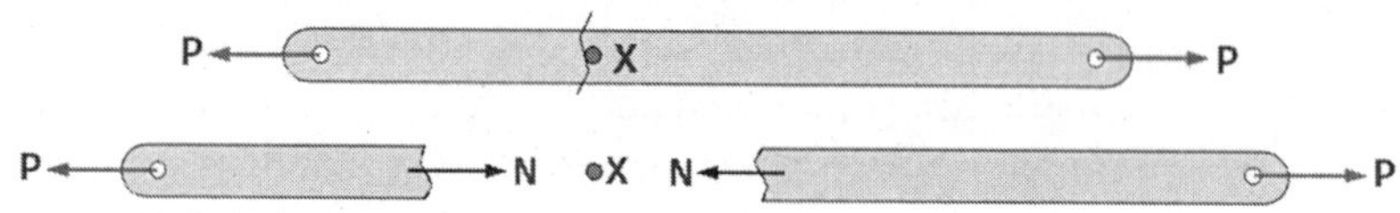

(a) Internal loads in bar under axial force

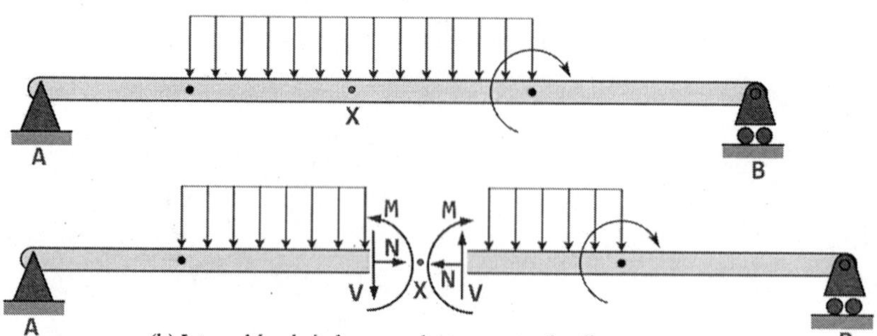

(b) Internal loads in beam under transverse loading

Fig. 6. 5 Internal loading in bars and beams

To design a beam, prior knowledge of the locations and magnitude of maximum shear force as well as maximum bending moment is of utmost importance. Ultimately, for a beam, the deflections under the applied loads (Fig.6.3) induces strains, stresses and displacements. These induced quantities vary along the beam's length and height. These could be easily established when one has the equations depicting the variation of shear force and bending moment along the span of the beam. So, the first step in calculating spatial variation of stresses and deflections is to calculate the shear forces and the bending moments across the length of the beam. A more convenient approach is to visualize the distribution of shear force (SF) and bending moment (BM) by plotting the diagrams. Hence, shear force (SF) and bending moment (BM) diagrams are plots/graphs which report the distribution of internal SF and BM plotted along the length of the beam. These allow one to observe graphically the locations along the span where maximum loads occur thereby allowing to optimize the weight and cost of the structure and to prevent failures. So before plotting these diagrams, one must be aware of the definitions of shear force and bending moment. In a beam, for any section that is analysed, the algebraic summation of all the transverse forces is called the *Shear Force* acting on the either side of the said section whereas the algebraic summation of the moments of all the transverse forces is called the *Bending Moment* acting on the either side of the said section.

In order to have same magnitude and sign of SF and BM on either side of the chosen beam section, a convenient sign convention must be adopted for the calculations.

6.5 Sign Conventions

The standard sign conventions used in applied mechanics are sometimes not convenient for depicting the internal loads in cases of beams. Shear forces, acting along vertically upward direction, are not always positive since these acts in pairs on the cut section and at any given point in the cut, the shear force acts in both, up and down directions. Hence for depicting internal loads, a different sign convention is needed. The selection of this is somewhat arbitrary and may differ from text to text, however, the choice of these sign conventions does not affect the standard sign convention used for equilibrium equations.

To maintain the consistency of the solution, the readers must persist with a particular sign convention of internal loads for a given problem. For this text, the sign conventions of SF and BM are shown in Fig.6. 6. One must note that though the sign conventions are arbitrarily chosen and could be reversed. The conventions chosen in this text result in providing mathematically correct interpretations (signs) of slopes and deflections of beams and, hence, are strongly endorsed.

In this text, shear force that moves the left section up or right section down are taken positive shear force whereas, shear force that moves the left section down or the right section up is regarded as negative shear force. In other words, shear force, upon breaking the section, if rotates the element clockwise, is a positive shear force, otherwise, is a negative shear force.

A bending moment that causes sagging by bending the beam thus producing concavity upwards is considered positive, while conversely, a bending moment that tends to hog the beam by bending it to produce convexity upwards is considered negative. In this unit, mainly the shear force and bending moment calculations of simply supported, overhanging and cantilever beams are discussed under different load conditions.

In case of SSB or Overhanging beams, the end reactions at beam supports are to be first calculated using the equilibrium equations (Eq. 1), as applicable. Once the end reactions are calculated, depending upon the beam loading, one must ascertain the total number of equations to be derived for calculating SF and BM each. This is simply determined by the loads and their locations on the beam such that the applications of loads/moments divide the beam span into sub-spans. To aid the reader in grasping this concept, an example of SSB under a combination of different loadings is explained in example 2.

Once the SF and BM equations for each sub-span are derived, the SF and BM diagrams are plotted by varying the independent variable, x in each equation. However, one must note that the SF and BM must be calculated at the start and end of each sub-span because point (concentrated) loads, and external moments always serve as singularity functions in the equations of SF and BM respectively thereby adding/subtracting their respective values by 'jumps' at their respective locations. Cantilever beams, being free at one end, are comparatively easier to solve since the necessity of calculating the end reactions in case of cantilever beams do not exist and one may straightaway initiate the solution by deriving then SF and BM equations from the free end. The unbalanced SF and BM calculated at the fixed end yield the balancing end reactions and end moments in such cases.

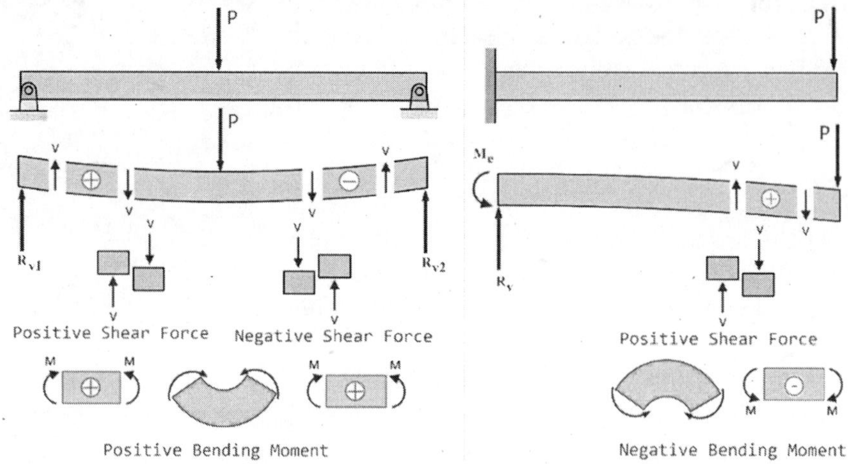

Fig. 6. 6 Sign Conventions for shear force and bending moment

Example 1: Determine the distributions of internal loads for the SSB shown in Fig.6. 7.

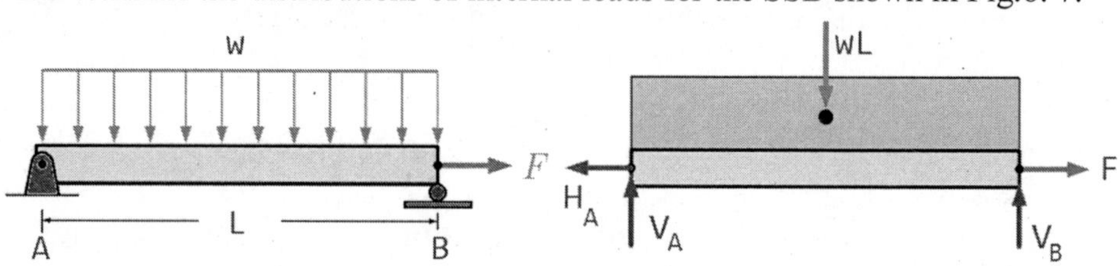

Fig. 6. 7 SSB carrying UDL

Solution: The beam carries an axial load ***F*** at point B and a uniformly distributed load of intensity ***w*** over its entire span. The free body diagram of the beam is shown subsequent figure. The total transverse load acting on the beam along its span is measured by representing the distributed load as a rectangle of sides ***L*** (as the load intensity spans over ***L***) and ***w*** (the load intensity). So, the total transverse load acting on the beam is ***wL*** (area of the rectangle) and is supposed to act from centroid of the rectangle. Support A, being a hinge/pin, shall have two reactions, H_A and V_A as shown in the figure and support B, being roller, shall have only one reaction,

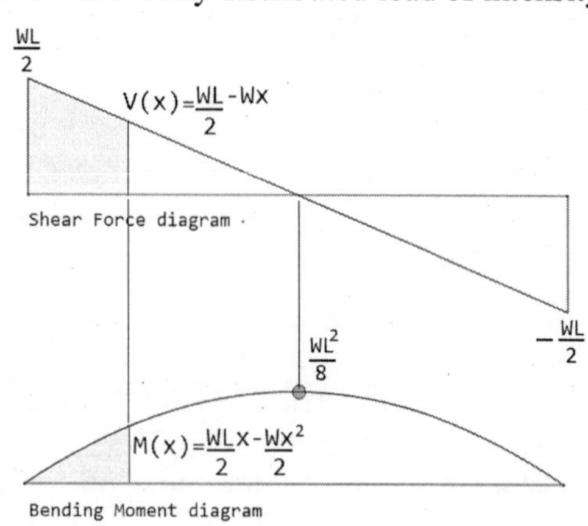

V_B. 1. First, the end reactions are calculated. Using the equilibrium equations

$$\sum F_x = 0; \quad -H_A + F = 0; \quad H_A = F$$

$$\sum F_y = 0; \quad V_A + V_B - wL = 0; \quad V_A + V_B = wL$$

$$\sum M_B = 0; \quad (wL).\frac{L}{2} - V_A.L = 0;$$

$$V_A = \frac{wL}{2}, \quad V_B = \frac{wL}{2}$$

2. Next, the internal loads are calculated by assuming a cut section of the beam at distance x from A. As shown in figure, the transverse forces acting on the cut section of the beam, when added algebraically, gives the shear force at C and the algebraic summation of moment of all the forces acting at C results into the bending moment at C.

Hence the summation of transverse forces in the section is:

$$V = V_A - wx$$

Summation of moments of forces at C is

$$M = V_A x - wx.\frac{x}{2}$$

And the axial force in the section is

$$N = F$$

3. Using the equations of V and M derived above, the shear force and bending moment diagrams are plotted.

6.6 RELATIONSHIPS BETWEEN LOAD INTENSITY, SHEAR FORCE AND BENDING MOMENT

Let a simply supported beam of span L, transversely loaded with UDL of intensity w (N/m) between its two supports, as shown in Fig.6. 8(a), be considered for deriving the relationship amongst the load intensity, w, with shear force and bending moment. Let V and M be the values of shear force and bending moment of an element of beam located at position x from the left support. Also, let the SF, and BM at a section located at x+dx from the left end support be V+dV and M+dM respectively (Fig.6. 8(b)). The element carries a total load of w.dx (N) through its centre.

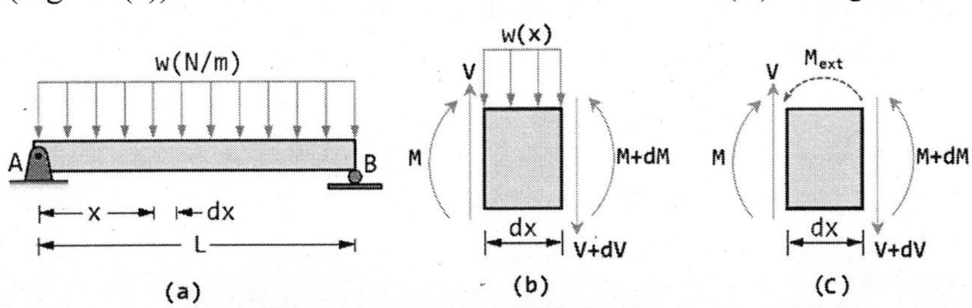

Fig. 6. 8 Simply supported beam with uniformly distributed load

Taking equilibrium of moments for the elementary length, **d*x***, one arrives at the bending moment at section ***x*+d*x*** as follows:

$$M_{x+\mathrm{d}x} = M + \mathrm{d}M = M + V\mathrm{d}x - w\mathrm{d}x.\frac{\mathrm{d}x}{2}$$

$$M_{x+\mathrm{d}x} = M + \mathrm{d}M = M + V\mathrm{d}x \text{ (Ignoring the last term)}$$

$$\frac{dM}{dx} = V(x) \qquad (6.3)$$

$$M_B - M_A = \triangle M = \int_A^B V(x)\mathrm{d}x \qquad (6.4)$$

Hence, from equation (6.3), it is established that the first derivative of BM with respect to ***x*** yields the SF at a given location in the beam. Further, when this attains a zero value at a location, then shear force at that location changes sign and bending moment attains a maximum value. If the bending moment diagram is plotted, then equation (6.3) also implies that the slope of the bending moment diagram at a given location along the beam span gives the value of shear force at the same location. Further, in equation (6.4), the RHS is equal to the area under the shear force diagram and is equal to the variation in bending moment ($\triangle M$) between any two points on the beam span. These relations are valid for concentrated forces too.

Similarly considering the equilibrium of forces for the elementary length, **d*x***, one arrives at the shear force at ***x*+d*x*** as follows:

$$V_{x+\mathrm{d}x} = V + \mathrm{d}V = V - w\mathrm{d}x$$

$$\frac{dV}{dx} = -w(x) \qquad (6.5)$$

$$V_B - V_A = \triangle V = -\int_A^B w(x)\mathrm{d}x \qquad (6.6)$$

Here, equation (6.5) implies that first derivative of the SF with ***x*** is yields the intensity of the distributed load and it can be inferred from equation (6.6) that the area under the load-span diagram (RHS of eq. 6.6) yields the change in shear force for a given length between any two points on the beam span. From equations (6.3) and (6.5):

$$\frac{d^2M}{dx^2} = -w(x) \qquad (6.7)$$

Interestingly, the relationship between shear force and bending moment presented in equations (6.3) and (6.4) does not remain valid for external moments/couple acting between A and B. From Fig.6. 8(c), the equilibrium of forces results into **d*V***=0. This means that shear force does not change at the point of application of external moment/couple. However, the equilibrium of moments of the section is as follows.

$$M_{x+\mathrm{d}x} = M + \mathrm{d}M = M + V\mathrm{d}x - M_{ext}$$

$$dM = V\mathrm{d}x - M_{ext} \qquad (6.8)$$

From eq. (6.8), it becomes evident that, at the point of application of external moment, the bending moment undergoes an abrupt change.

Example 2: Determine and plot the SF and BM of the SSB as shown in the Fig.6. 9(a).

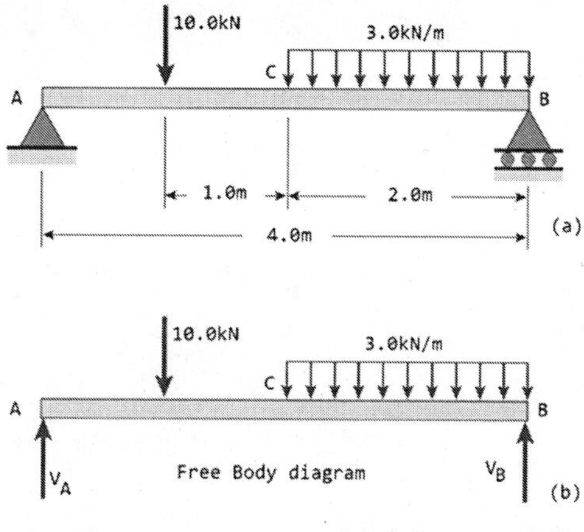

Fig. 6. 9 Simply Supported beam

Solution:

End reactions:

$\sum F_y = 0;$

$V_A + V_B - 10 - 3x2 = 0;$

$$\sum M_B = 0;$$

$(3x2).\frac{2}{2} - V_A x4 + 10x3 = 0;$

$V_A = 9\text{kN}, \quad V_B = 7\text{kN}$

To derive the equations of shear force and bending moment distributions, the beam subspans are shown in Fig.6. 9(c-e).

In the table below, as per the sign conventions, while calculating shear forces, transverse force is taken positive if it tends to shear the left portion (Fig.6.9(c-e)) in upward direction, otherwise negative. Similarly while calculating the bending moments, the moment of transverse forces that tend to sag the left portion of the beam are taken positive, otherwise negative. The distributions of shear force (SF) and bending moment (BM) are plotted above. In the remaining examples, too, same sign conventions are followed to arrive at the given solution.

S. No	Sub-span	Shear force at x, V(x)	Bending moment at x, M(x)
1	$0 \le x \le 1$	V_A	$V_A.x$
2	$1 \le x \le 2$	V_A-10	$V_A.x - 10.(x-1)$
3	$2 \le x \le 4$	$V_A - 10 - 3.(x-2)$	$V_A.x - 10.(x-1) - 3.\frac{(x-2)^2}{2}$

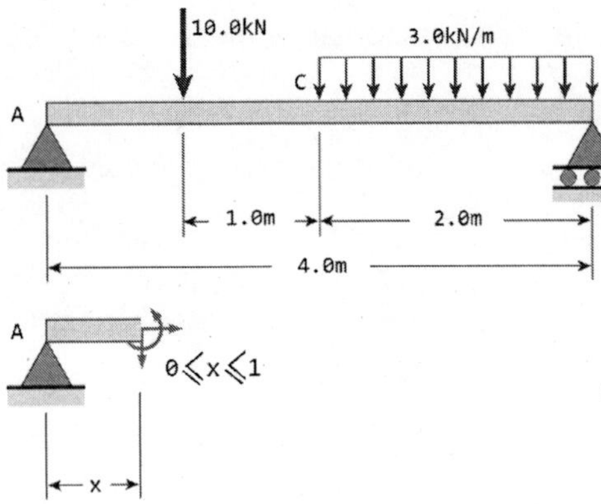

(c)

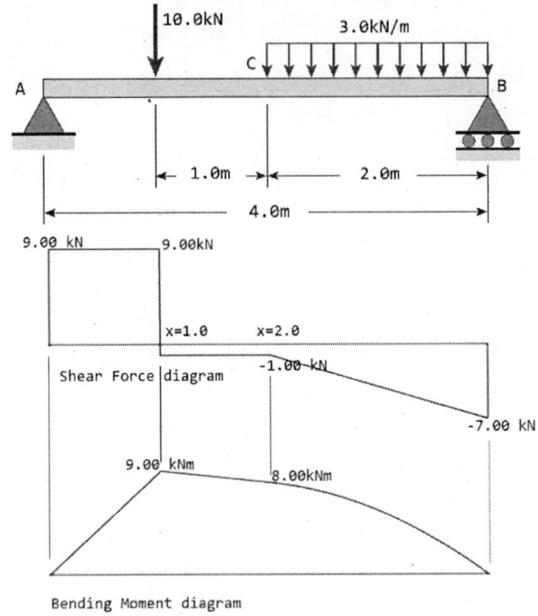

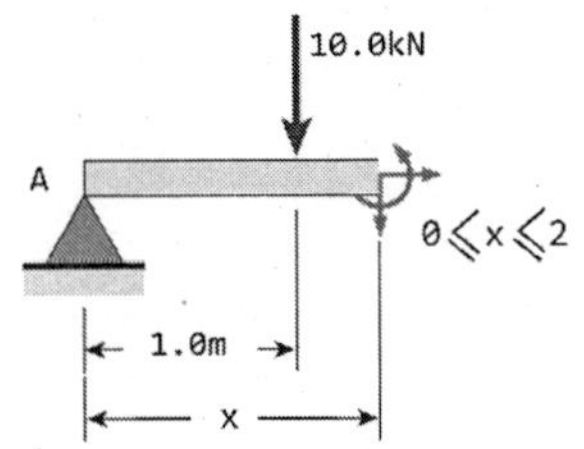

(d)

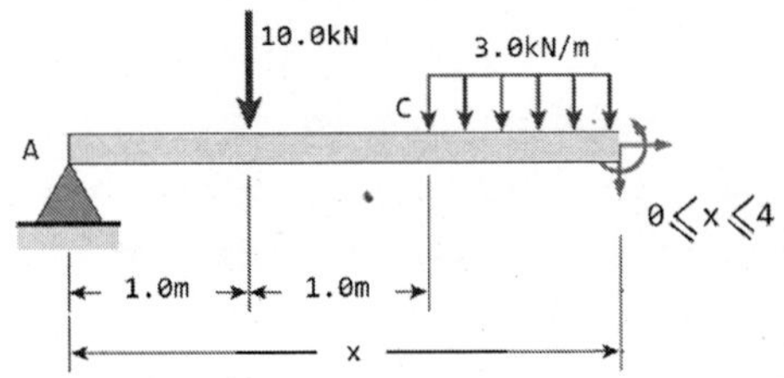

(e)

Example 3: Determine and plot the SF-BM of the overhanging as shown in the Fig.6. 10(a).

Solution:

End reactions: $\sum F_y = 0; V_A + V_B - 5x4 - 2x2 = 0;$

$\sum M_B = 0; (5x4).\left(\frac{4}{2} + 4\right) - V_A x8 - (2x2).(2 + \frac{2}{2}) = 0;$

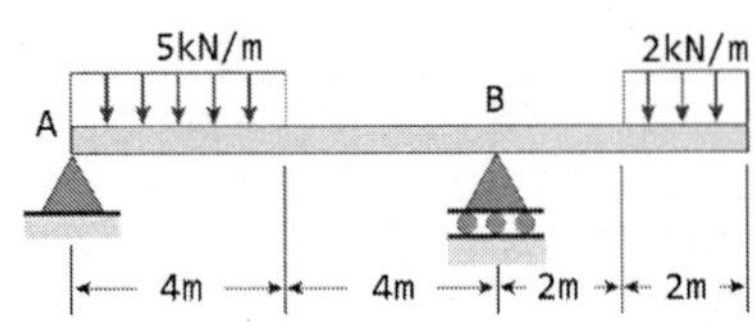

(a)

$V_A = 13.5\text{kN}, \quad V_B = 10.5\text{kN}$

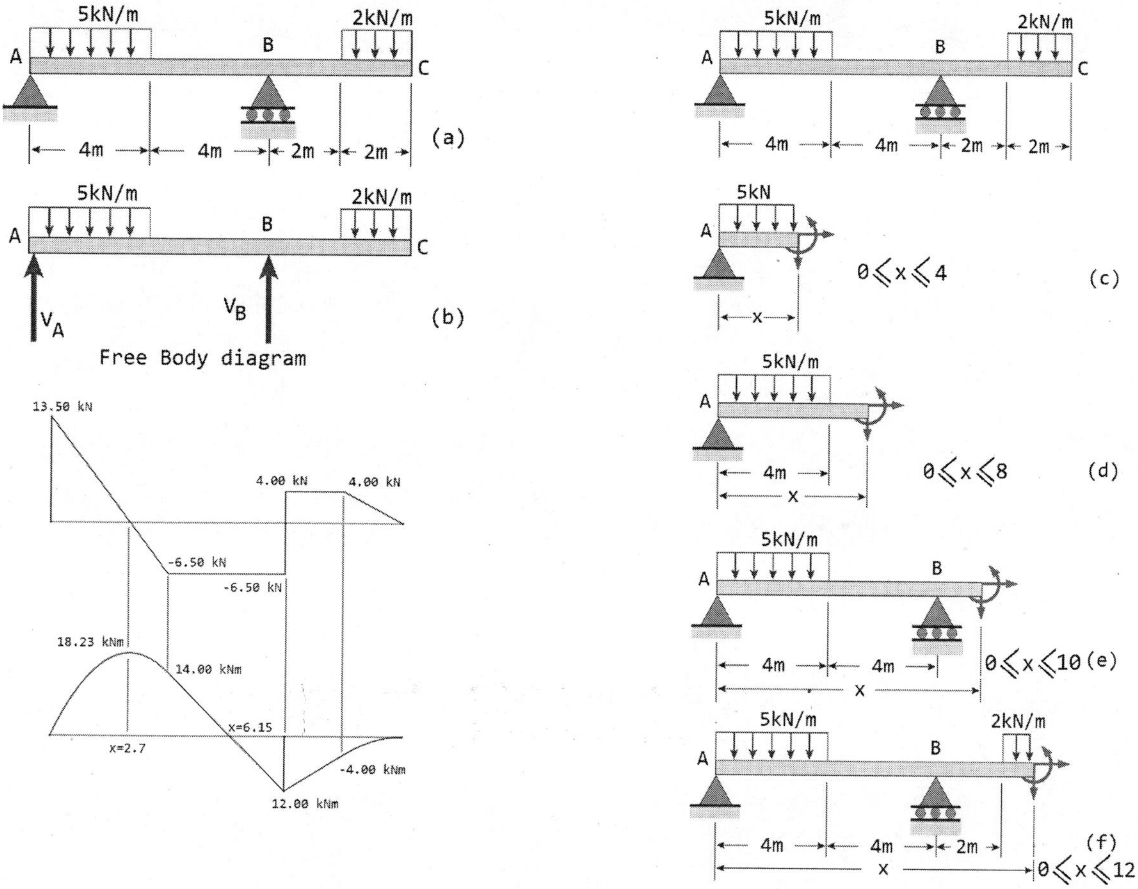

Fig. 6. 10 Overhanging Beam

To derive the equations of shear force and bending moment distributions, the beam subspans are shown. One must note here that the bending moment in case of overhanging beams (other type of beams too, depending upon the combination of external load and moments applied) may change sign along the beam span. Readers may recall from their courses in mathematics that when any function (plotted along y-axis, i.e., dependent variable) changes its sign along the x-axis (independent variable), then the corresponding x-value is called inflexion point.

As happens in case of bending moment, too, the point along the beam span (x=6.15 m, in this case) where bending moment changes sign, is the inflexion point and is popularly known as ***point of contraflexure***. The significance of point of contraflexure is that it is at this point, the tension side switches to be the compression side and vice-versa. Visually, one may say that at this point, the sagging flips to hogging and vice-versa and hence the beam is said to flip its curvature. In reinforced beams (concrete and steel, say), the design of the beam calls for lesser flexural reinforcement.

SF-BM equations:

S. No	Sub-span	Shear force at x, V(x)	Bending moment at x, M(x)
1	$0 \le x \le 4$	$V_A - 5x$	$V_A.x - 5.\frac{x^2}{2}$
2	$4 \le x \le 8$	$V_A - 20$	$V_A.x - 20.(x-2)$
3	$8 \le x \le 10$	$V_A - 20 + V_B$	$V_A.x - 20.(x-2) + V_B.(x-8)$
4	$10 \le x \le 12$	$V_A - 20 + V_B - 2.(x-10)$	$V_A.x - 20.(x-2) + V_B.(x-8) - 2.\frac{(x-10)^2}{2}$

Example 4: For the beam shown in Fig.6. 11, determine and plot the SF-BM distribution.

Solution: The end reactions are calculated first.

$\sum F_y = 0;$

$V_A + V_B - 5\text{x}10 - 4 - 6 = 0;$

$\sum M_B = 0;$

$(5\text{x}10).\left(\frac{10}{2}\right) + 6\text{x}10 + 4\text{x}20 - V_A\text{x}15 = 0;$

$V_A = 26\text{kN}, \quad V_B = 34\text{kN}$

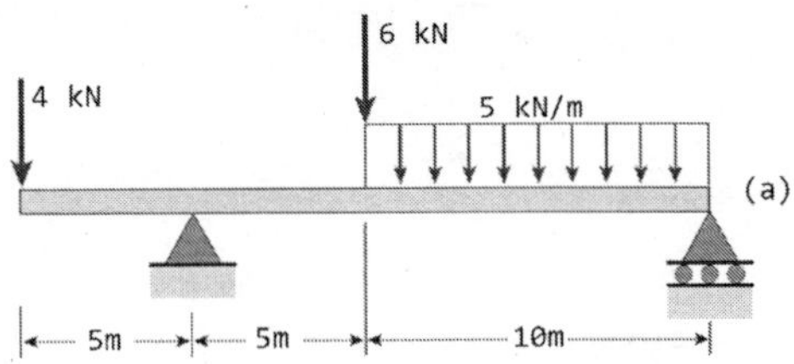

Fig. 6. 11 Overhanging beam

The beam subspans are shown in the corresponding figure and are used to derive the shear force and bending moment equations.

S. No	Sub-span	Shear force at x, V(x)	Bending moment at x, M(x)
1	$0 \le x \le 5$	-4	$-4x$
2	$5 \le x \le 10$	$-4 + V_A$	$-4x + V_A(x-5)$
3	$10 \le x \le 15$	$-4 + V_A - 6 - 5(x-10)$	$-4x + V_A(x-5) - 6(x-10) - 5\frac{(x-10)^2}{2}$

Point of contraflexure, maximum bending moment and its location are shown in the bending moment diagram.

Example 5: For the cantilever beam shown in Fig.6. 12, calculate and plot the shear force and bending moment diagrams.

Solution: In case of cantilever beams, the shear force and bending moments could be derived from the free end to do away with the necessity of calculating the end reaction forces and moments. The shear force and bending

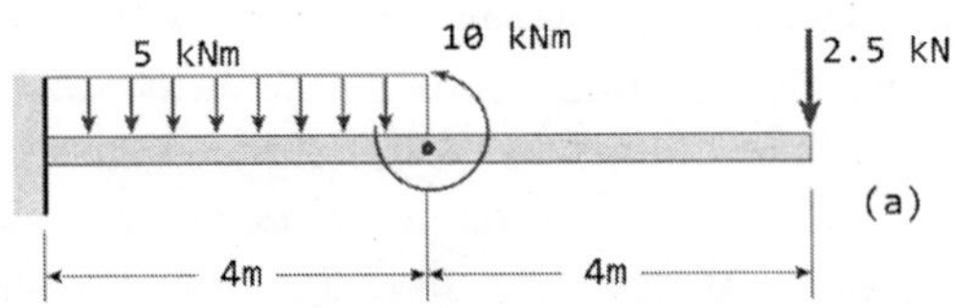

Fig. 6. 12 Cantilever beam

moment values at the fixed end yield the values of V_A and M_A.

The beam subspans are shown in the corresponding figure and are used to derive the shear force and bending moment equations.

S. No	Sub-span	Shear force at x, V(x)	Bending moment at x, M(x)
1	$0 \le x \le 4$	2.5 (+ve, since downward right section)	$-2.5x$ (-ve, due to hogging)
2	$4 \le x \le 8$	$2.5 + 5(x-4)$	$-2.5x + 5\frac{(x-4)^2}{2} + 10$ The third term is due to the external moment acting at x=4m, causing sagging of the beam

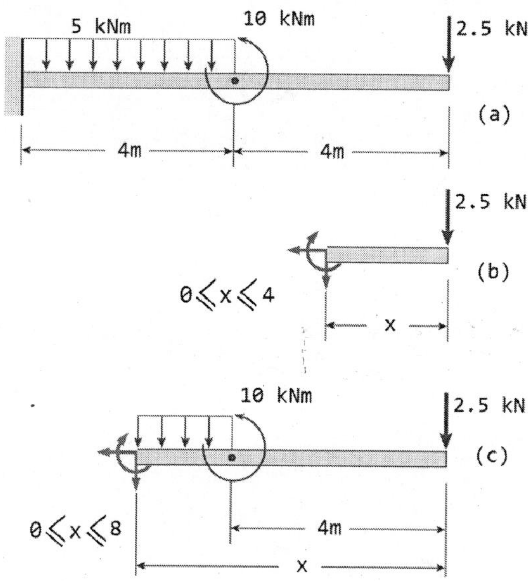

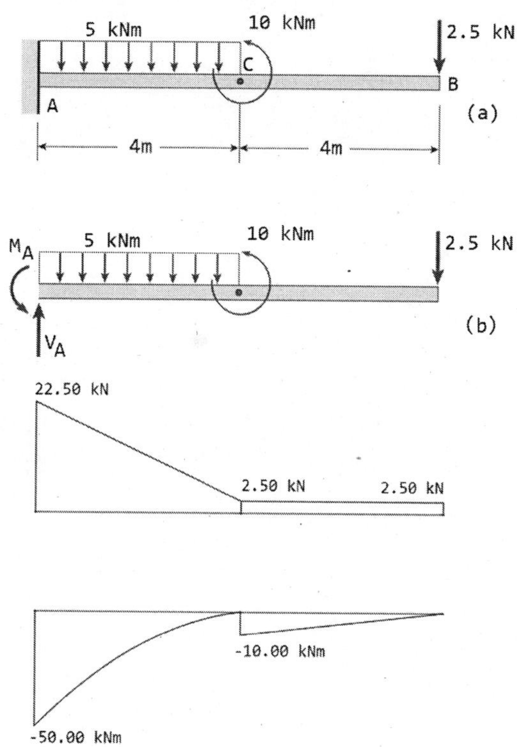

The end reactions at the fixed end are as follows:
$V_A = 22.5$ kN
$M_A = 50$ kNm
Readers must note here that the sign of the end reactions and end moments at the fixed end are not decided by the sign conventions applicable for SF and BM calculations but by the sign conventions that are standard for equilibrium equations.

Example 6: For the beam loading shown in Fig.6. 13, plot the SF-BM diagrams.

Solution: The end reactions are calculated first as shown in the figure.

$\sum F_y = 0; V_A + V_B = 0;$

$\sum M_A = 0; -10 - 20 + V_B x12 = 0;$

$V_B = 2.5\text{kN}, \quad V_A = -2.5\text{kN}$

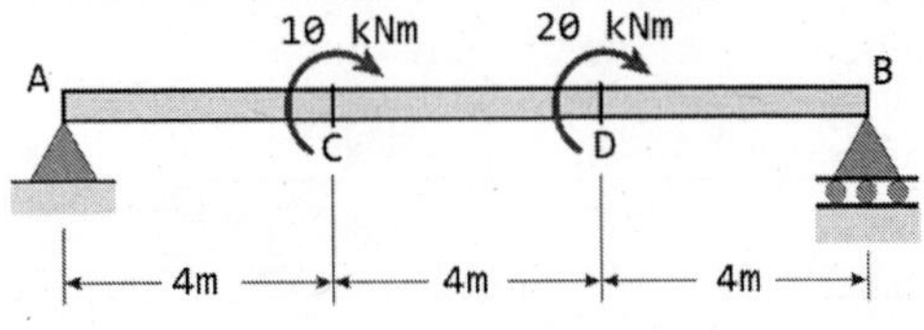

Fig. 6. 13 SSB with external moments

The beam subspans are shown in the corresponding figure and are used to derive the shear force and bending moment equations.

S. No	Sub-span	Shear force at x, V(x)	Bending moment at x, M(x)
1	$0 \le x \le 4$	-2.5	$-2.5x$
2	$4 \le x \le 8$	-2.5	$-2.5x + 10$
3	$8 \le x \le 12$	-2.5	$-2.5x + 10 + 20$

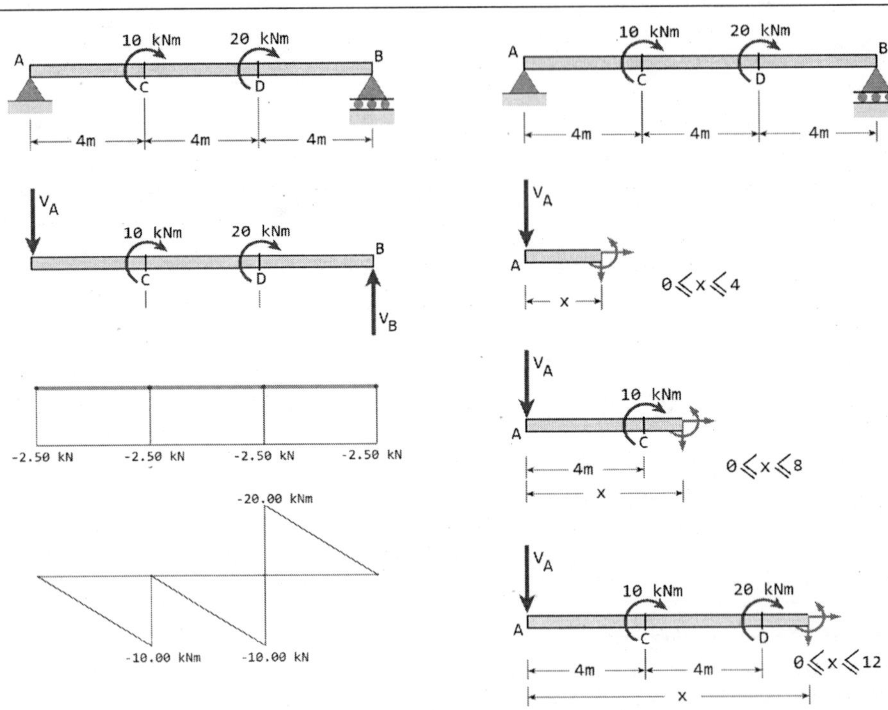

Readers must note that, point of contraflexure may appear in any type of beam depending upon the combination of load, moments and position of supports of the beam. In this, problem, though of simply supported beam, due to the application of external moments, the point of contraflexure appears at x=8m where the bending moment changes sign.

Example 7: For the beam resting on the ground and carrying the 8kN/m UDL as shown in Fig.6. 14, determine the BM at its mid-span and plot the SF-BM diagram.

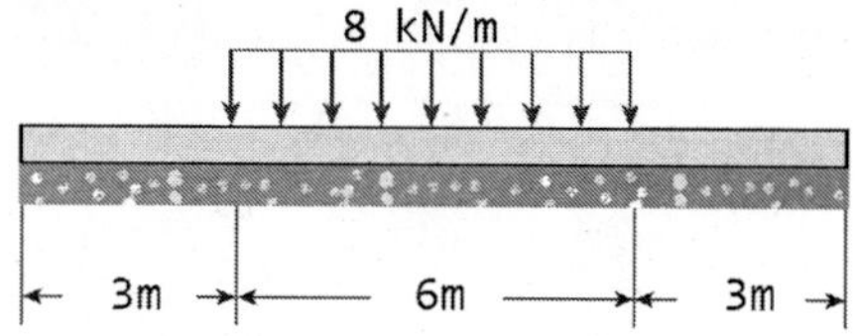

Fig. 6. 14 Beam with distributed reaction

Solution: The beam is supported on the ground entirely along its length. Hence, the support reaction acts along the entire length, too. Assuming the uniformly distributed reaction to be V kN/m, and balancing the forces in along Y-axis gives,

$$-8\text{x}6 + V\text{x}(3 + 6 + 3) = 0;\; V = 4\ \text{kN/m}$$

The beam sub-spans are shown in figure and are used to derive the equations of shear force and bending moment as follows.

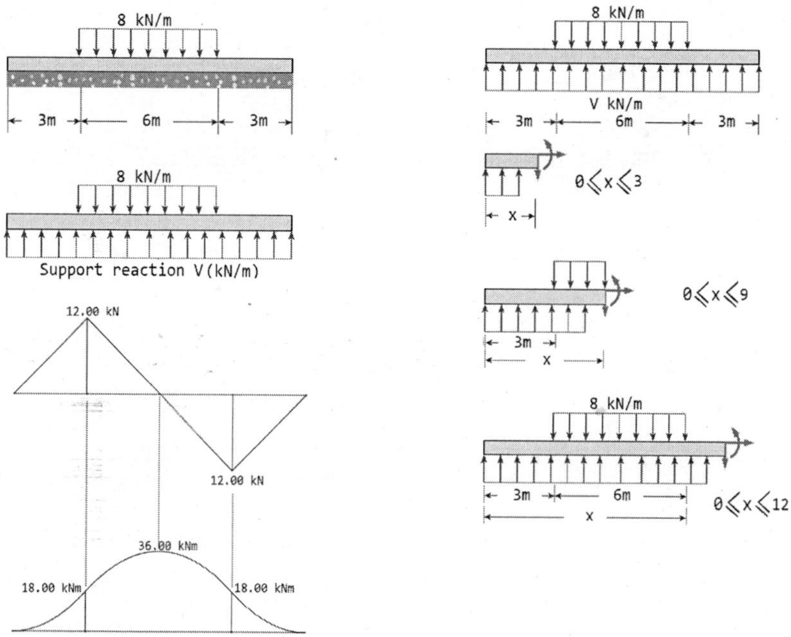

S. No	Sub-span	Shear force at x, V(x)	Bending moment at x, M(x)
1	$0 \le x \le 3$	Vx	$V\frac{x^2}{2}$
2	$3 \le x \le 9$	$Vx - 8(x-3)$	$V\frac{x^2}{2} - 8\frac{(x-3)^2}{2}$
3	$9 \le x \le 12$	$Vx - 8\text{x}6$	$V\frac{x^2}{2} - 8\text{x}6\text{x}\left(x - \left(3 + \frac{6}{2}\right)\right)$

The maximum bending moment is 36 kNm at the mid-span of the beam.

Example 8: The distributed load acting on the wing-span of the aeroplane under cruising condition is idealized as shown in Fig.6. 15. Derive the equations of shear force and bending moment along the wing span.

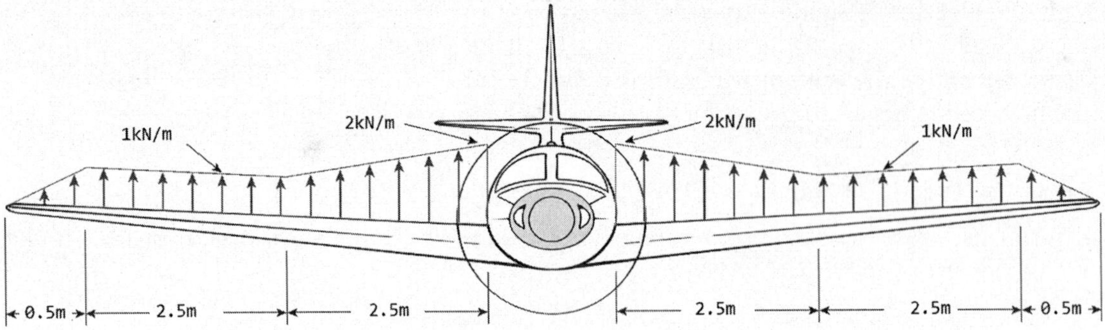

Fig. 6. 15 Loading on the wing-span of a cruising aeroplane

Solution: Considering the symmetry of the plane and idealizing the wingspan as a cantilever beam carrying variable loading, the entire wing span is divided into three sub-spans as shown in the figure. The variable load acting on the sub-spans are determined by calculating the area under the load diagram. The load diagrams are shown numbered in the figure as I for sub-span $0 \leq x \leq 0.5$m, II for sub-span $0 \leq x \leq 3$m , III and IV for sub-span $0 \leq x \leq 5.5$m. It is observed from the loading that for area I, the height of the triangle is 1 kN and the width is 0.5 m. So, the intensity of varying load for I is 2 kN/m. Similarly for area IV, the height of the triangle is 1 kN and the base of the triangle is 2.5 m. Hence the intensity of variable load under IV is 0.4 kN/m. Thus, the total load acting in each sub-span is calculated cumulatively as follows:

Section Number	Type of load	Dimensions		Area (total load)
		Width	Height at x m	
I	UVL, triangle	x m	$2x$	$(2x^2)/2$
II	UDL, rectangle	x-0.5 m	1	$(x-0.5)$
III	UDL, rectangle	x-3.0 m	1	$(x-3.0)$
IV	UVL, triangle	x-3.0 m	$0.4x$	$0.4x(x-3.0)/2$

The cumulative total load obtained from the table above for each sub-spans is used to derive the equations of shear force and bending moment as follows.

S. No	Sub-span	Shear force at x, V(x)	Bending moment at x, M(x)
1	$0 \leq x \leq 0.5$	x^2	$x^2.(\frac{x}{3})$
2	$0.5 \leq x \leq 3$	$\left(\frac{0.5x1}{2}\right) + 1.(x-0.5)$	$\left(\frac{0.5x1}{2}\right).(x - \frac{2}{3}x0.5) + 1.\frac{(x-0.5)^2}{2}$
3	$3 \leq x \leq 5.5$	$\left(\frac{0.5x1}{2}\right) + 1x2.5 + 1.(x-3.0)$ $+0.4x\frac{(x-3.0)}{2}$	$\left(\frac{0.5x1}{2}\right).(x - \frac{2}{3}x0.5 + 1x2.5.(x-1.75)$ $+1.\frac{(x-3.0)^2}{2} + 0.4x\left(\frac{(x-3.0)}{2}\right).\left(\frac{(x-3.0)}{3}\right)$

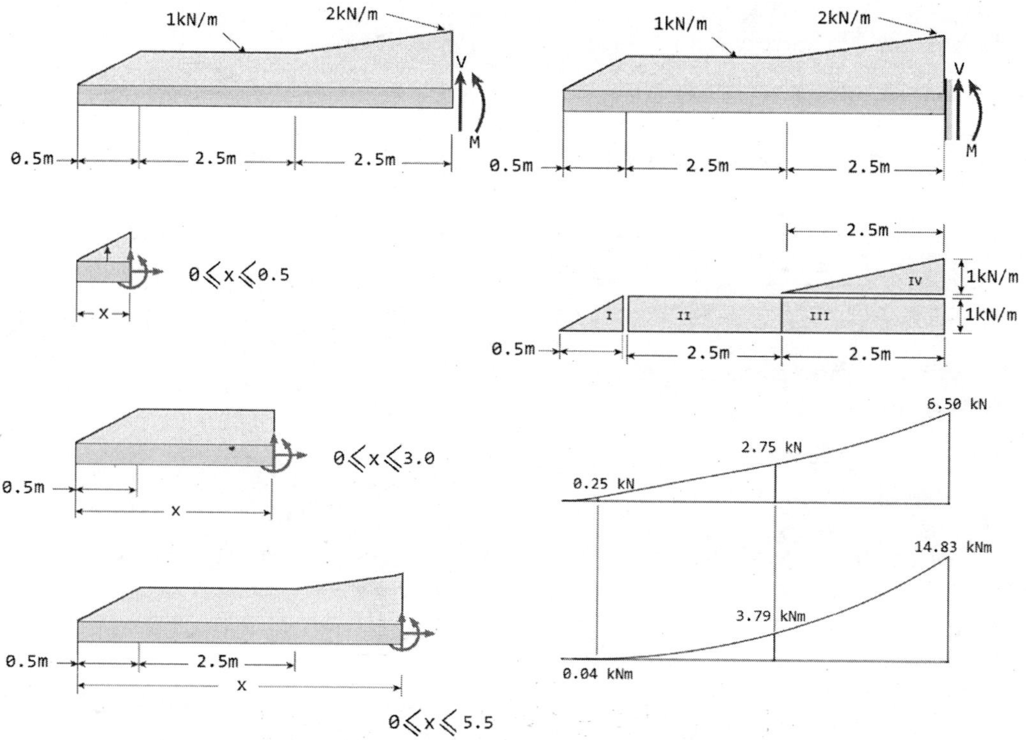

Using the equations, the fixed end SF and BM of the idealized cantilever beam is obtained as follows:

V = 6.50 kN

M = 14.83 kNm.

UNIT SUMMARY

In the present unit, different types of beams and beam loadings are discussed. The present unit also explains the method to calculate shear force and bending moment at any beam section. It also teaches to identify location of maximum bending moment so as to provide efficient design of the structural member. Beams having points of contraflexures are also discussed. Various problems of different combinations of beam support and loading are discussed.

Exercises

Multiple Choice Questions

1. Change of bending moment per unit length of the beam section is equal to

(a) shear force of the section

(b) deflection of the section

(c) Load intensity at the section

(d) slope at the section

2. In a beam, at any point, bending moment is the algebraic sum of
(a) vertical forces at the point
(b) all horizontal forces at the point
(c) all forces on either side
(d) moments on either side

3. For a beam fixed at one end and free at the other with a load at its centre, maximum bending moment appears:
(a) between centre and fixed end
(b) under the load
(c) at fixed end
(d) at free end

4. For SSBs, bending moment at the end supports is:
(a) Greater than unity
(b) Zer0
(c) Unity
(d) Less than unity

5. A 1 m long SSB carries a UD load of 0.4 kN/m. Maximum bending moment (kN-m) is:
(a) 1.0
(b) 0.1
(c) 0.05
(d) 0.025

6. For the beam shown here, maximum bending moment (N-m) is
(a) 25
(b) 30
(c) 35
(d) 60

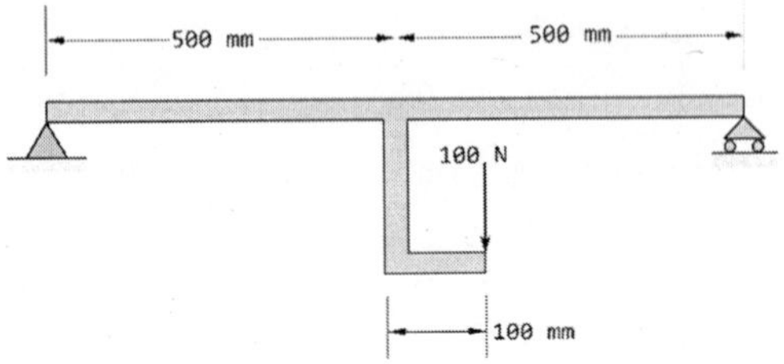

7. If shear force is zero in a simply supported beam, then the bending moment shall be:
(a) zero
(b) maximum
(c) minimum
(d) zero or minimum

8. Point of contra-flexure normally occurs in
(a) SSB
(b) beams with load changing from zero to maximum from one end to other
(c) cantilever beams
(d) overhanging beams

9. Point of inflexion or contraflexure is the location where:
(a) SF changes in sign
(b) BM changes in sign
(c) SF is maximum

(d) SF is minimum

10. Bending moment in a simply supported beam carrying concentrated load at midspan shall be:
(a) minimum at the support
(b) minimum at the centre
(c) maximum at the support
(d) None of the above

11. For a section having zero shear force, bending moment shall be:
(a) Minimum
(b) Maximum
(c) Either minimum or maximum
(d) Zero

12. Shear force at the centre of a simply supported beam with UDL of intensity w N/m distributed over its entire length is:
(a) $\frac{wl^2}{8}$
(b) $\frac{wl}{4}$
(c) zero
(d) $\frac{wl}{2}$

13. Irrespective of type of load, the maximum BM and maximum SF in a cantilever beam is obtained at:
(a) fixed end
(b) free end
(c) middle
(d) any point

14. Bending moment diagram for SSB with a load at the mid-span is a:
(a) parabola
(b) hyperbola
(c) rectangle
(d) triangle

15. Maximum bending moment is reported in a beam section where shearing force is:
(a) Minimum
(b) Maximum
(c) Zero
(d) Changing sign

16. Shear force for the section of a beam where the bending moment is constant is:
(a) Zero
(b) Increasing
(c) Decreasing
(d) Constant

17. A simply supporting beam having two end supports reports maximum bending moment:
(a) usually on the supports
(b) always at mid span
(c) where there is no shear force
(d) where the deflection is maximum

18. A beam of length L is simply supported at ends with a distributed load sin (3πx/L) N/m, distance x measured from the left support. The vertical reaction (N) at the left support is:
(a) zero
(b) L/3π
(c) L/π
(d) 2L/π

19. A cantilever beam of length 5 m carries 10 kN at mid-span and 10 kN at its free end. The bending moment at beam's fixed end is:

(a) −25 kN m

(b) −50 kN m

(c) −75 kN m

(d) None of these

20. A beam of length 10 m has equal overhang of 2 m at both ends and carries concentrated loads each of 40 kN at its ends and a point load of 80 kN at its middle; The location of inflexion point is:

(a) 4 m

(b) 3 m

(c) 2 m

(d) None of these

Answers of Multiple-Choice Questions

1	2	3	4	5	6	7	8	9	10
a	d	c	b	c	b	b	d	b	a
11	12	13	14	15	16	17	18	19	20
c	c	a	d	c	a	c	b	c	a

Short and Long Answer Type Questions

1. What are the different types of beams? What are the different types of loads acting on a beam?
2. State the necessity of drawing the beam SF-BM diagrams.
3. What shall be the reactions provided by a flat surface upon which a beam with concentrated loads at different points is uniformly supported on a flat surface?
4. Define the terms axial force, shear force, bending moment, and point of contraflexure.
5. The BM in a beam is maximum or minimum where the SF is zero. Is the converse true? Why?
6. What is meant by upward concave and upward convex bending moment?

Numerical Problems

1. A 3 m long cantilever beam has a 4 kN/m UDL over a length of 1 m from the fixed end in addition to a concentrated load of 0.5 kN at 1 m from the free end. Calculate and draw the S.F. and B.M. diagrams.
2. A cantilever beam of 4 m length supports a 5kN/m UDL for a length of 2 m from the free end. Additionally, it also bears a concentrated load of 3000 N and an anticlockwise moment of 2kNm at 1 m from the free end. Calculate and plot the shear force and bending moment.
3. A 6 m long SSB carries gradually varying load of 0.45 kN/m from its left support to 0.9 kN/m to its right support. Calculate and plot the shear force and bending moment and determine the magnitude, nature and position of the maximum bending moment over the beam.
4. For Figs. P.6.1 to P.6.7, calculate and plot S. F. and B.M.

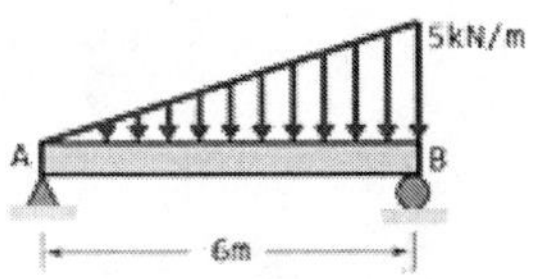

Fig. P.6.1

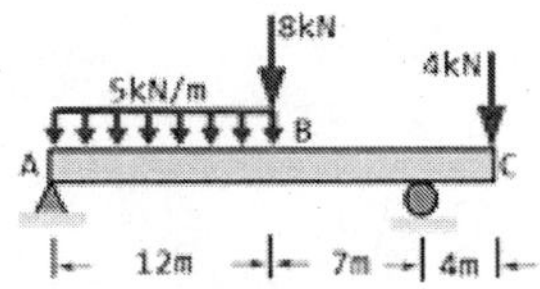

Fig. P.6.2

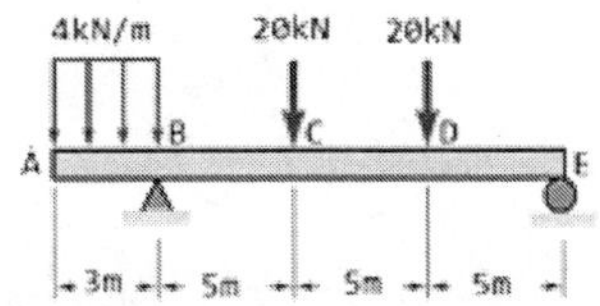

Fig. P.6.3

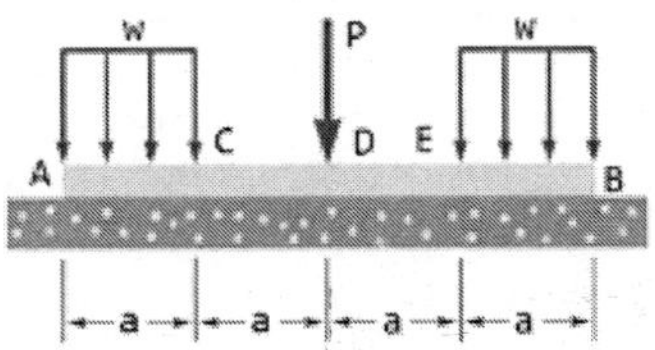

Fig. P.6.4

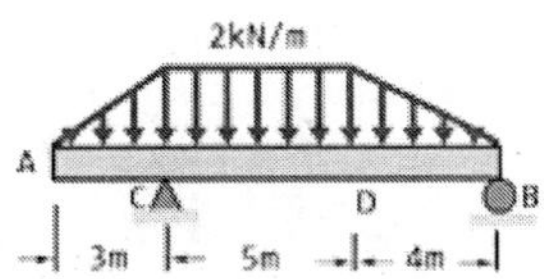

Fig. P.6.5

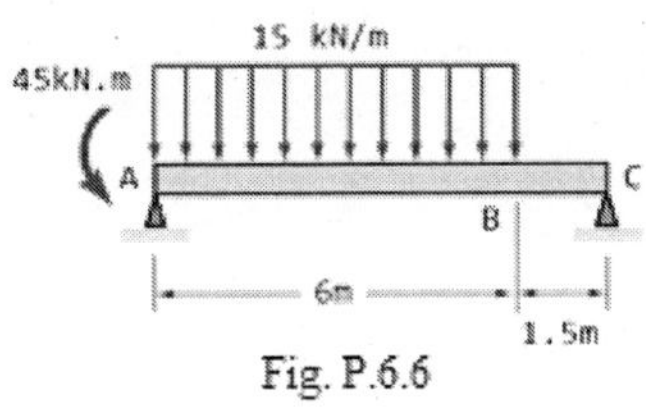

Fig. P.6.6

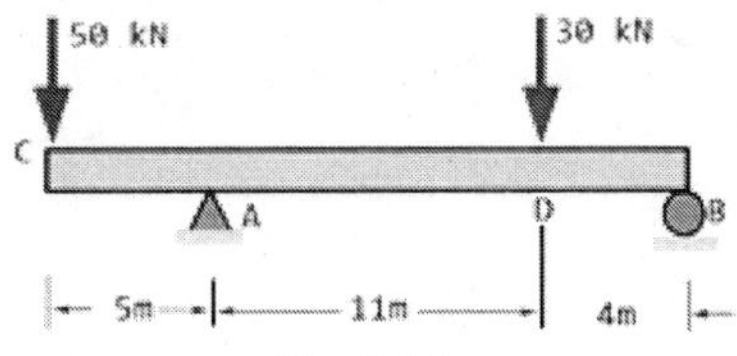

Fig. P.6.7

KNOW MORE

Bending Moment in Beams Experiment

REFERENCES AND SUGGESTED READINGS

Strength of Materials, R Subramaniam, Oxford University Press

1. Mechanics of Materials, Ferdinand P. Beer, E. Russell Johnston Jr., John T. DeWolf, David F. Mazurek, McGraw Hill.
2. Mechanics of Materials, Barry J. Goodno, James M. Gere, Cengage Learning.

7 Torsion

UNIT SPECIFICS

This unit discussed the following concepts:

- *Deformations of circular shafts;*
- *Shear stress in a circular and torsion equation derivation;*
- *Angle of twist in circular and hollow shafts;*
- *Power transmission of circular shafts.*

RATIONALE

This unit on mechanics helps students to understand the structural members behaviour subjected to torsion. The torsional motion of circular shafts is explained in this unit to understand the practical applications. The unit also explores the stresses and strains in circular cross-sectional members that have been exposed to twisting couples, or torques and power transmission in more detail.

PRE-REQUISITES

- *Nil*

UNIT OUTCOMES

The following are the outcomes of this unit:

U7-O1: Describe the torsion of circular shafts

U7-O2: Explain the torsional stresses and strains

U7-O3: Apply torsion equation to solve practical scenarios or problems

Unit - 7 Outcomes	EXPECTED MAPPING WITH COURSE OUTCOMES (*1- Weak Correlation; 2- Medium Correlation; 3- Strong Correlation*)					
	CO-1	**CO-2**	**CO-3**	**CO-4**	**CO-5**	
U7-O1	2	3	2	2	3	
U7-O2	3	2	2	2	3	
U7-O3	2	3	2	2	3	

7.1 INTRODUCTION

In many engineering applications, members in torsion are encountered. A revolving machine component with a circular cross section is a transmission shaft, commonly known as a shaft. The usage of a shaft allows for the transmission of rotating motion from one place to another. Transmission shafts offer the most widespread application. Many machine components are subject to torsional loading, which can occur either to transmit power, as in the case of driveshafts or axle shafts in vehicles, or to support dynamic loads, such as coil springs or torsion bars. Circular solid shafts and circular hollow shafts are commonly used for power transmission due to their ease of construction, balance, and the ability of the outermost material to bear the majority of the stress. A circle's complete surface has access to more material for a given maximum size than a square's four corners have. Because of this, understanding the behavior of these types of shafts under torsion is crucial in the design and analysis of machine components.

Consider about the setup in Fig. 7.1a, which consists of a transmission shaft AB connecting an electric generator A and a steam turbine B. By disassembling the system into three parts, as depicted in Figure 8.1b, it becomes possible to observe how the turbine applies a twisting couple, or torque T, to the "shaft," which in turn applies an equal amount of torque to the generator. The turbine is subjected to a torque T_A applied by the shaft and an equal and opposite torque T_A applied by the generator in response.

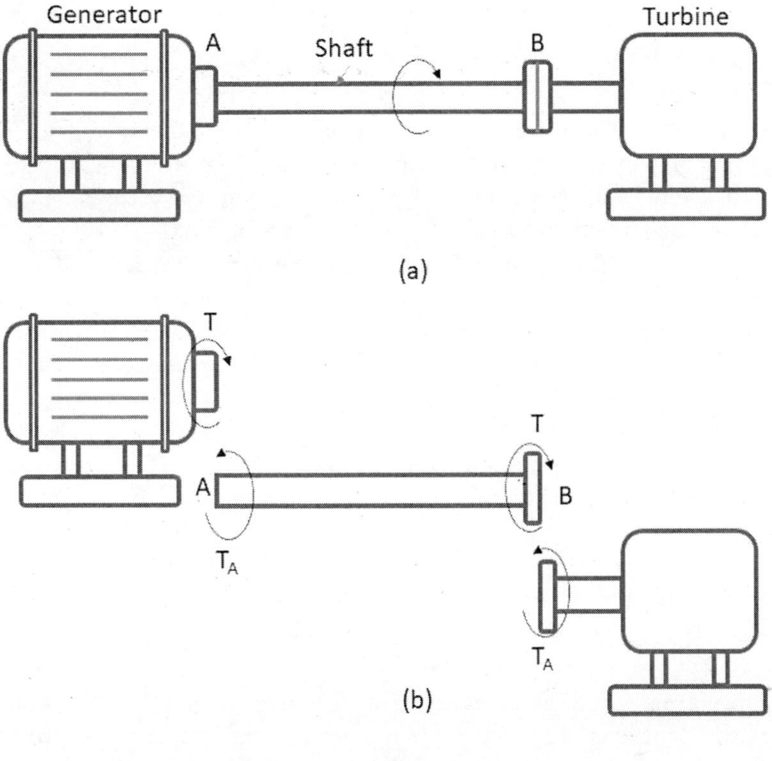

Fig. 7. 1 Transmission of shaft

To comprehend how circular shafts behave when subjected to torsion, it is crucial to analyze the strains and deformations that take place within them. It has been observed that each cross section of a circular shaft remains plane and unaltered during torsion, rotating like a "solid rigid slab" while the many cross sections along the shaft rotate at varied angles. This allows for the identification of the distribution of shearing strains within the shaft, which varies linearly with distance from the shaft axis. Additionally, the distribution of shearing stresses in a circular shaft can be determined using the elastic torsion equation, which takes into account deformations in the elastic range and applies Hooke's law for shearing stress and strain. These concepts are crucial for analyzing the behavior of machine components that are loaded in torsion and for designing structures that can withstand these types of loads.

7.2 DEFORMATIONS OCCURING IN A CIRCULAR SHAFT

As shown in Fig. 7.1a, imagine a circular shaft that has one end secured to a support. If a torque T is exerted on the opposite end of the shaft, with its free end spinning through an angle called the angle of twist (symbolized by φ), as depicted in Figure 7.1b. According to observation, the ϕ is proportional to T within a given range of values of T. Additionally, it demonstrates that ϕ is proportional to the shaft's length L. This section shows the relation between ϕ, L, and T.

Imagine a small square section that is formed by two straight lines and two adjacent circles drawn on the surface of a cylinder having a radius of ρ. This is depicted in Figure 7.2a, and the section is removed from the shaft before it is subjected to any external load. After the application of a torsional load to the shaft, the previously square element is deformed into a rhombus shape as shown in Figure 7.2b. The amount of shearing strain γ in an element can be calculated by measuring the change in the angles formed by its sides. The term shearing strain γ is used to describe any variation in the angle between two perpendicular planes. In the case of the circular element being considered, the shearing strain γ is equivalent to the angle between the lines AB and A'B since the two circles forming two of the sides of the element remain unchanged. The value of the shearing strain is typically measured in radians. The equation AA' = $L\gamma$ can be used to express the arc length AA'. Additionally, AA' = $\rho\phi$ can be used to indicate the same arc length. i.e.

$$\gamma L = \rho\phi \Rightarrow \gamma = \frac{\rho\phi}{L} \tag{7.1}$$

Where γ and ϕ are in radians. The Eq. (7.1) shows that the relationship between the shearing strain γ and the twist angle ϕ at a particular location in a shaft under torsion is linear. Moreover, the equation also signifies that the shearing strain (γ) is directly proportional to the radial distance (ρ) between the axis of the shaft and the point being analyzed. As a result, the shearing strain in a circular shaft change in a straight-line manner with the shaft's radial distance. Since Eq. (7.1) has proven that the surface of the shaft, where ρ = c, experiences the most shearing strain, Eq. (7.1) becomes,

$$\gamma_{max} = \frac{c\phi}{L} \tag{7.2}$$

From the equations (7.1) and (7.2), $\gamma = \frac{\rho}{c}\,\gamma_{max}$ (7.3)

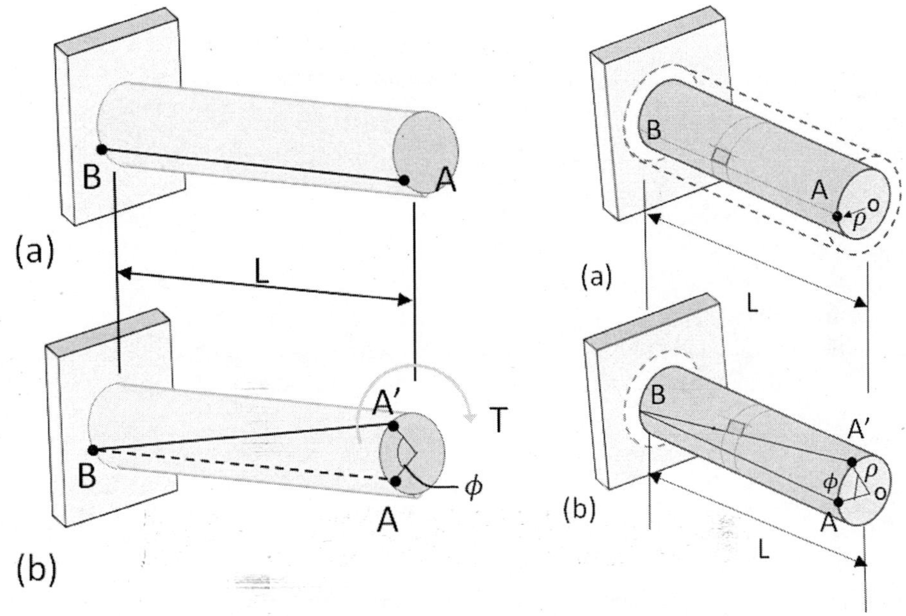

Fig. 7. 2 Shaft with fixed support and Shear strain.

7.3 SHEAR STRESSES OF CIRCULAR SHAFT

Shear stress is a force that causes deformation in a material by inducing slippage along one or more planes parallel to the applied stress. In other words, it is the stress that acts in a direction parallel to the surface or the plane of the material. A shaft experiences an internal torque in proportion to the external torque applied to it. In this section, we will develop an equation that relates the distribution of shear stress acting on the cross-section of

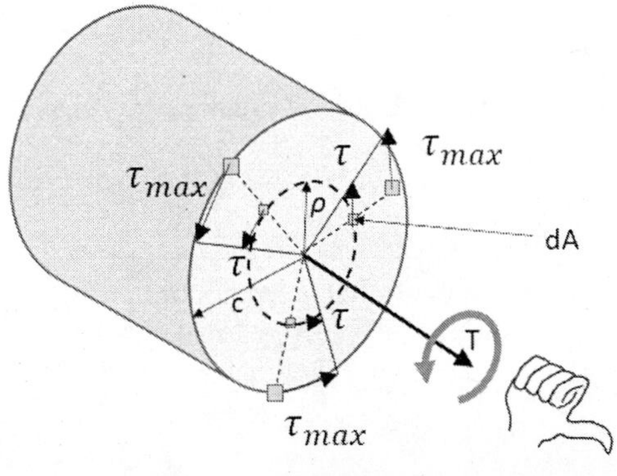

Fig. 7. 3 Shear stress variation

the shaft to the "internal torque". A round "shaft" will twist through an angle when you apply a torque T to it (as shown in Fig. 7.3). Twisting is a sign that a material is deforming, indicating that there is strain present.

The strain is at its maximum towards the surface of the shaft, while it is absent or zero at the center of the shaft. From the middle of the shaft to its surface, the strain changes linearly. If a material has linear elastic properties, Hooke's law can be applied, represented by $\tau = G\gamma$ or $\tau_{max}=G\gamma_{max}$. As a result of this, when there is a linear increase in shear strain, a corresponding linear change occurs in shear stress along any radial line. Therefore, the shear stress τ varies linearly from zero at the longitudinal axis of the shaft to its maximum value, τ_max, at its outer surface, as indicated in Fig. 7.3.

$$\tau = \left(\frac{\rho}{c}\right) \tau_{max} \tag{7.4}$$

If a circular shaft remains within its yield strength or proportional limit, then the equation indicates that the shearing stress within the shaft will have a linear relationship with the distance r from the axis of the shaft. This means that the shearing stress will be at its maximum at the shaft's outermost surface and will decrease linearly toward the center of the shaft. In Figure 7.4a and 7.4b, the stress distribution in solid and hollow circular shafts with inner and outer radii of c_1 and c_2, respectively, of radius c is depicted. According to Eq. (7.4), in the later scenario,

$$\tau_{min} = \left(\frac{c_1}{c_2}\right) \tau_{max} \tag{7.5}$$

As depicted in Fig. 7.3, every element of area positioned at distance ρ experiences a force of $dF = \tau\, dA$ due to the applied shearing stress. Consequently, the corresponding induced torque is $dT = \rho\ (\tau\ dA)$. Summing up all the contributions of torque from each element across the entire cross section leads to the following expression:

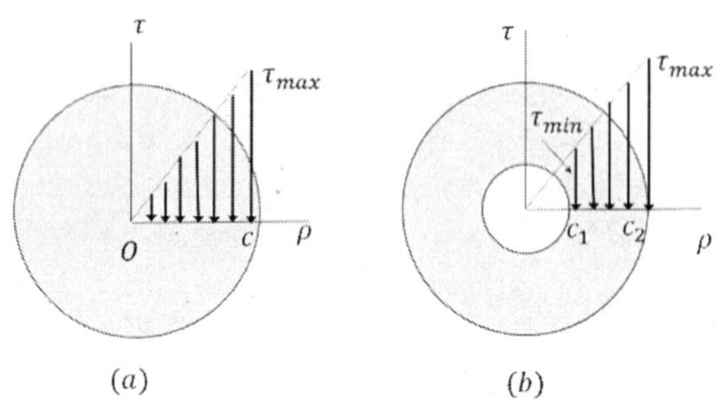

Fig. 7. 4 Shear stress distribution

$$T = \int_A \rho(\tau dA) = \int_A \rho \left(\frac{\rho}{c}\right) \tau_{max} dA \tag{7.6}$$

Since $\frac{\tau_{max}}{c}$ is constant, Eq. (7.6) becomes

$$T = \frac{\tau_{max}}{c} \int_A \rho^2 dA \tag{7.7}$$

The equation (7.7) demonstrates that the integral is related to the polar moment of inertia J of the cross section with respect to its center O. In light of this,

$$T = \frac{\tau_{max}}{c} J \Rightarrow \tau_{max} = \frac{Tc}{J} \tag{7.8}$$

Similarly, the shearing stress is expressed as the following for any distance ρ from the shaft axis:

$$\tau = \frac{T\rho}{J} \tag{7.9}$$

The formulas given by Eq. (7.7) and Eq. (7.9) are commonly known as the elastic torsion formulas. These equations rely on Hooke's law and are only valid if the shaft has a circular cross section, the material is uniform, and it displays linear elastic behavior.

Examples: 7.1: The transmission of "torque" from one "shaft" to another requires a 100 mm diameter circular shaft. If the shaft must transfer a safe amount of torque and the shear stress cannot be greater than 60 MPa, determine the safe torque.

Solution 7.1: Diameter, d = 100 mm, radius c=50 mm; $\tau_{max} = 60$ MPa;

Polar moment of inertia of circle, $J = \frac{\pi}{2}c^4$

From Eq. (7.8), $\tau_{max} = \frac{Tc}{J} \Rightarrow T = \frac{\tau_{max}J}{c}$

Therefore, the torque $T = \frac{60}{c}\frac{\pi}{2}c^4 = 30\pi\ 50^3 N.mm = 11.78\ kN.m$

Example 7.2: An 800 Nm torque is applied to a hollow shaft with inner and outer dimensions of 100 mm and 120 mm, respectively. What is the inside and exterior surfaces' shear stress? Give your response in MPa.

Solution 7.2: T = 800 Nm; d_i = 100mm; d_o = 120mm;

The distance from the inner surface to the centroid,

$$r_i = \frac{d_i}{2} = \frac{100mm}{2} = 50mm.$$

The distance from the exterior surface to the centroid,

$$c = \frac{d_o}{2} = \frac{120mm}{2} = 60mm$$

A hollow circle's "polar moment of inertia" equals $J = \frac{\pi(d_o^4 - d_i^4)}{32}$.

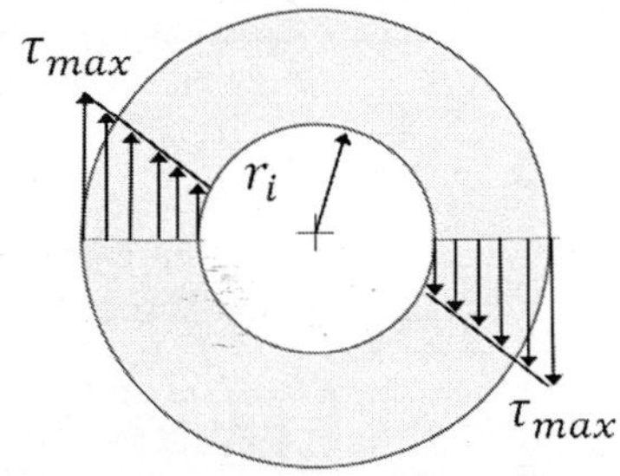

Fig. E7.1: Shear stress distribution

The internal and external shear stresses can be calculated using the same polar moment of inertia J. The resulting shear stresses on the inside and outside surfaces are given by

$$\tau_i = \frac{Tr_i}{J} = 800N.m50mm \frac{32}{\pi\left[(120mm)^4 - (100mm)^4\right]} \left| \frac{MPa \cdot m^2}{10^6 N} \right| \frac{(10^3 mm)^3}{m^3} = 3.80MPa.$$

$$\tau_o = \frac{Tr_o}{J} = 800N.m\ 60mm \frac{32}{\pi[(120mm)^4 - (100mm)^4]} \left| \frac{MPa \cdot m^2}{10^6 N} \right| \frac{(10^3 mm)^3}{m^3} = 4.56MPa.$$

Example 7.3: A hollow cylindrical steel shaft of 2.0 meters in length, has an inner diameter of 60 mm and an outer diameter of 80 mm (as shown in Figure E7.2). The maximum torque that can be applied to the shaft is determined based on (a) maximum allowable shearing stress of 120 MPa, and (b) calculate the minimum shearing stress on the shaft.

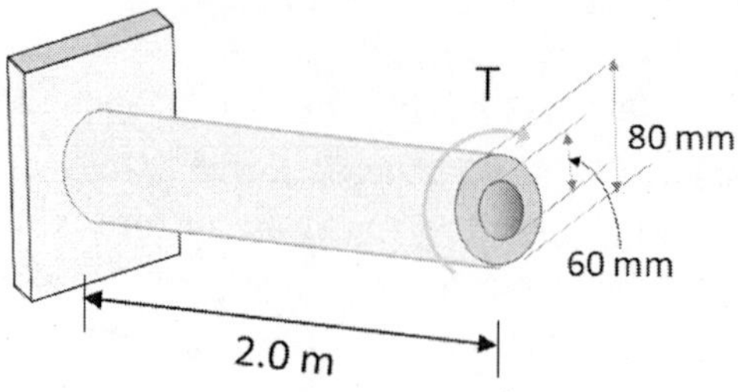

Fig. E7.2

Solution 7.3: $d_i = 60mm$; $d_o = 80mm$; $\tau_{max}=120MPa$.

(a) The polar moment of inertia $J = \frac{\pi(d_o^4-d_i^4)}{32} = \frac{\pi(80^4-60^4)}{32} = 2.75 \times 10^6 mm^4$

c = 80mm/2 = 40mm at maximum shear stress

We can apply Eq. (7.8) because the given maximum shear stress is lower than steel's yield strength. Then torque T is as follows:

$$T = \frac{\tau_{max}}{c} J = \frac{120N/mm^2}{40mm} \times 2.75 \times 10^6 mm^4 = 8.25 \times 10^6 N.mm = 8.25\ kN.m$$

(b) The smallest value of shearing stress on the shaft can be found on its inner surface. This can be obtained using Eq. (7.5) which relates τmin to c1 and τmax to c2 in a proportional manner:

$$\tau_{min} = \left(\frac{c_1}{c_2}\right) \tau_{max} = \frac{30mm}{40mm} \times 120MPa = 90MPa$$

7.4 ANGLE OF TWIST

Assuming that the shaft remains in the elastic region, we will analyze a shaft with a uniform cross-section of radius c and length L that is being subjected to a torque T at its free end. This is illustrated in Figure 7.5. In the elastic range, apply Hooke's law, Eq. (7.2), and Eq. (7.8) can be written as follows,

$$\gamma_{max} = \frac{c\phi}{L} = \frac{\tau_{max}}{G} = \frac{Tc}{JG} \quad (7.10)$$

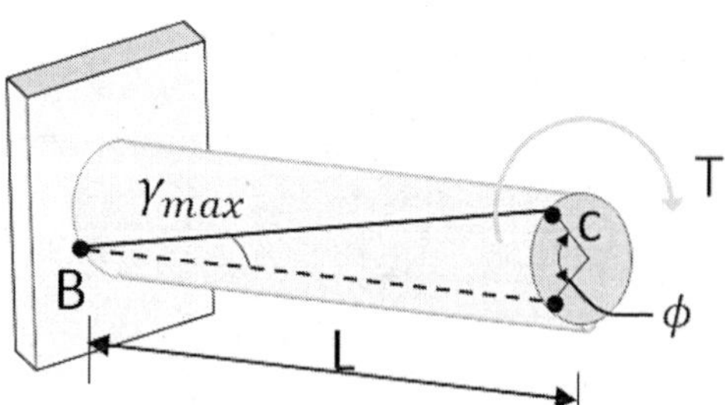

Fig. 7. 5 Angle of twist

The expression given in Equation (7.10) can be used to calculate the twist angle, the twist angle ϕ in radians can be denoted as

$$\phi = \frac{TL}{JG} \quad (7.11)$$

The equation (Eq. (7.11)) shows that the amount of twisting angle (measured in radians) in a shaft within the elastic limit is directly proportional to the torque T that is applied to the shaft.

Multiple Torques: The figure shown in Figure 7.6 depicts that if the shaft is exposed to variety of torques, or if there is a sudden change in the cross-sectional area or the shear modulus from one part to the another of the shaft, then Eq. 7.11 should be applied individually to each part of the shaft where these quantities remain constant. The angle of twist for one end of the shaft relative to the other can be determined by adding the algebraic sum of the angle of twists of each part. In this scenario,

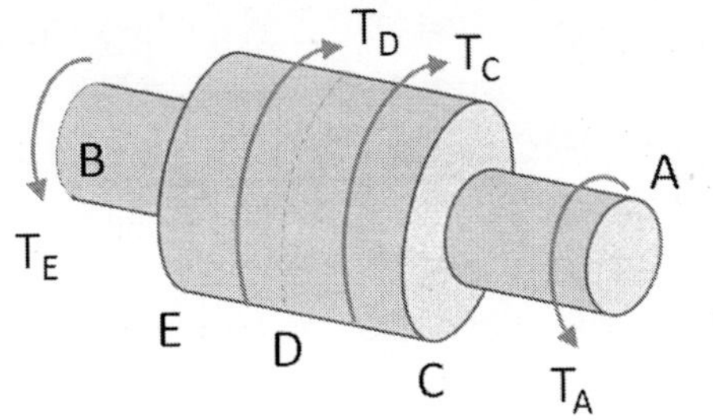

Fig. 7. 6 Multiple torques with multiple sections

$$\phi = \Sigma \frac{TL}{JG} \tag{7.12}$$

In order to use the equation effectively, a sign convention should be employed for both the internal "torque" and the "angle of twist" of one end of the shaft relative to the other. For this purpose, the right-hand rule is used, where the thumb is pointed outward from the shaft and the fingers are curled in the direction of the torque, as depicted in Figure 7.7. If this convention is followed, both the torque and the angle will be positive.

When a shaft has a varying circular cross section, such depicted in Fig. 7.8, a disc of thickness dx may be subjected to the Eq. (7.11). The angular relationship between the rotation of one disc face and to the other is

$$d\phi = \frac{Tdx}{JG}$$

The total "twisting angle" of the shaft is calculated by integral in x from 0 to L:

$$\phi = \int_0^L \frac{Tdx}{JG} \tag{7.13}$$

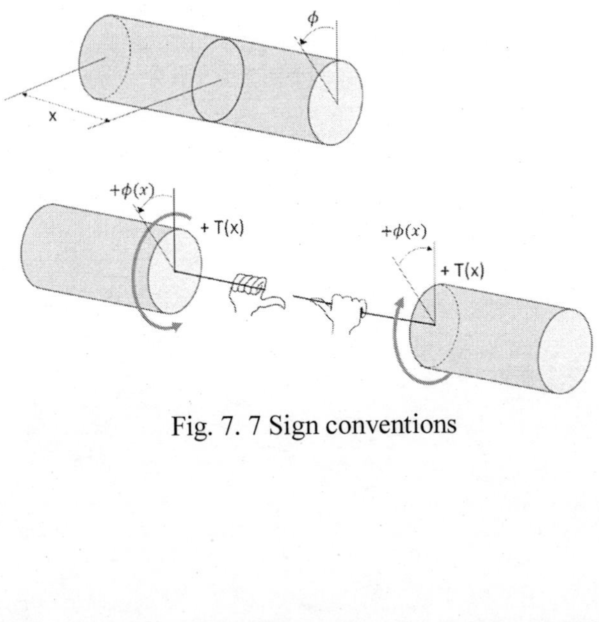

Fig. 7. 7 Sign conventions

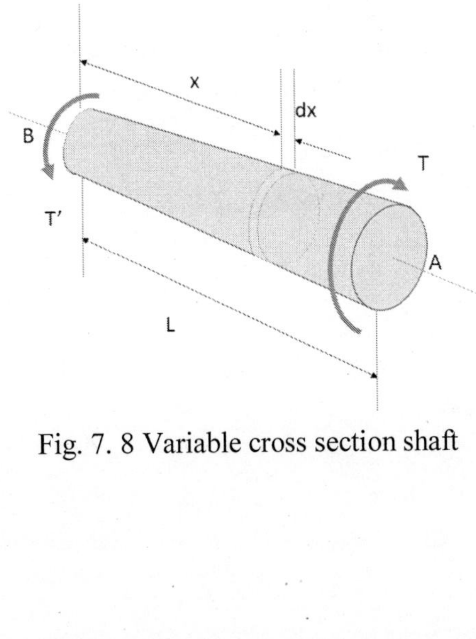

Fig. 7. 8 Variable cross section shaft

Example 7.4: A 2.0 m long steel hollow shaft has 50 mm inner and 100 mm outer diameter respectively. To produce a twist of 2^0, what amount of torque must be applied to the shaft's end? Use the shear modulus of steel shaft, G= 77 GPa.

Solution 7.4: $\phi = 2^0 = 2^0 \left(\frac{2\pi\, rad}{369^0}\right) = 34.9 \times 10^{-3}\, rad$; L = 2.0m = 2000 mm;

Polar moment of inertia $J = \frac{\pi(d_o^4 - d_i^4)}{32} = \frac{\pi(100^4 - 50^4)}{32} = 9.2 \times 10^6 mm^4$

From Eq. (7.11), $T = \frac{\phi JG}{L} = \frac{(34.9 \times 10^{-3}\, rad)(9.2 \times 10^6 mm^4)(77 \times \frac{10^3 N}{mm^2})}{2000mm} = 12.36 \times 10^6 N.mm$

Torque applied to the end of shaft is 12.36 kN.m

Example 7.5: What degree of twisting will cause the inside of the hollow steel shaft, which has a length of 2.0 m and diameters of 50mm and 100mm, to experience a shearing stress of 70 MPa?

Solution 7.5: d_i = 50mm; c_i = 25mm

Using Hooke's law, $\gamma_{min} = \frac{\tau_{min}}{G} = \frac{70 \times 10^6 Pa}{77 \times 10^9\, Pa} = 0.909 \times 10^{-3}$

From Eq. (7.1), shear strain at interior of hollow shaft can be written as

$$\gamma_{min} = \frac{c_i \phi}{L} \Rightarrow \phi = \frac{\gamma_{min} L}{c_i} = \frac{0.909 \times 10^{-3} \times 2000mm}{25mm} = 0.072\, rad$$

Angle of twist in degrees, $\phi = (0.072\, rad)\left(\frac{360^0}{2\pi rad}\right) = 4.125^0$

Example 7.6: The torques shown in Fig. E7.3 are applied to a horizontal shaft AD that is mounted to a base at D. Part CD of the shaft has a 44 mm-diameter hole bored into it. Find the twisting angle at end A, keeping in mind that the shaft is entirely constructed of steel with a G of 77 GPa.

Solution 7.6: The shaft is composed of three pieces, AB, BC, and CD, each with a uniform cross section and a constant internal torque, therefore Eq. (7.12) can be used. Using a free body diagram displayed in Fig. E7.4a, a section through the shaft between A and B, we establish

$\sum M_x = 0$: (300 N.m) – T_{AB}= 0, T_{AB} = 300 N.m

Now consider the section between B and C as shown in Fig. E7.4b,

$\sum M_x = 0$: (300 N.m) +(2500 N.m) – T_{BC}= 0,

T_{BC} = 2800 N.m

As C is not subjected to torque, $T_{CD} = T_{BC}$ = 2800 N.m

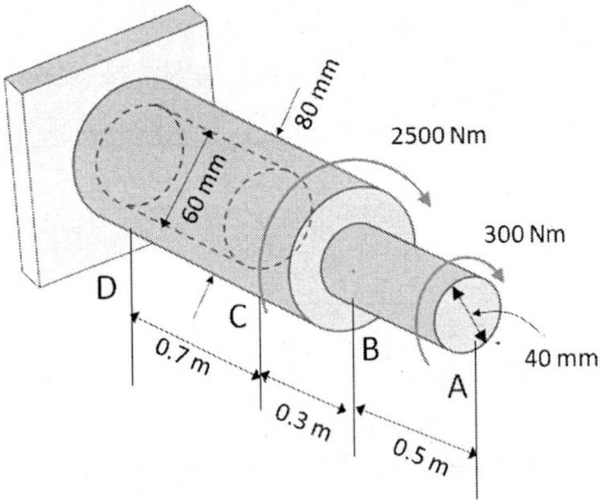

Fig. E7.3

Polar moment of inertia,

$$J_{AB} = \frac{\pi}{2}c^4 = \frac{\pi}{2} \times 0.02^4$$
$$= 0.251 \times 10^{-6} m^4$$

$$J_{BC} = \frac{\pi}{2}c^4 = \frac{\pi}{2} \times 0.04^4$$
$$= 4.021 \times 10^{-6} m^4$$

$$J_{CD} = \frac{\pi}{2}(c_2^4 - c_1^4) = \frac{\pi}{2} \times (0.04^4 - 0.03^4)$$
$$= 2.749 \times 10^{-6} m^4$$

$$\phi_A = \sum_i \frac{T_i L_i}{G J_i} = \frac{1}{G}\left(\frac{T_{AB}L_{AB}}{J_{AB}} + \frac{T_{BC}L_{BC}}{J_{BC}} + \frac{T_{CD}L_{CD}}{J_{CD}}\right)$$

$$\phi_A = \frac{1}{77\times10^9 Pa}\left(\frac{300N.m\times0.5m}{0.251\times10^{-6}m^4} + \frac{2800N.m\times0.3m}{4.021\times10^{-6}m^4} + \frac{2800N.m\times0.7m}{2.749\times10^{-6}m^4}\right) = (7.761 \times 10^{-3} rad + 2.713 \times 10^{-3} rad + 9.259 \times 10^{-3} rad) = 0.019733\, rad$$

$$\phi_A = (0.019733) \times \left(\frac{360^0}{2\pi\, rad}\right) = 1.13^0$$

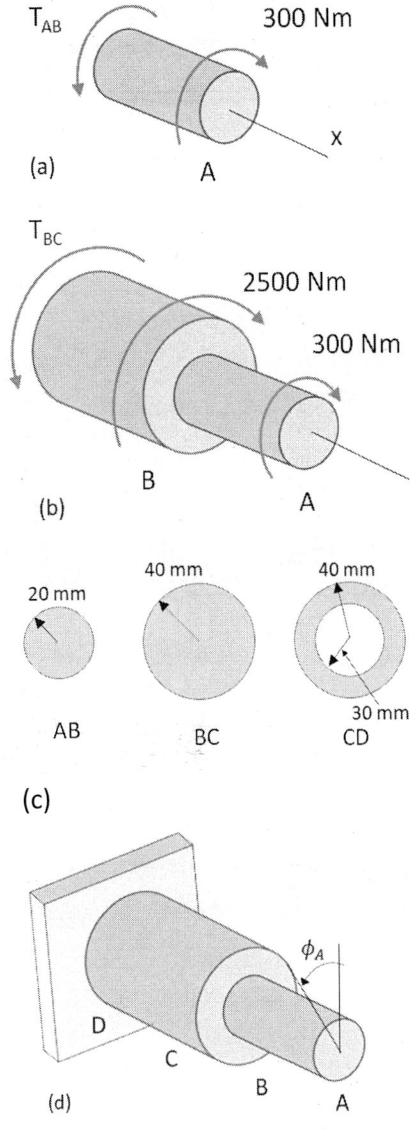

Fig. E7.4

7.5 POWER TRANSMISSION

It is common practise to use shafts and tubes having circular cross sections to transmit the power produced by a machine. They experience a torque when utilised for this purpose, that depends on both machine's power output and the shaft's rotational speed. The work done in a unit of time is referred to as *power*. Additionally, the torque applied multiplied by the rotational angle equals the work delivered by a rotating shaft. Therefore, if a shaft rotates dθ as a result of a torque T applied for a short period of time dt, the work done is Tdθ, and the instantaneous power (P) is

$$P = \frac{Td\theta}{dt} \quad (7.14)$$

Since the angular velocity of the shaft $\omega = \frac{d\theta}{dt}$, the Eq. (7.14) becomes

$$P = T\omega \quad (7.15)$$

In the SI system, power is expressed in watts W when torque is measured in newton-meters (N.m) and ω is measured in radians per second (rad/s). Although **horsepower** (hp) is frequently employed in engineering, 1 hp = 746 W

For machinery, the frequency, f, of a shaft's rotation is usually provided. The number of revolutions, or "cycles," the shaft makes in a second is measured and expressed in hertz, where 1 Hz is equivalent to one cycle in a second. Since ω = 2πf and 1 cycle equals 2π rad, the power equation previously mentioned can also be written as

$$P = 2\pi f T \quad (7.16)$$

Example 7.7: To transmit 300kW at 100 rpm, a hollow shaft is used. If the internal diameter is 0.7 of the exterior diameter of the shaft and the shear stress isn't to exceed 60MPa, determine the shaft's diameters.

Solution 7.7: P = 300 kW, N=100rpm, $d_i = 0.7d_o$, τ=60MPa, $c_2 = d_o/2 = 0.5d_o$, $c_1 = d_i/2 = 0.35d_o$

Polar moment of inertia, $J = \frac{\pi}{2}(c_2^4 - c_1^4) = \frac{\pi}{2}[(0.5d_o)^4 - (0.35d_o)^4] = 0.0746 \times d_o^4 m^4$

From Eq. (7.8), $T = \frac{\tau_{max}}{c}J = \frac{60\times10^6 Pa}{0.5d_o m} \times 0.0746 \times d_o^4 m^4 = 8.952 \times 10^6 \times d_o^3\ N.m$

Frequency, f = 100/60 = 1.667 revolutions per second

From Eq. (7.13), $300 \times 10^3 W = 2\pi \times 1.667 rps \times 8.952 \times 10^6 \times d_o^3\ N.m$

$d_o^3 = 3.199 \times 10^{-3} m^3 \Rightarrow d_o = 0.147m$=147 mm

$d_i = 73.65\ mm$

UNIT SUMMARY

The study and design of circular shafts subject to torques or twisting couples were the focus of this unit. The deformation and shear strains of the shafts subject to torque, shear stresses, twisting angle and power transmission of circular shafts were discussed.

EXERCISES

Multiple Choice Questions

1. The range of the shear stresses caused in a shaft by applied torque is as follows:
   a) Greatest at the circumference and zero at the centre
   b) Zero at the circumference and maximum at the centre
   c) Maximum at the circumference and minimum (non-zero) at the centre
   d) Uniform from centre to circumference of the cross section

2. When a solid shaft diameter (d) is under shear stress (τ), the torque delivered by the shaft is equal to--------------
   a) $\frac{\pi}{16}\tau d^2$ b) $\frac{\pi}{16}\tau d^3$ c) $\frac{\pi}{32}\tau d^2$ d) $\frac{\pi}{32}\tau d^3$
3. What should a shaft's new diameter be if its speed is doubled while still transferring the same amount of power and the maximum shear stress that the shaft is subjected to stays the same?
   a) one-half of its initial diameter
   b) 2 times its initial diameter
   c) $\sqrt{2}$ times its initial diameter
   d) $1/2^{1/3}$ of the initial diameter
4. Another name for torsional sectional modulus is----------------
   a) Torsion rigidity
   b) Sectional modulus
   c) Polar modulus
   d) Torsion modulus
5. ________ is a measure for shaft rotation strength.
   a) Torsional stiffness
   b) Polar modulus,
   c) Sectional modulus
   d) Torsional modulus
6. Shear stress variation in shaft torsion is-------------
   a) parabolic
   b) linear
   c) cubical
   d) quadradic

Answers of Multiple Choice Questions

1. (a) 2. (b) 3. (d) 4. (c) 5. (b) 6. (b)

Short and Long Answer Type Questions

1. How do you determine a circular solid shaft's torsional stiffness?
2. Derive the equation for twisting angle of hollow circular shaft?
3. Derive the torsion equation for hollow circular shaft?
4. Draw the shear stress distribution of hollow circular shaft?
5. A solid shaft in an engine transmission will be replaced with a hollow shaft made of superior quality steel, increasing the allowed stress by 24%. The replacement shaft will match the existing solid shaft's outside diameter in order to preserve the bearings that are currently in place. Calculate: (a) the hollow shaft's outer diameter in relation to the bore diameter.

   (b) the % weight savings, assuming that the steel densities of both shafts are equal.

Numerical Problems

1. In a manufacturing, a circular shaft diameter of 100 mm is needed to deliver torque. Calculate the greatest torque it can produce, if the permissible shear stress is 60 MPa.

2. To transmit 8.0 kN.m of torque, a solid steel shaft is used. What should be the shaft's minimum diameter if the shear stress at its highest is 50 MPa?
3. 7.65 kW is transmitted using a steel tube (Fig. P7.1) with 80 mm outer diameter and a speed of 30 revolutions per minute. Calculate the inner diameter d of the tube, if the permitted shear stress is 60 MPa.
4. Calculate the torque T that will cause 520 MPa maximum shearing stress in the hollow shaft illustrated in Fig. P7.2, knowing that d is 30mm.
5. Find the maximum shearing stress that would result from a torque of T = 0.9 kN.m. The hollow shaft of 23 mm inner diameter as illustrated in Fig. P7.2.
6. Internal torque of 20 kN-m is imparted to the hollow circular shaft as illustrated in Fig. P7.3. Determine the shear stress between A and B. A volume element should be used to represent each stress state.

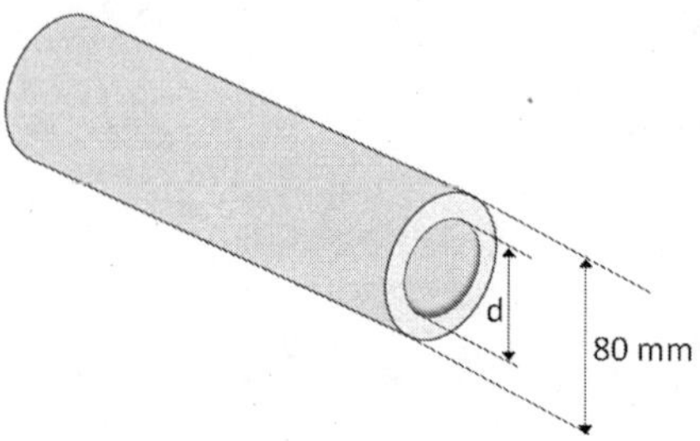

Fig. P7.1

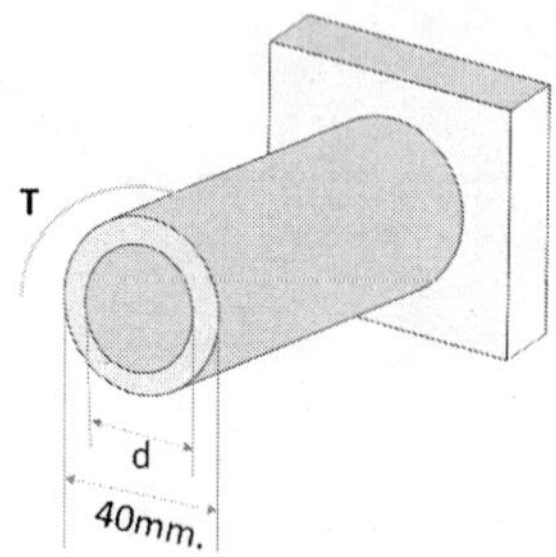

Fig. P7.2

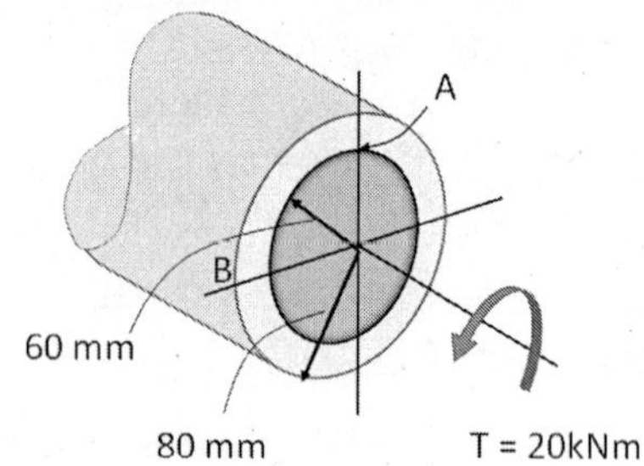

Fig. P7.3

7. The shaft is solid from B to C and hollow from A to B as shown in Fig. P7.4. Find the maximum shear stress on the shaft. The wall of the hollow portion of the shaft is 15 mm thick, and the shaft's outside diameter is 100 mm.
8. Find the shaft's as illustrated in Fig. P7.5 maximum shear stress, which has a 60 mm diameter.

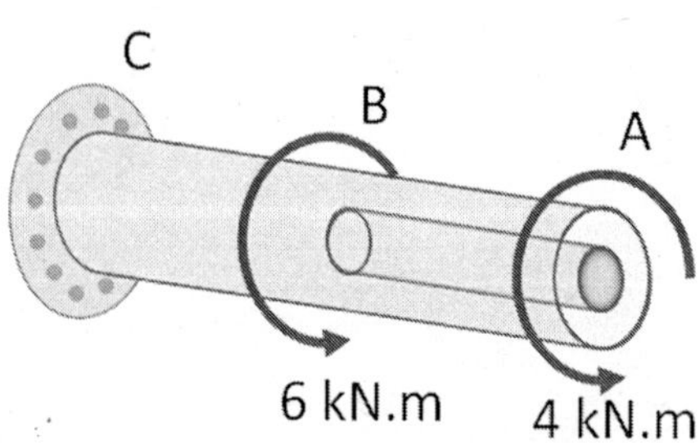

Fig. P7.4

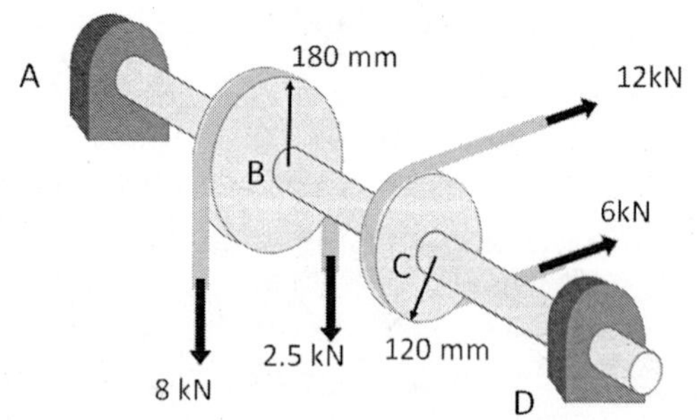

Fig. P7.5

9. The torques imparted to the gears are transferred to the solid 60 mm diameter shaft as illustrated in Fig. P7.6. Find the shaft's maximum possible shear stress.
10. According to the cross section as shown in Fig. P7.7, a rigid disc is attached to a fixed support by a steel shaft and an aluminium tube. If the initial stresses are zero and the allowable stresses for the steel shaft and aluminium tube are 120 MPa and 70 MPa, respectively, what is the maximum torque T_0 that can be applied to the disc? Use G = 77 GPa for steel and G = 27 GPa for aluminium.

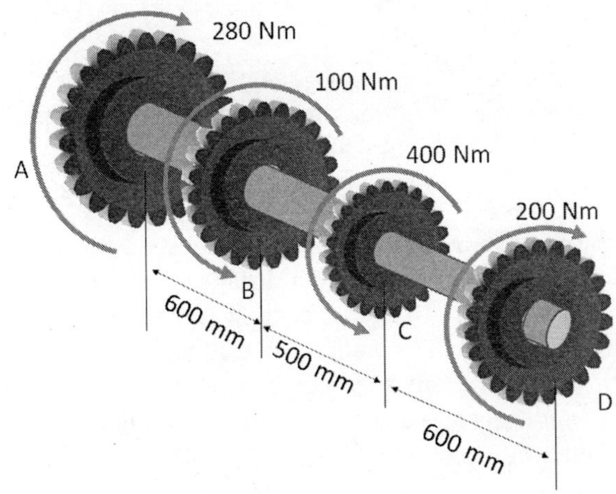

Fig. P7.6

11. For the aluminium shaft depicted in Fig. P7.8 (G = 27 GPa), calculate (a) the torque T that causes an twisting angle of 4^0, and (b) the twisting angle produced by a given torque T applied to a solid cylindrical shaft of equal length and cross-sectional area.
12. The hollow aluminium shaft of 6061-T6 grade has depicted in Fig. P7.8. Establish the twisting angle of end A. $T_B = k_B \phi_B$, where k_B is the torsional stiffness of 100 kN m/rad, because the support at B is adjustable like a torsional spring.
13. One end the steel shaft is fixed and other end is free has a 100 mm diameter and 1000 mm long. If the free end is being subjected to uniformly distributed torque, calculate the twisting angle there. Consider about G = 75 GPa.
14. For power transmission at 150 rpm, a circular shaft of 100 mm in diameter is needed. Calculate the shaft's power output if the shear stress cannot be greater than 60 MPa.

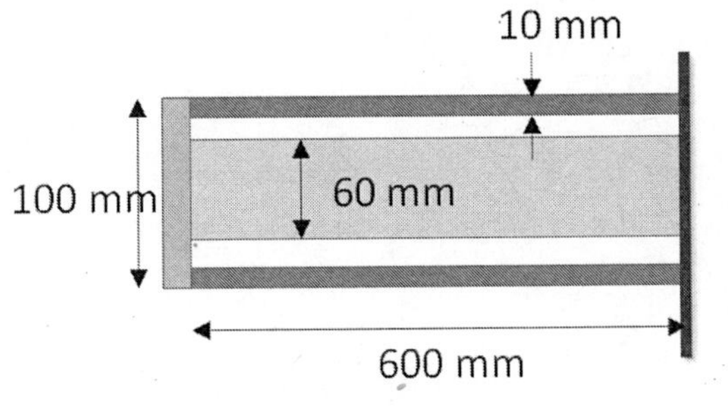

Fig. P7.7

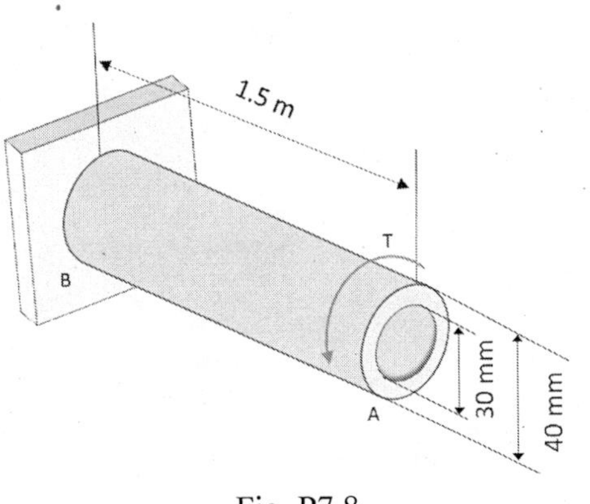

Fig. P7.8

PRACTICAL

Aim: To establish a connection between the clamping length (L), torque (T) and twist angle (ϕ), as well as the "shear modulus of elasticity (G)" for circular shafts made of steel, aluminum, and brass.

Materials Required: Torsion apparatus (Fig. L7.1) and measuring dial.

Theory:

The shaft that is fastened between two holding clamps will receive a torque from the device as shown in Fig. L7.1. By applying a force of F through a distance spindle arm via the load handle above the system, torque is created. The shaft should be tightly gripped between the end holding clamps as much as is practical. Twisting will cause "shearing stresses" to be applied to the test specimen, and the specimen with a higher rigidity modulus will be more rigid or resistant to twisting. Once more, the modulus of rigidity is a property of the material and has a constant value for every material under normal environmental conditions. In this experiment, two or three specimens will be put to the test.

These samples will have the same geometrical dimensions and just differ in terms of the kind of material. Steel, aluminium, and brass among the different materials tested. By twisting the test specimen and computing the slope of the shear stress vs. shear strain graph, the modulus of rigidity can be ascertained. In order to determine the shear stress at the limit of proportionality, the torsion test specimen will also be twisted until it breaks. The highest value of shear stress at which a material will act elastically is the shear stress at the limit of proportionality.

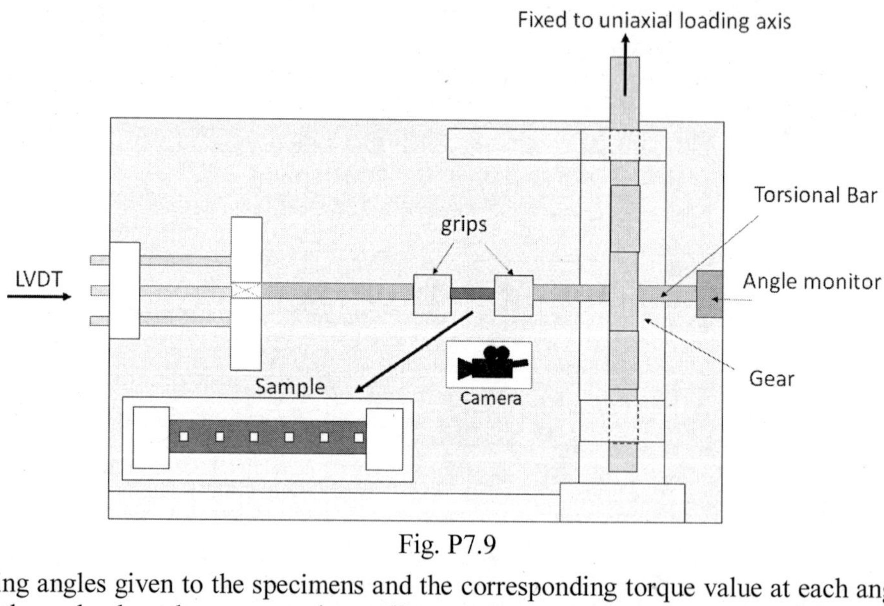

Fig. P7.9

The twisting angles given to the specimens and the corresponding torque value at each angle of twist are the real values that are experimentally recorded as the specimens are being twisted. The 6^0, 360^0, and torque scale locations should be made clear by the instructor. For converting twist and torque measurements to the equivalent shear stress and shear strain values, equations are given below.

$$\tau = \frac{T\rho}{J}; \gamma = \frac{\rho\phi}{L}; \phi = \frac{TL}{JG} \Rightarrow G = \frac{TL}{J\phi}$$

Procedure:

1. Slide the rod into the shaft after loosening the tightening screws on the test equipment holding chucks. By fastening the sliding arm and tightening screws, tighten the rod.
2. Determine a fixed length with a 200 mm spread. The rod's diameter should be measured.
3. Calculate the distance between the rod's centre and the point where the weight is applied. Comparing this measurement to the value on the lab poster will help.
4. The distance between the rod's centre and the vernier measuring point should be measured. In light of equipment worth, compare this value.
5. Set the Vanier reading to zero or save the position of the zero load.
6. As you gradually apply loads, note the deflection at the vernier point. Keep track of how many times the gauge reader has been turned, if it has been more than one complete rotation.
7. Increase the load by a factor of 100 gm, apply a force of 0 to 2000 gm, and note the deflections. Continue for every specimen.
8. Repeat the measurements with a length change from 150 mm to the longest span, a steep angle of 100 mm, and 450 mm.

Observation:

Use the suggested table below to collect data.

Material of the specimen:

Length of specimen:

Diameter of circular shaft:

Applied Load (gm)	Measured Deflection	Angle of Twist Radian	Calculated Angle of Twist	Modulus of Rigidity
100				
200				
300				
.				
.				
.				
2000				

Calculations:

1. For each of the three materials, plot a graph of the Toque verse twist angle.
2. Draw a graph showing the twist angle in relation to length L for each of the three materials.
3. Using information from job 1 and formulas for all materials, calculate the shear modulus of elasticity (G) and torsional shear stress on the top surface.
4. Quantify the inaccuracies by comparing the calculated value of G with the theoretical value.

KNOW MORE

Torsion constant: The torsion constant is a geometric property of a linear-elastic, homogeneous bar that determines the connection between the applied torque and the angle of twist along the axis of the bar. The torsion constant, along with material properties and length, determines the torsional stiffness of the bar. The torsion constant is typically expressed in SI units as m^4.

A French engineer named A. Duleau, in 1820, made an analytical calculation by making assumptions that a plane section will remain planar after twisting and a diameter will remain a straight line. According to his calculation, the torsion constant of a beam is equal to the second moment of area normal to the section J, which has a precise analytical equation. Unfortunately, such assumption is false for every other shape where warping occurs and only holds true for beams with circular cross-sections.

There aren't any precise analytical equations for calculating the torsion constant for cross-sections that aren't circular. For many shapes, however, approximative solutions have been discovered. There are always warping deformations in non-circular cross-sections, hence numerical techniques are needed to determine the torsion constant precisely. If the end sections of non-circular beams are prevented from warping, for instance, through the use of rigid end blocks, the beams' torsional stiffness is significantly enhanced.

REFERENCES AND SUGGESTED READINGS

- Ferdinand P. Beer, E. Russell Johnston, Jr., John T. Dewolf, David F. Mazurek, Mechanics of Materials, McGraw Hill, New York, 2012.
- Hibbeler, R.C., Mechanics of Materials, Pearson Prentice Hall, 2018.

8 Friction

UNIT SPECIFICS

Through this unit we have discussed the following aspects:

- *History and types of friction;*
- *Five laws of friction;*
- *Resultant reaction at the contact surface;*
- *Angle of friction and angle of repose;*
- *Wedge friction;*
- *Slipping vs tipping conditions in friction;*

RATIONALE

This unit on friction shall help the students to understand about the laws of dry friction, its applications and assist the students in solving friction problems. These are relevant to the students since the knowledge gained shall be applied to solve the practical cases arising from applications of friction in advanced course of Mechanisms, Machine Design, Robotics and Tribology.

PRE-REQUISITES

None

UNIT OUTCOMES

List of outcomes of this unit is as follows:

U8-O1: Students shall be able to grasp the fundamentals of friction

U8-O2: Students shall be able to analyse different applications of friction

U8-O3: Students shall be able use the concepts in advanced courses involving topics of friction

U8-O4: Students shall be solve numerical problems of friction and its applications

Unit - 8 Outcomes	EXPECTED MAPPING WITH COURSE OUTCOMES (*1- Weak Correlation; 2- Medium Correlation; 3- Strong Correlation*)					
	CO-1	**CO-2**	**CO-3**	**CO-4**	**CO-5**	**CO-6**
U8-O1	3	3	3	-	3	1
U8-O2	1	1	2	2	1	-
U8-O3	2	1	3	1	2	1
U8-O4	-	-	3	1	2	2

8.1 FRICTION

Resisting force to any relative motion between two contacting surfaces results in friction. In mechanics, initially, problems are discussed by safely ignoring friction, however, many interesting situations arise in statics and dynamics wherein friction must be accounted for, such as wedges, screws, bearings, transmission elements etc. In case of elementary problems ignoring friction, each surface only exerts normal force to the other surface and the two surfaces can move freely relative to each other. However, if the surfaces are rough with asperities or have an interface of adhesion in between, then, apart from the normal force, tangential forces are also developed that tend to prevent the motion of one surface with respect to the other. These tangential forces give rise to dry friction and are called friction forces. A friction force is unlike the conventional applied force that causes a body to accelerate. Instead, friction force is an internal force that actually opposes the externally applied force.

In general, friction can be categorised as dry friction wet friction. Dry friction is also known as Coulomb friction, and wet friction is synonymous to fluid friction or viscosity. In addition to this, from the viewpoint of material science, there is one more type of friction (called internal friction) that materials experience intrinsically due to the resisting forces present inside the bodies that resist deformation. In essence, then, friction can be categorized mainly as:

1. Dry friction: the friction that opposes one solid surface sliding across another solid surface. Dry friction is further categorised as static friction or stiction on stationary surfaces and kinetic friction on moving surfaces.

2. Fluid friction: the friction between layers of a viscous fluid in motion. When different layers of fluid move with different velocities, fluid friction develops between such layers. Such situations arise in cases of flow over surfaces (both internal and external flow) and are dealt within the subject of Fluid Mechanics.

3. Internal friction: the force resisting internal deformation of a solid material.

In the present unit, readers are introduced to the concept of dry friction only wherein, two solid surfaces touch each other without the presence of any fluid in between.

8.2 HISTORY

The ancients Greeks were aware of friction, its causes and ways of mitigation, difference between static and kinetic applications so much so that Themistius, in 350 AD is known to state that "it is easier to further the motion of a moving body than to move a body at rest". The first methodical investigations of friction were made and recorded by Leonardo da Vinci (1452-1519) on wood-wood and wood-iron interface combinations wherein he measured the friction force needed to slide a mass across a surface. It is known that he noted two vital observations:

a. The friction force gets doubled when the weight is doubled, and
b. The friction force was independent of their positioning on the surface (i.e., F was observed to be independent of the area of contact).

Though many of the fundamental rules of friction are believed to have been formulated by Leonardo da Vinci, for generations, these observations remained in obscurity and subsequently were being rediscovered by Guillaume Amonton in 1699. Amonton was the first to propose the friction's nature in terms of irregularities at contacting surfaces. Bernard Forest de Bélidor and Leonhard Euler, in 1750, took ahead the work of Amonton and provided an expression of angle of repose of a weight on an inclined plane. Bélidor and Euler also were the first to distinguish between static and kinetic friction. It was Coulomb, in 1781, who, further analysed and verified these laws despite a lot of scepticism and noted the non-dependence of friction force on velocity of sliding. Coulomb explored the impact of four key factors on friction:

i. the nature/type of the materials in contact, coatings on such surfaces, if any.
ii. the surface area;
iii. the normal load;
iv. the length of time that the surfaces remained in contact (time of repose).

Coulomb also investigated the effect of velocity of sliding, operating temperature and prevailing humidity, to elaborate upon the different explanations that were proposed before his time on the nature of friction.

In 1794, the role played by adhesion phenomenon in friction was reported by John Theophilus Desaguliers. Desaguliers proposed that friction was necessary to tear those surfaces apart that got adhered due to microscopic adhesion forces. The concept of sliding versus rolling friction was first proposed by Arthur Jules Morin in 1833. The classic empirical model of friction that encompasses static, kinetic, and fluid friction as is commonly used engineering today was made complete when Osborne Reynolds, in 1866, reported the mathematics behind viscous flow. In the 20th century, the main interest behind the research on friction was to understand the physical mechanisms behind it. In 1950, Bowden and Tabor reported that the actual contact area between surfaces is a very small fraction of the apparent area at the microscopic level. They also reported that as the contact pressure increases, the actual contact area of the asperities also increases. With the advent of atomic force microscope in 1986, the researchers started gaining better insights into the underlying physics of friction and contact at the atomic scale are subsequently identified that, at atomic scale, dry friction is basically the product of the shear stress at the interface and the contact area.

8.3 THE FIVE LAWS OF FRICTION

Amonton's laws of friction that generally help to understand the characteristics and decipher the practical applications are as follows:

i. According to Amonton's first law, for any object in motion, force of friction is proportional and perpendicular to the normal force exerted between the surfaces in contact. A simple and practical demonstration of this law is observed when one finds it difficult to push heavy objects in comparison to pushing the lighter ones.
ii. Amonton's second law illustrates that friction depends on the surface characteristics of the contacting surfaces and is affected by the nature/type of contact surfaces, i.e.it depends on material of which the contact surfaces are made of, roughness of the contacting surfaces, and whether the surfaces in contact are dry or have fluid within.
iii. The third law states that friction force is independent of the area of contact. If one assumes two objects of same weight but different sizes resting on a ground, then the weight of the objects, being same, shall exert same force at the contact surface on the ground, however, different pressure over different contact areas compensates for the unequal area of contact of the both objects with the ground.
However, there are many situations where this law is not valid. For example, in case of an automobile, the traction force between the road and the tyre can be increased by deflating the tyre thereby increasing the area of contact.
iv. According to the fourth law, coefficient of static friction is always larger than coefficient of kinetic friction as more forces are acting on an object to keep it at rest than to maintain its motion.
v. The fifth law is the Coulomb's law of friction. According to this, velocity has no effect on kinetic friction. As long as the normal force and the type of material remain constant, the speed of an object does not change the amount of friction between two surfaces.

8.4 DRY FRICTION

Dry friction always arises between two contacting surfaces that are devoid of any other medium in between them. The friction force, distributed over the contacting area, always acts in the direction which opposes the relative motion of the contact surfaces. While studying friction, this distributed friction force is assumed to be acting from a point on the area of contact and, hence, is mathematically considered as a concentrated friction force. For mathematical simplicity, hence, the actual reaction force obtained from the ground/support is decomposed into normal reaction and friction force acting at a point in the surface of contact (Refer Fig. 8.1).

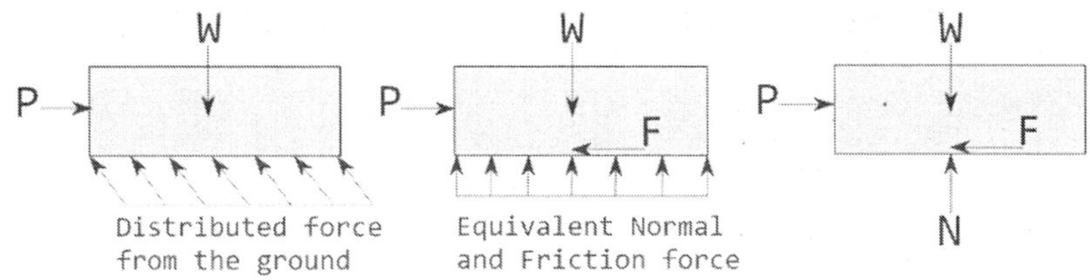

Fig 8. 1 Response at contact interface of two bodies

Depending on the situation, the friction force may hold the object in equilibrium or may cause it to accelerate/retard. Consider the example of a book resting on a table as shown in Fig. 8.2(a). Prior to its use, the weight of the book is supported by a normal force directly underneath its centre of gravity, and the friction force at the contact area of book-table is zero. If someone pushes the book to slide to the left, dry friction will come into play and a friction force shall develop along the bottom surface of the book in order to prevent it from sliding to the left. This frictional force is represented as a point normal force, offset from the centre of gravity of the book. If the pushing force is large enough, equilibrium will break and the book shall begin to slide towards left and hit the vase. On the other hand, the friction forces acting on a rolling object, say a bike as shown in Fig. 8.2(b), are unlike the sliding objects and are more complicated. As shown in the figure, both wheels, if not braked, rotate in the clockwise direction, but the rear wheel is driven by chain, whereas the front wheel is rotated due to the road friction. The bike maintains its speed due to the friction force acting rightwards on the rear wheel while the friction force arising due to the dry friction on the front wheel acts towards left and retards the motion of the motorcycle.

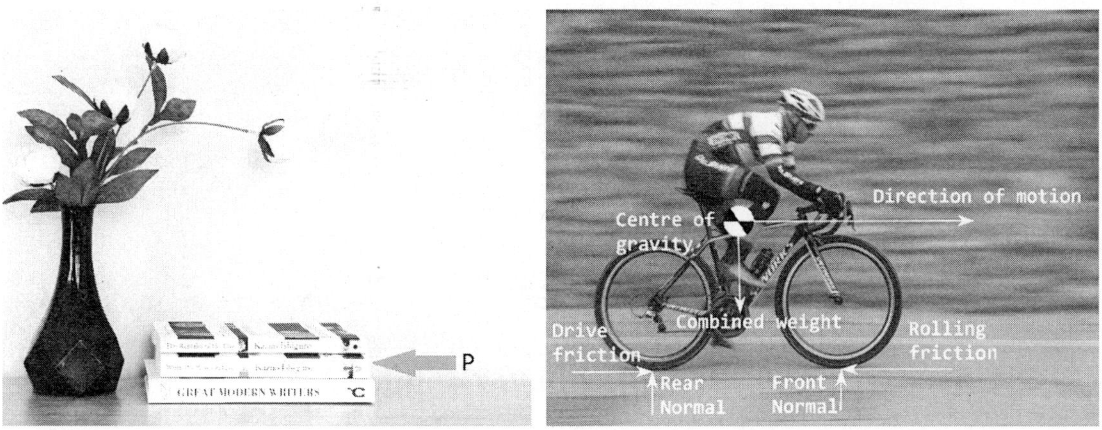

Fig 8. 2 Dry friction between (a) book-table depicting sliding friction and (b) wheels and road depicting rolling friction

Commonly used model for dry friction is Coulomb friction, in which different phases are involved as shown in Fig. 8.3. Coulomb friction relies on a friction coefficient μ to serve as the proportionality constant between the normal and friction forces.

For static friction:

$F_l = \mu_s N$; μ_s is the coefficient of static friction (8.1)

For kinetic friction:

$F_k = \mu_k N$; μ_k is the coefficient of kinetic friction (8.2)

Measurement of coefficient of friction is done empirically through experiments. Rougher surfaces tend to have higher coefficient of friction. One must note that coefficient of friction is better categorized as a system property instead of being a material property since it depends on system variables such as velocity, temperature, humidity in addition to surface roughness and material deformation characteristics at contact interface. The surface roughness parameterized by fractality of surfaces also plays an important role in determining the magnitude of the static friction. Coefficient of friction, symbolized by the Greek alphabet μ, is a non-dimensional scalar that defines the ratio of friction force and the normal force between two bodies. As stated earlier, coefficient of friction depends on the materials at the interface of two contacting bodies. The friction coefficient μ is always greater than 0 and commonly less than 1. The coefficient of friction normally varies between 0.3 to 0.6 for most of the material combinations under dry conditions, however, exceptions do exist (Teflon has coefficient of friction value of 0.04). Zero value of coefficient of friction is an elusive property and signifies absence of friction. It is noted that coefficient of friction amongst the surfaces of same metals is greater than that between the surfaces of different metals. Usually, in most applications, coefficient of friction is observed to be less than unity, however, a value equal or more than 1 (as observed in cases of silicone rubber, acrylic rubber-coated surfaces or glued surfaces) is normally found in cases of materials that exhibit positive adhesion to each other and implies that the friction force or the force required to push an object along such surfaces is larger than the normal force at the contact surface. In other words, in such circumstances, the friction force is dominant over the normal force in terms of calculation of the resultant force at the contact interface.

To illustrate the laws of dry friction as shown in Fig.8. 3, assume a block of weight ***W*** is resting on a horizontal surface. The block is acted upon by its weight, ***W*** and the normal reaction at the contact surface. Due to absence of any horizontal external force no horizontal reactive forces are produced. This situation is depicted by point O in the figure. Now, if a force ***P*** is applied horizontally to the block, then to maintain equilibrium, three situations may arise.

Firstly, if ***P*** is small, then the block remains in rest and consequently this indicates that a reactive horizontal force of equal magnitude but opposite in direction must have come to existence. This other/reactive force is called static-friction force ***F***. This static friction force is actually a resultant of large number of contact forces at the surface irregularities or asperities between two contacting surfaces i.e., block and plane. This state is represented by A-A in Fig.3 and is known as ***static friction state***.

Secondly, if the magnitude of ***P*** is further increased, then ***F*** also increases until its magnitude reaches a maximum value, F_L, denoted by B-B in Fig.8.3. The friction force at its maximum is called limiting friction force. This state is known as ***limiting friction state*** or ***static-but-impending state***.

Thirdly, if ***P*** is further increased, the block starts to slide as the resulting friction force at the contact surface cannot counter-balance the external force ***P***. At this instant, as the motion impends, the magnitude of F falls from F_L to F_k, since during the relative motion, less interpenetration occurs within the asperities of the contacting surfaces and F_k is known as kinetic friction force. This situation is represented by C-C in Fig. 3 and this state is known as ***kinetic friction state***. Hereupon, the block keeps sliding and F_k remains nearly constant.

The maximum friction possible is reached at impending motion (B-B in Fig.8. 3) and the resulting limiting friction force is given as $F_L = \mu_s N$.

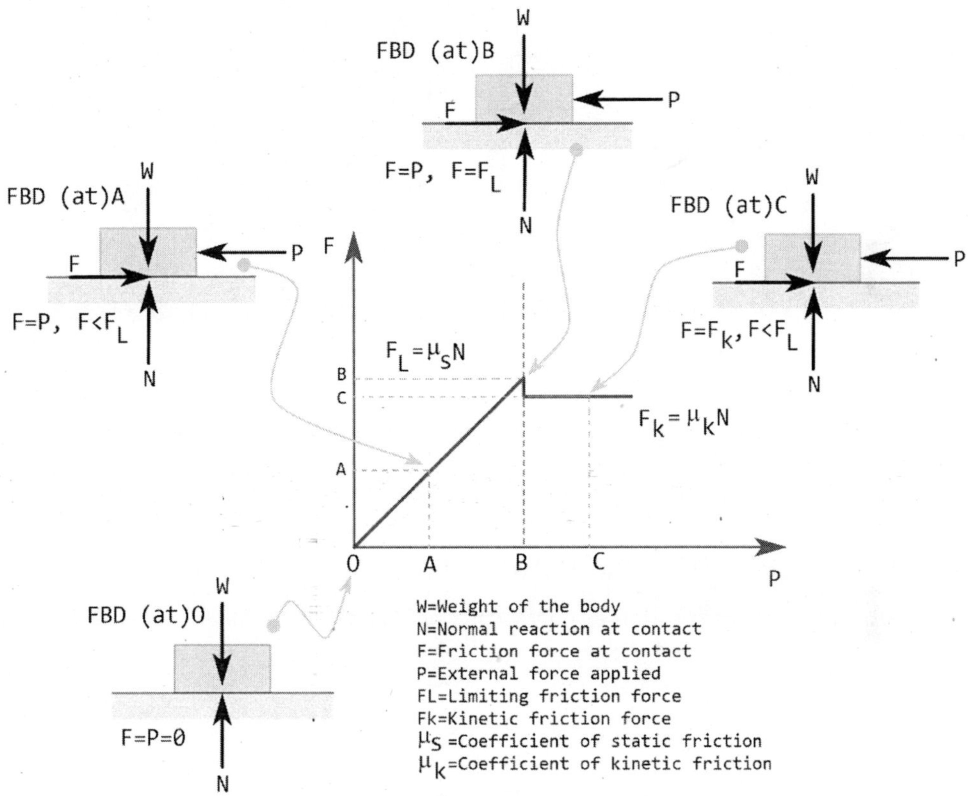

Fig 8. 3 Coulomb friction graph for a block-plane interaction

When the pushing force is increased beyond impending motion and the block starts to slide over the plane due to the influence of kinetic friction, then kinetic friction force is given as $F_k = \mu_k N$.

One must note that the limiting friction force at impending motion is always greater than kinetic friction force since coefficient $\mu_s > \mu_k$ for most materials. This is because, as stated earlier, during motion, the interpenetration within the irregularities or the asperities of the contacting surfaces are much less than these are during static condition.

8.5 RESULTANT REACTION AT THE CONTACT SURFACE

For 2-D problems, the normal forces are seldom concentrated at one point and are mostly distributed across the contact area, however these are usually shown at the centre of the contacting surfaces and in the free body diagrams, pertaining to friction problems, the force vector is applied at a point on the contact surface and is equivalent to the distributed force. This is represented by means of figure below (Fig.8. 4).

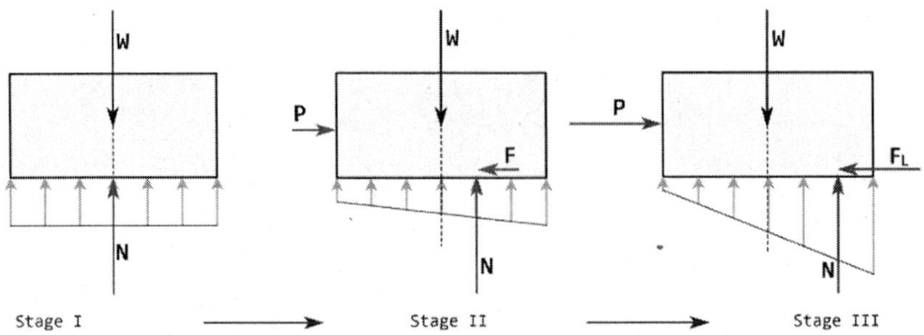

Fig 8. 4 Free Body Diagram for friction problem

(Stage I) An object without any push and at rest. The normal reaction is distributed uniformly at the contact surface but is represented by a point reaction N directly below the weight force at the centre of the contact surface. (Stage II) When a push P is applied to the object, the distributed resultant reaction changes shape resultant normal reaction N shifts towards right to maintain equilibrium. (Stage III) When the push P causes the friction force to be equal to F_L, the final shape of distributed resultant reaction, normal reaction (N) and limiting friction force (F_L) and their final point of application.

8.6 FRICTION ANGLE, FRICTION RESULTANT

The friction force and normal reaction at contacting surface can be represented in any of the two following but equivalent ways (also represented in Fig.8. 5):

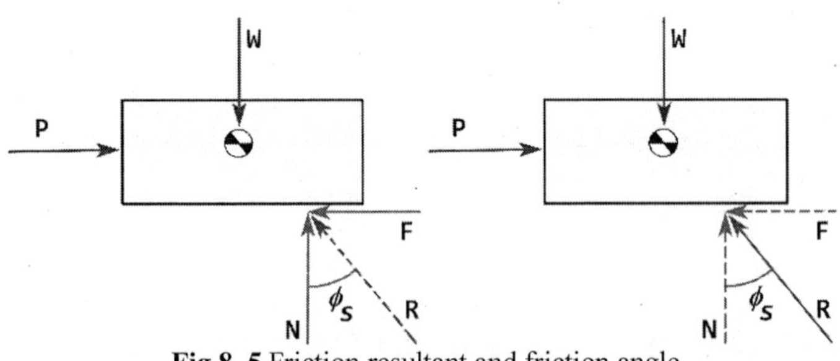

Fig 8. 5 Friction resultant and friction angle

a. **In terms of friction force and normal reaction.** In many applications of friction such as block and wedge problems, the orthogonal vectors representing friction force and normal reaction are most convenient to apply and are valid for all three phases of friction (static friction, limiting friction and kinetic friction).

b. Resultant force and friction angle. Since resultant force is the sum of two or more vectors, the resultant of orthogonal vectors representing friction force and normal reaction is termed as friction resultant and is given as

$$\vec{R} = \vec{F} + \vec{N}$$

$$\text{Magnitude, } R = \sqrt{F^2 + N^2}$$

$$\text{Direction, } \emptyset = \tan^{-1}\left(\frac{F}{N}\right)$$

At $F = F_l$, $\emptyset = \emptyset_s$ and is called the friction angle between the friction resultant and normal forces and is only applicable at impending motion. For the cases when $\emptyset \leq \emptyset_s$, the system is under equilibrium and $\emptyset = \emptyset_s$, indicates that limiting friction is reached and sliding takes place hereafter.

$$\emptyset_s = \tan^{-1}\left(\frac{F_l}{N}\right) = \tan^{-1}\left(\frac{\mu_s N}{N}\right) = \tan^{-1}(\mu_s)$$

Beyond the limiting case, once the object begins to slide, $\emptyset = \emptyset$, angle of kinetic friction

$$\emptyset_k = \tan^{-1}\left(\frac{F_k}{N}\right)\tan^{-1}\left(\frac{\mu_k N}{N}\right) = \tan^{-1}(\mu_k)$$

The use of friction resultant and friction angle is convenient for solving problems of screw, flexible belt, and journal bearing type problems.

8.7 ANGLE OF REPOSE

Roughly speaking, to define angle of repose, one must consider a rough horizontal surface with a block resting on it without any external force except its weight and a corresponding normal reaction at the contact surface (Fig.8. 6(a)). Now the reader must visualize the gradual and incremental rotation of this plane such that as the plane is rotated by an angle θ, the weight of the block gets resolved into two components, one along the incline and other orthogonal to the incline as shown in Fig.8. 6(b). Along these two orthogonal directions, since the object does not move, when forces are resolved, equilibrium condition prevails.

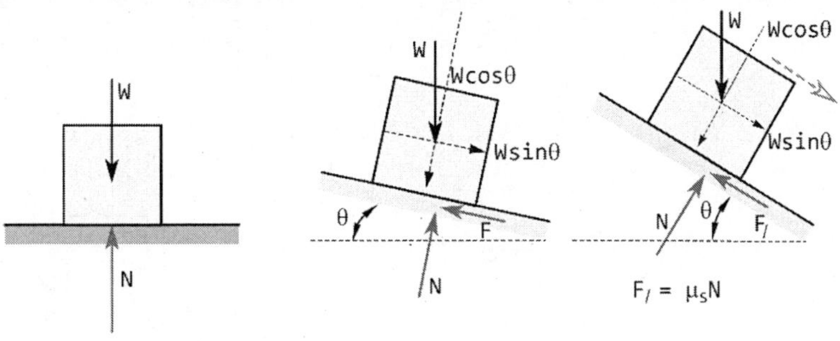

Fig 8. 6 Angle of Repose

$\sum F_y = 0$; N = Wcosθ

$\sum F_x = 0$; F = Wsinθ

However, one must note that since motion does not impend, $F < F_l$.

Upon further gradual increment of the inclination, a situation shall arise when, with increasing angle θ, Wsinθ also increases thereby increasing F. As F reaches the limiting friction force value, the block just impends to slide as shown in Fig.8. 6(c). Hence, the corresponding equilibrium equation shall read as follows:

$\sum F_x = 0$; $F_l = W\sin\theta$

Also under impending sliding,

$F_l = \mu_s N = \mu_s W\cos\theta$

Hence, using these equations;

$W\sin\theta = \mu_s W\cos\theta$

$\tan\theta = \mu_s$

$\tan\theta = \tan\phi_s$

$\theta = \phi_s$

Hence, though mathematically it is proved that angle of repose and angle of friction are identical for a block resting on a horizontal rough surface, but physically, angle of repose is the angle, that a rough inclined plane makes with the horizontal, when a block placed on it is on the verge of impending sliding motion.

8.8 PROBLEM CLASSIFICATION AND ANALYSIS

The inequality in Coulomb's law does not provide the friction force; instead, it only gives the largest possible or limiting friction force. The equality (Eq.1) is used only if slipping impends. Since the induced friction force at the contact surface (F) is not necessarily same as limiting friction force (F_l), a single approach to solve friction problems is not always valid. Hence, problems of application of friction are categorised into three different types, each with a different way of solving:

Type I, Problems without the mention of impending motion:

In such problems, the normal reaction (N) and the friction force (F) are the unknowns and using equilibrium equations, N and F are calculated. Hence first equilibrium is assumed and once the values of N and F are obtained, using Eq. 1, limiting friction force (F_l) is also calculated. Now, if $F \leq F_l$, assumption of equilibrium is correct. But if $F > F_l$, then the assumption is wrong and it becomes evident that the problem belongs to that of kinetic friction. Under such circumstances, using Eq. 2, F_k is calculated. The balancing of equation of forces along the contact surface, then provides the magnitude and direction of force required to establish equilibrium.

Type II, Problems of impending motion:

In such problems, it is implied that sliding is about to impend, thereby meaning that the system is under static equilibrium and $F = F_l$, thus allowing the substitution of $F_L = \mu_s N$. The sense of F_l should be taken opposite to the direction of impending sliding.

Type III, Problems of impending motion without prior knowledge of the surfaces experiencing sliding:

These are the most typical problems to analyse as these involve trial and error approaches. In such cases, the foremost act is to ascertain the possible ways in which sliding may impend. The next step is to set $F = F_l$ at any one of these surfaces and ensure that the sense of limiting friction force is taken correctly and solve the equilibrium equations. Subsequently, after obtaining F at other surfaces using equilibrium equations, verify the solution at each of these surfaces by comparing the calculated F with F_l at these remaining surfaces. If $F \leq F_l$, then the solution is correct elseif $F > F_l$, then the obtained solution is not valid and entire process is to be repeated at any other surface till the correct solution is obtained.

Friction problems are similar to equilibrium problems of rigid bodies with additional friction forces at all contact surfaces.

8.9 NUMERICAL PROBLEMS

Example 1: For the problem shown in Fig.8. 7, (i) determine the horizontal force P required to cause impending motion. (ii) If P is applied downward at 30 degrees (with horizontal) at same location, what shall be P? (iii) Also determine the minimum force required to start motion. Coefficient of static friction is 0.40

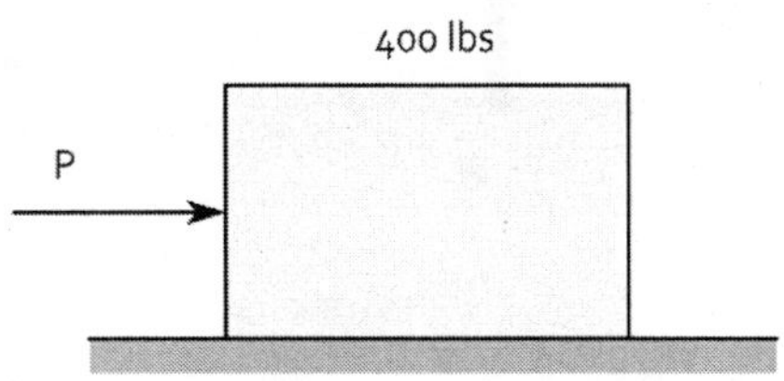

Fig 8. 7 Block on a horizontal surface

Solution:

(i) Using equilibrium equations,

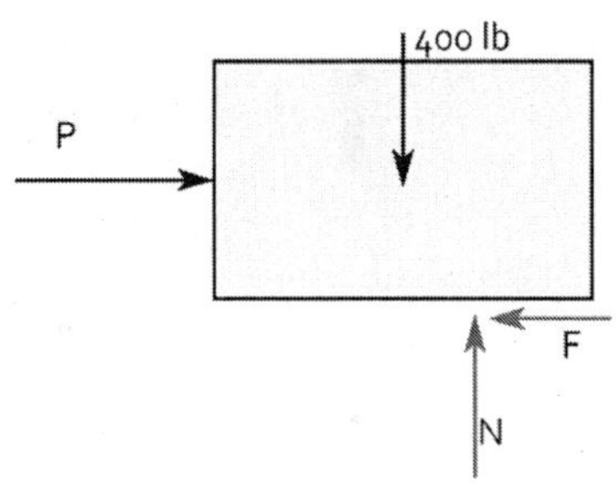

$$\sum F_y = 0;$$

N = 400 lb

$$\sum F_x = 0;$$

F = P

However, since the problem is of impending motion, F = μ_sN

Hence

P= μ_sN = 0.4 (400) = 160 lb

(ii) When P is applied at 30 downward as shown in Fig.

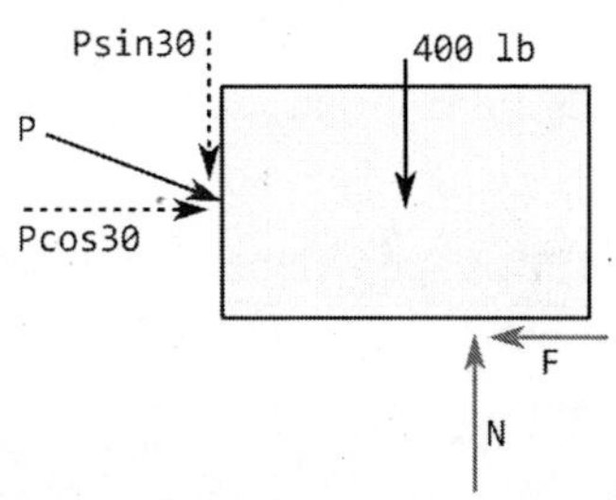

$$\sum F_y = 0;$$

N = Psin30 + 400

Also, since the problem is of impending motion, F = μ_sN, hence

F = 0.4(Psin30 + 400)

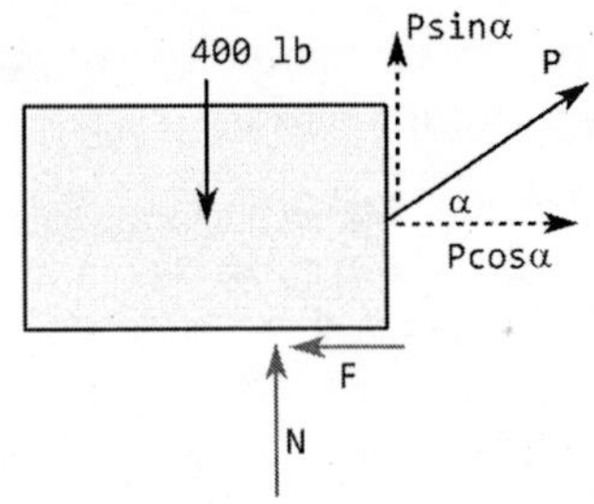

F= 0.2P+160

$$\sum F_x = 0;$$

Pcos30 = F

Pcos30 =0.2P+160

P=240 lb

(iii)

$$\sum F_y = 0;$$

N + Psinα = 400

Also, since the problem is of impending motion, F = μ_sN, hence

F = 0.4(400-Psinα)

F=160-0.4Psinα

$$\sum F_x = 0;$$

Pcosα = F

Pcosα = 160-0.4Psinα

(cosα+0.40sinα) P=160

For P to be minimum,

$$\frac{dP}{d\propto} = \frac{-160(-sin \propto +0.4cos \propto)}{(cos \propto +0.4sin \propto)^2} = 0$$

Upon solving, α=21.80° and P = 148.5 lb

Example 2: For the block shown in Fig.8. 8, determine the value of horizontal force (P) necessary to

(i) impend motion of the block up the incline

(ii) just prevent the block from sliding down.

(iii) If P = 1.78 kN, determine the friction force?

Coefficient of static friction is 0.40

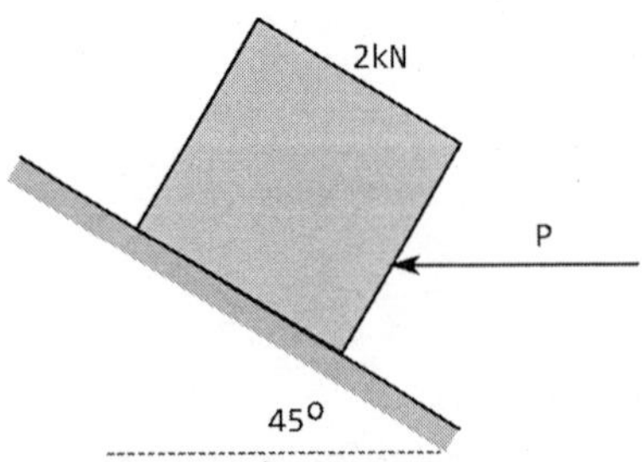

Fig 8. 8 Block on an inclined surface

Solution: Assuming the direction along the plane to be along x-axis under local reference plane and using corresponding orthogonal coordinates,

(A) Force required to move the block up the incline:

In this case, the friction force shall be the limiting friction force and shall act in downward direction as shown in the FBD.

Using equilibrium equations,

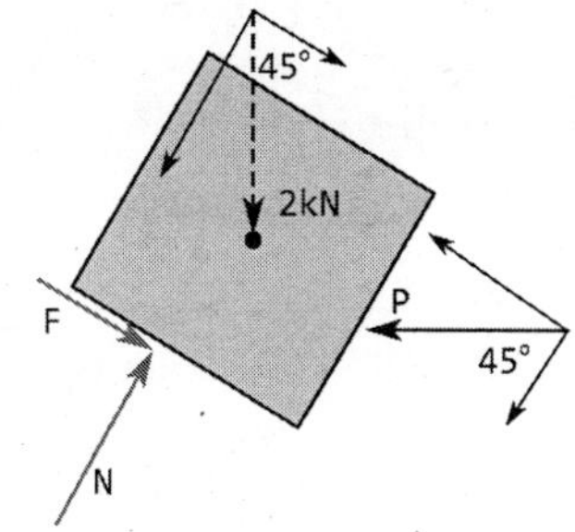

$\sum F_x = 0;$

F+2cos45=Psin45

Also, since the problem is of impending motion, F = μ_sN, hence

μsN+2cos45=Psin45

0.4(N)+1.414=0.707P

N=-3.535+1.7675P

$\sum F_y = 0;$

N=2sin45+Pcos45

N=1.414+0.707P

Upon solving the two equations simultaneously,

-3.535+1.7675P=1.414+0.707P

P= 4.66 kN

(B) Force required to prevent the block from moving down the incline:

In this case, the friction force shall again be the limiting friction force but to prevent the block from slipping down, must act in upward direction to maintain static equilibrium. So the FBD shown in (i) shall only change to show F acting in upward direction.

Using equilibrium equations,

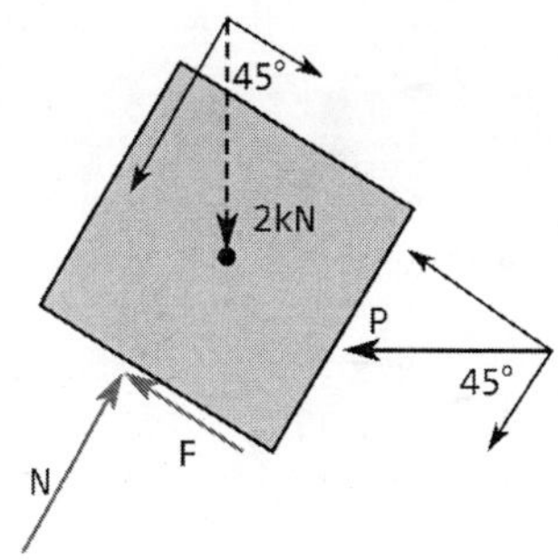

$\sum F_x = 0;$

2cos45=Psin45+F

Also, since the problem is of impending motion, F = μsN, hence

2cos45=Psin45+ μsN'

1.414=0.707P+0.4(N)

N=3.535-1.7675P

$\sum F_y = 0;$

N=2sin45+Pcos45

N=1.414+0.707P

Upon solving the two equations simultaneously,

3.535-1.7675P=1.414+0.707P

P= 0.85 kN

(C) It is evident from the FBDs shown above that F shall case to exist if the components of P and 2 kN along the plane (x-axis) are same. These components are P_x (=Psin45) and 2cos45 (=1.414

kN). So, if P_x >1.414 kN, then F shall act opposite to Px, thereby resisting upward motion and shall be represented by 1st FBD. If P_x <1.414 kN, then F shall act in the same direction as P_x to prevent the block from sliding down under the action of its weight and shall be represented by 2nd FBD. For the given problem, P=1.780 kN and hence P_x = 1.2586 kN (<1.414 kN). So, the correct representation of block, under the circumstances, is made by FBD shown under (ii) and the friction force acts upwards along the incline. Physically, this means that the friction force assists the external force P to prevent the block from sliding down. By balancing the forces along the incline (x-axis), F can be determined as follows.

$$\sum F_x = 0;$$

2cos45=Psin45+F

2cos45=1.78sin45+ F

1.414=1.258+F

F=3.535-1.7675P

F=0.156 kN

Example 3: For the problem shown the Fig.8. 9, determine the coefficient of friction at the contact interface. The impending motion is up the incline.

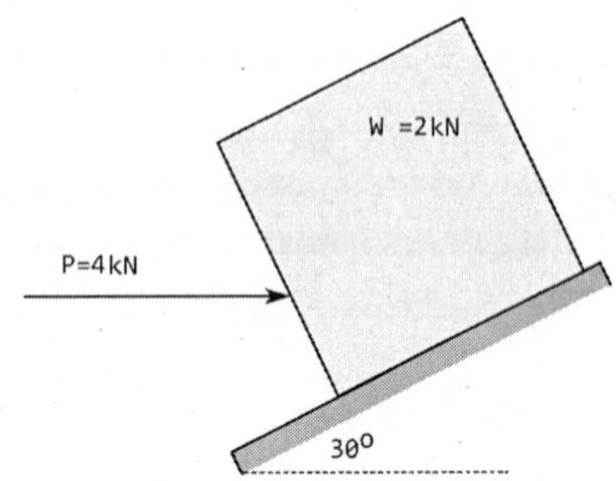

Fig 8. 9 Block on an inclined surface

Solution: The free body diagram is shown below:

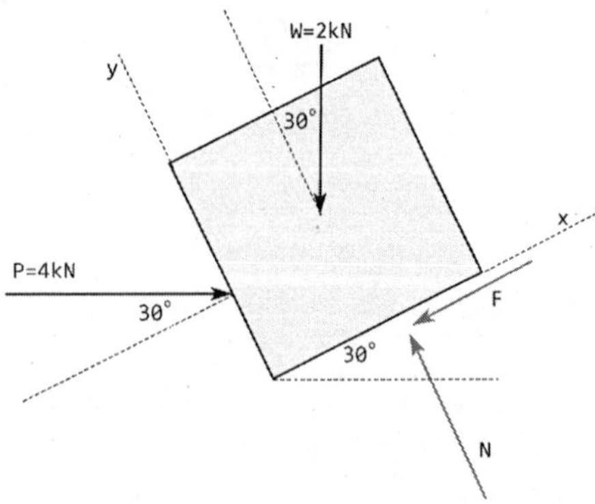

$$\sum F_y = 0;$$

N=2cos30+Psin30

N=1.414+0.707P

N=1.732+0.5P

$$\sum F_x = 0;$$

Pcos30=2sin30+F

0.866P=1.0+ μ_sN

In the problem, P = 4 kN. So,

N =1.732+0.5(4)

N=3.732 kN

And

0.866(4) = 1.0+ μ_s(3.732)

Upon solving the equation, one obtains:

μ_s=0.66

8.10 WEDGE FRICTION

A wedge is a simple device used to create mechanical advantage. It is a tapered device used to either to force objects apart (known as low friction wedges) or to prevent impending motion (high friction or self-locking wedges), these should be slippery so as to be easily driven into the crack in one hand and on the other, must have sufficient friction to stay in place upon removal of the driving force (Refer Fig.8. 10).

The difference in the free body diagrams of low friction and high friction wedges lies in ascertaining the direction of impending motion in the former case (when the wedge is driven to split or move objects apart) and the supposed direction of motion that may take place if the wedge is removed (or fails to exhibit self-locking nature). In wedge related problems, the inclined surfaces cause additional force components in equilibrium equations.

Fig 8. 10 Splitting firewood using a wedge

One must remember that once the directions of friction and normal forces are ascertained for a wedge, for the adjacent body, both friction and normal forces must be taken in opposite direction. As shown in Fig.8. 11, the FBD represents the state of forces for the block to be raised under the action of external force P acting on the wedge. The reader must note the direction of the friction forces on the surfaces on the wedge. These are so, since the wedge is pushed in order to raise the block. At the contact surface between the wedge and the block, on the block side the direction of normal reaction (N_2) and friction force (F_2) is taken reverse to the one on the wedge side. One must remember that if the problem statement is changed to determine the force P to be applied to hold the block in equilibrium (or prevent the block from falling down, then the directions of friction forces (F_1 and F_2) shall be reversed since under this condition, if the block impends to fall, the wedge shall be driven out and subsequently the direction of friction force on all surfaces of the edge shall be opposite to the direction of impending sliding. Accordingly, the friction force on the block shall also change.

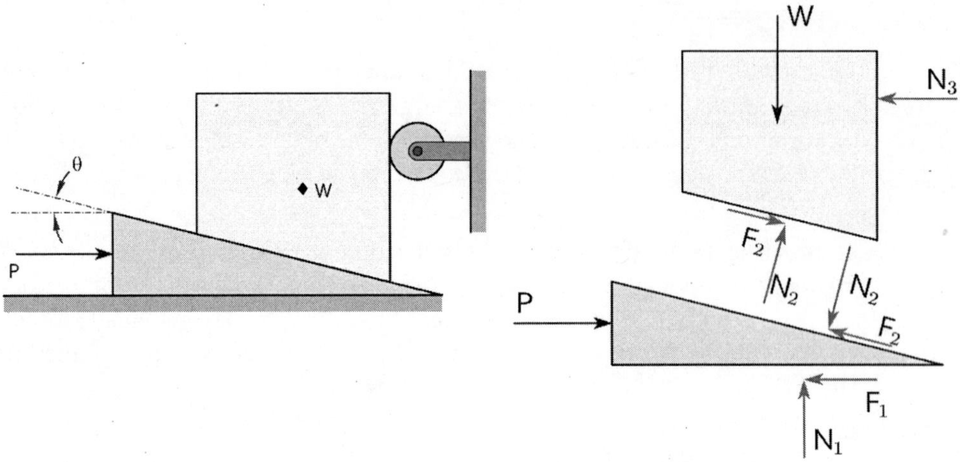

Fig 8. 11 Free-body diagrams to find the minimum P_1 to raise the block.

Example 4: For splitting of logs, a wedge of cone angle (α) is used as shown in Fig 12. The angle of friction between the wedge and the log is ϕ, Determine the maximum value of wedge angle, α, for which the wedge remains embedded in the log.

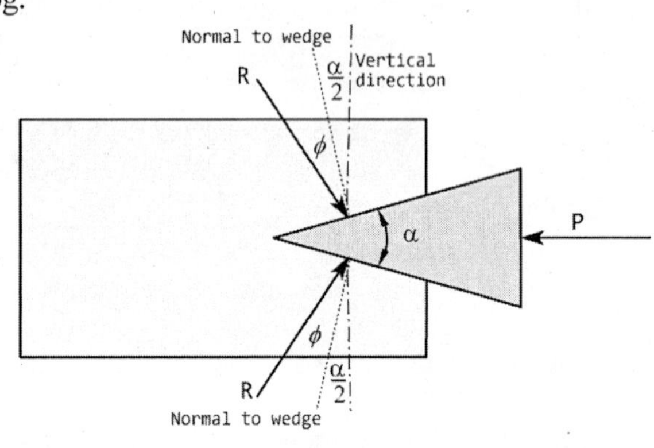

Fig 8. 12 Wedge of cone angle

Solution: The line of action of normal reaction makes an angle of $\alpha/2$ with the vertical. The angle of friction between the wedge and the log is ϕ, which is the angle between the normal reaction and the resultant of friction and normal, as shown if the Fig.

$$\sum F_x = 0;$$

$$P = 2R\sin\left(\frac{\propto}{2} + \emptyset\right)$$

Wedge in a log

$$\frac{\partial P}{\partial \propto} = \frac{1}{2}R\cos\left(\frac{\propto}{2} + \emptyset\right) = 0$$

$$\cos\left(\frac{\propto}{2} + \emptyset\right) = 0 = \cos 90$$

$$\left(\frac{\propto}{2} + \emptyset\right) = 90$$

$$(\propto + 2\emptyset) = 180$$

$$\propto = 180 - 2\emptyset$$

$\sin(\propto) = \sin(180 - 2\emptyset) = \sin(2\emptyset)$

$\propto = 2\emptyset$

Example 5: For the wedge shown in Fig.8. 13, determine the force (P) required to start the wedge. For all contact surfaces, friction angle (Ø) is 15°.

Solution: As the problem is of impending motion of the wedge, at all friction surfaces the respective friction forces are:

$F_1 = \mu_s N_1$; $F_2 = \mu_s N_2$; $F_{13} = \mu_s N_3$

$\mu_s = \tan(\emptyset) = \tan(15) = 0.27$

As there are three unknown variables in the FBD of wedge (P, N_2 and N_3) whereas for the block, there are two unknown variables (N_1 and N_2), and each has two equilibrium equations of force balances along the two perpendicular axes (x-y), hence the block is solved first to obtain N_2.

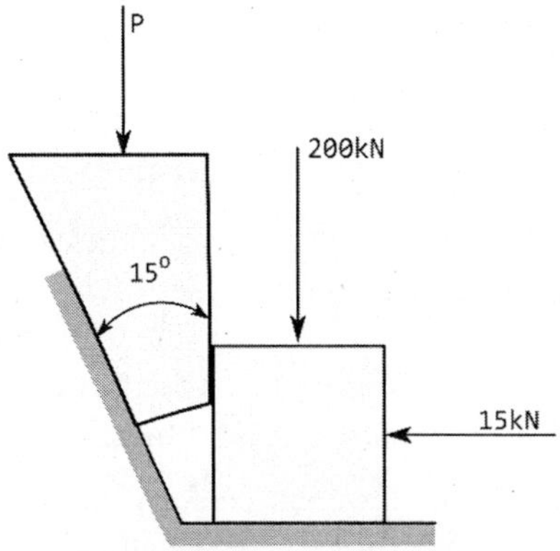

Fig 8. 13 Wedge and block

This shall serve to reduce the three unknowns of wedge to two unknown variables (P and N_3). Using the two equilibrium equations of wedge, then, P shall be obtained.

Solving for block:

$\sum F_y = 0;$

$200 + 0.27N_2 = N_1$

$N_1 - 0.27N_2 = 200$

$\sum F_x = 0;$

$15 + 0.27N_1 = N_2$

$-0.27\,N_1 + N_2 = 15$

Upon solving the two equations;

$N_1 = 220.09$ kN and $N_2 = 74.42$ kN

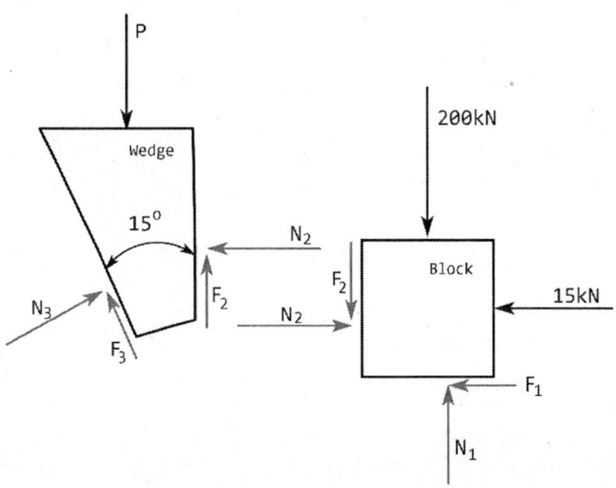

Solving for wedge:

$\sum F_x = 0;$

$N3\cos15 = N_2 + 0.27N3\sin15$

$0.96N_3 - 0.069N_3 = 74.42$

$0.891N_3 = 74.42$

$N_3 = 83.52$ kN

$\sum F_y = 0;$

$P=0.27N_2+0.27N_3\cos15+N_3\sin15$

$P-0.52N_3=20.09$

$P-0.52(83.52) =20.59$

$P=64.02$ kN

8.11 SLIPPING VS. TIPPING

As already known by now that the instant at which an object starts to move is known as the instant of impending motion. When a block rests on a rough surface and is acted upon by an external force, friction comes into play and till friction force reaches a maximum/limiting value, the block remains in equilibrium. As the force is increased further, two possibilities arise; either the box starts to slide, or the external force and the friction force creates a couple causing the block to tip on its corner.

For a block, of dimensions b, h, and d, resting on a rough surface, and pushed by a force P, as shown in Fig.8. 14, assuming that W is the weight of the block, it is required to find force P that will shall cause impending motion of the block, which may either be impending sliding or impending tipping. In the figure, x is the distance that locates the line of action of normal reaction N.

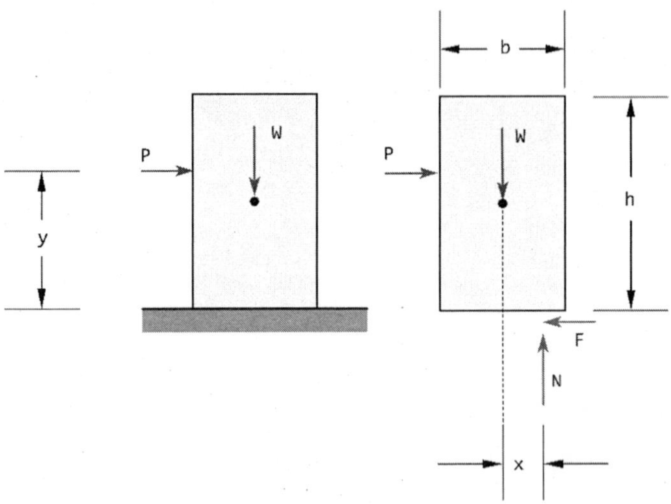

Fig 8. 14 Slipping vs. tipping

A ready approach to solve the problem of tipping or sliding is to solve for the pushing force once assuming slipping and then expecting tipping. The solution that provides the least pushing force is the answer to the impending motion being slipping or tipping. One must observe that, in order to determine the mode of impending motion, there are 4 unknown variables (P, N, F and x) while there are only three independent equilibrium equation. To solve this degeneracy, one must take the help of an additional equation. The approach is as follows: If one assumes impending slipping, then the additional equation is

$$F = F_l = \mu_s N$$

But, if one assumes the impending motion to be that of tipping, then, as slipping does not occur, F does not reach the limiting value, instead the line of action of normal reaction must reach the extreme corner along the width and the additional required equation is taken as x = b/2.

Using both the approaches one by one iteratively, one can assuredly determine the mode of impending motion.

Example 6: For the Fig.8. 15 shown here, determine the greatest height h where force P may be applied parallel to the incline so as to cause sliding of the block up the incline without tipping over.

Coefficient of static friction is 0.30. Dimensions in cm.

Solution: Using equilibrium equations, since it is mentioned that the block slides up the incline,

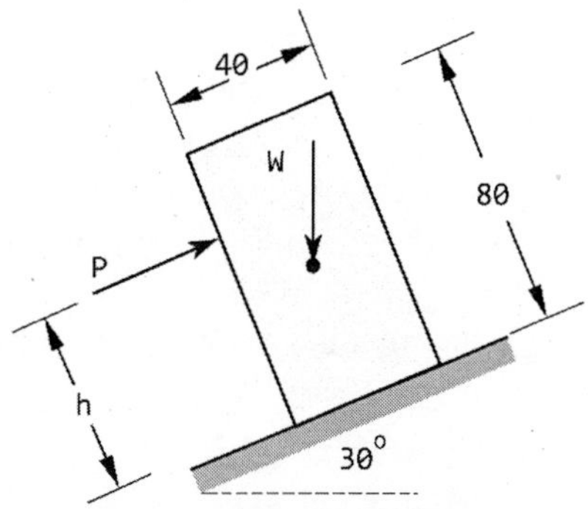

Fig 8. 15 Sliding block

Block on incline

$F = F_l = \mu_s N = 0.3N$

$\sum F_y = 0;$

N=Wcos30

N=0.866 W

$\sum F_x = 0;$

P=Wsin30+F

P=Wsin30+0.3(0.866W)

P=0.5W+0.26W=0.76W

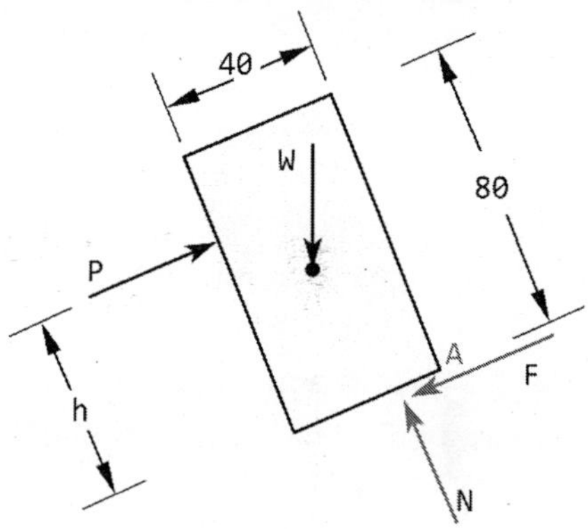

$\sum M_A = 0;$

Moment of N and F at corner A would be non-existent as the normal reaction and friction force under tipping shall be acting at A only. Hence,

P. h= (Wcos30).(20) + (Wsin30).(40)

Upon Solving, h=49.10 cm.

UNIT SUMMARY

The present unit provides a detailed explanation of the types of friction, its laws, and its applications. The approaches to be taken by students in solving the problems of dry friction are detailed and demonstrated using several examples.

EXERCISE

Multiple Choice Questions

1. A 1 kg block is resting on a surface with coefficient of friction μ = 0.1. A force of 0.8 N is applied to the block. The friction force is
   (i) 0 (ii) 0.8 N (iii) 0.98 N (iv) 1.2 N
2. A block weighing 981 N is resting on a horizontal surface. The coefficient of friction between the block and the horizontal surface is μ = 0.2. A vertical cable attached to the block provides partial support as shown in Fig.8. A man can pull horizontally with a force of 100 N. What will be the tension, T (in N) in the cable if the man is just able to move the block to the right?

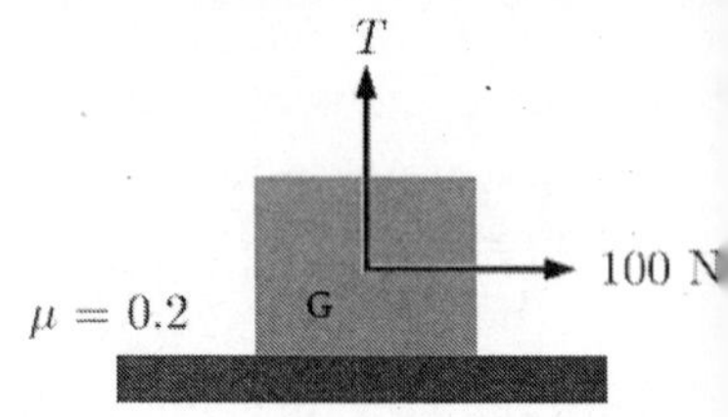

   (i) 176.2 N (ii) 196 N (iii) 481 N (iv) 981 N
3. Two books of mass 1 kg each are kept on a table, one over the other. The coefficient of friction on every pair of contacting surfaces is 0.3. The lower book is pulled with a horizontal force F. The minimum value of F for which slip occurs between the two books is
   (i) zero (ii) 1.06 N (iii) 5.74 N (iv) 8.83 N
4. Find the incorrect statement:
   (i) the tangent of the angle of friction is equal to coefficient of friction
   (ii) the angle of repose is equal to angle of friction
   (iii) the tangent of the angle of repose is equal to coefficient of friction
   (iv) the sine of the angle of repose is equal to coefficient to friction
5. If a ladder rests on smooth ground and rests on a rough vertical wall, the friction will be
   (i) towards the wall at its upper end
   (ii) away from the wall at its upper end
   (iii) upwards at its upper end
   (iv) downwards at its upper end
6. Find the incorrect statement about friction force for dry surfaces from the following.
   (i) Friction force is proportional to normal load between the surfaces
   (ii) Friction force is dependent on the materials of contact surface
   (iii) Friction force is proportional to velocity of sliding
   (iii) Friction force is independent of the area of contact surfaces
7. A body is resting on a plane inclined at angle 30^0 to horizontal. What force would be required to slide it down, if the coefficient of friction between body and plane is 0.3?
   (i) Zero (ii) 1 kg (iii) 5 kg (iv) 9.8 kg
8. A ladder of weight 250 N is placed against a smooth vertical wall and a rough horizontal floor (μ = 0.3). If the ladder is on the point of sliding, the reaction at A will be
   (i) 250 N (ii) 261 N (iii) 125 N (iv) 500 N
9. A horizontal force of 400 N is applied on a body of weight 1200 N, placed on a horizontal plane. If the body is just on the point of motion, the angle of friction would be
   (i) 20^0 (ii) 18^0 (iii) 10^0 (iv) 25^0

10. If the normal force is doubled, the coefficient of friction is—
    (i) Not changed (ii) Halved (iii) Doubled (iv) Triple

Answers of Multiple Choice Questions

1. (ii) 2. (iv) 3. (iv) 4. (iv) 5. (iii) 6. (iii) 7. (i) 8. (ii) 9. (ii) 10. (i)

EXERCISE PROBLEMS

1. Is the block shown in Fig. P8.1 in equilibrium? For P = 400N, find the magnitude and direction of the friction force. Coefficient of static and kinetic friction are 0.3 and 0.25 respectively.
2. What is the range of P for which the system of two blocks shall remain in equilibrium as shown in Fig. P8.2? The wall is smooth.

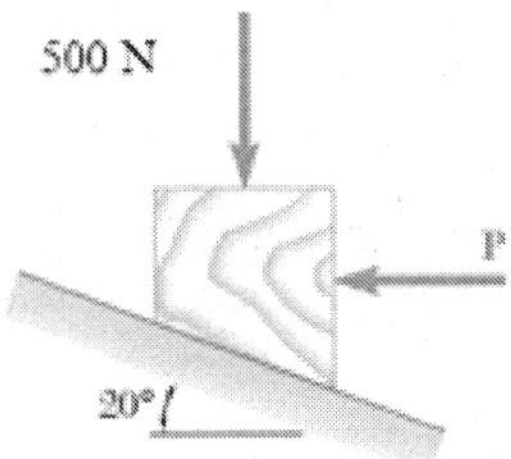

Fig. P8.1

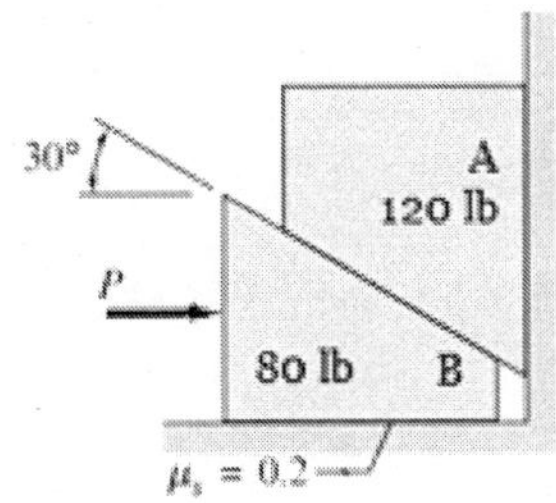

Fig. P8.2

3. The contact surface between the 36-lb block and 20-lb homogenous cylinder is frictionless as shown in Fig. P8.3. Can the system be in static equilibrium on the rough inclined plane?
4. Determine the largest angle θ of the incline, as shown in Fig. P8.4, for which the homogeneous block remains at rest.
5. Two 200-lb blocks as shown in Fig. P8.5, are pushed apart by a wedge (negligible weight). At all contact surfaces, angle of static friction is 12°. Calculate the force P required to impend sliding of blocks.

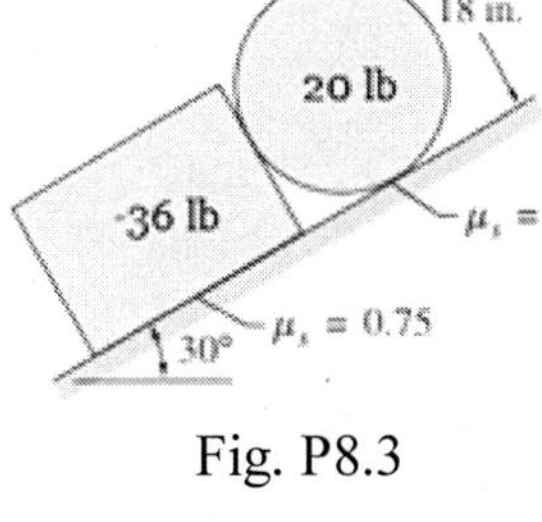

Fig. P8.3

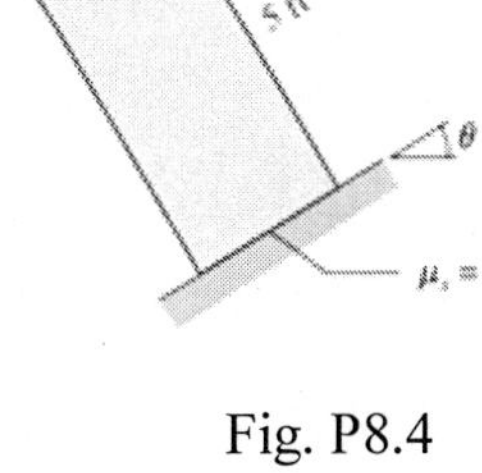

Fig. P8.4

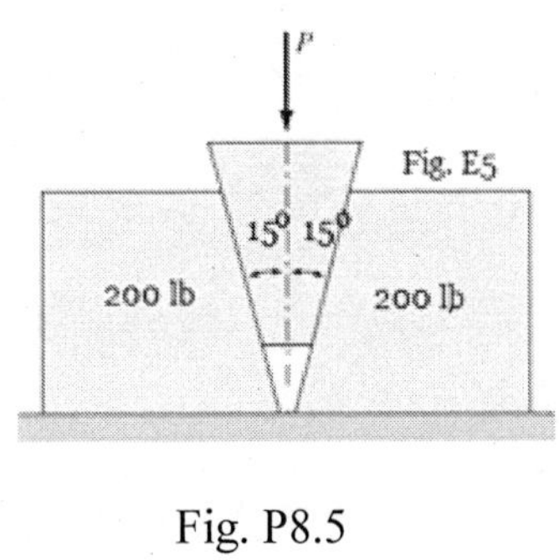

Fig. P8.5

KNOW MORE

Leonardo da Vinci

Leonardo da Vinci (15 April 1452 – 2 May 1519) was an Italian polymath from the Renaissance period, renowned for his paintings. He was also a scientist, mathematician, engineer, inventor, anatomist, sculptor, architect, botanist, musician, and writer. Driven by a desire to understand nature and discover how everything functioned, Leonardo excelled in studying, designing, and creating various inventions. Art historian Helen Gardner noted that no one has ever been quite like Leonardo. His vast array of interests made him seem like a giant in terms of intellectual curiosity, yet his personal character remains an enigma.

REFERENCES AND SUGGESTED READINGS

1. Engineering Mechanics, Manoj K Harbola, Cengage Learning.

2. Vector Mechanics for Engineers, Statics and Dynamics, 11th Edition, Beer, Johnston, Mazurek, Cornwell & Self, McGraw-Hill Education.

3. Engineering Mechanics, Statics, Pytel & Kiusalaas, Cengage Learning.

References for Further Learning

1. Millard F. Beatty, Principles of Engineering Mechanics Volume 1 Kinematics- The Geometry of Motion, Springer Science Business Media, LLC, 1986.
2. Millard F. Beatty, Principles of Engineering Mechanics Volume 2 Dynamics, Springer Science Business Media, LLC, 2006.
3. BERLINSKI, D., Newton 's Gift, Free Press, New York, London, 2000.
4. Ferdinand P. Beer, E. Russell Johnston, Jr., David F. Mazurek, Phillip J. Cornwell, Elliot R. Eisenberg, VECTOR MECHANICS FOR ENGINEERS Statics and Dynamics, 9th Edition, McGraw-Hill, New Delhi, 2009
5. DUGAS, R., A History of Mechanics, Central Book Co., New York, 1955.
6. Fox, E. A., Mechanics, Harper and Row, New York, 1967.
7. Greenwood, D.T. Principles of dynamics. Englewood Cliffs, New Jersey: Prentice-Hall, 1965.
8. MACMILLAN, W. D., Dynamics ofRigid Bodies, Dover, New York, 1960.
9. Meriam, J.L. and Kraige, L.G., Engineering Mechanics: Dynamics, Wiley, 2011.
10. Hibbeler, R.C., Engineering Mechanics: Dynamics, Pearson Prentice Hall, 2010.
11. N. H. Dubey, ENGINEERING MECHANICS STATICS AND DYNAMICS, McGraw Hill Education (India) Private Limited, New Delhi, 2015.
12. HERTZ, IS., The Principles of Mechanics. Dover, New York, 1956.
13. ROSCOE, T., Pig Boats, Bantam Books, New York, 1982.
14. Dietmar Grossn Wolfgang Ehlers Peter Wriggers Jörg Schröder, Engineering Mechanics 1, Statics – Formulas and Problems, Springer, 2017
15. Dietmar Grossn Wolfgang Ehlers Peter Wriggers Jörg Schröder, Engineering Mechanics 3, Dynamics – Formulas and Problems, Springer, 2017
16. I. H. Shames, Engineering Mechanics: Statics and dynamics, 4th Ed, PHI, 2002.
17. Andy Ruina and Rudra Pratap, Introduction to Statics and Dynamics, Oxford University Press, 2011.
18. E. Nelson, Charles Best, William Mclean, Schaum's Outline of Engineering Mechanics, McGraw-Hill Education; 5th edition, 1997.
19. A. Nelson, Engineering Mechanics Statics and Dynamics, McGraw Hill Education; 1st edition, 2017.
20. Andrew Pytel and Jaan Kiusalaas, Engineering Mechanics: Dynamics, Cengage Learning EMEA; 2nd edition, 1999.
21. Stephen Thornton, Jerry Marion, Classical Dynamics of Particles and Systems: Scott's Fatal Antarctic Expedition, Brooks/Cole; 5th edition, 2003.
22. H. Benaroya and S. L. Nagurka, Fundamentals of Engineering Mechanics, Second Edition: Basic Concepts in: Statics and Dynamics, Independently Published, 2022.
23. Daniel Kleppner and Robert Kolenkow, An Introduction to Mechanics, McGraw Hill Education; 1st edition, 2017.

CO and PO Attainment Table

Course outcomes (COs) for this course can be mapped with the programme outcomes (POs) after the completion of the course and a correlation can be made for the attainment of POs to analyze the gap. After proper analysis of the gap in the attainment of POs necessary measures can be taken to overcome the gaps.

Table for CO and PO attainment

Course Outcomes	Attainment of Programme Outcomes *(1- Weak Correlation; 2- Medium Correlation; 3- Strong Correlation)*											
	PO-1	PO-2	PO-3	PO-4	PO-5	PO-6	PO-7	PO-8	PO-9	PO-10	PO-11	PO-12
CO-1												
CO-2												
CO-3												
CO-4												
CO-5												
CO-6												

Index